W0235273

JAPANESE PHENOMENOLOGY

ANALECTA HUSSERLIANA

THE YEARBOOK OF PHENOMENOLOGICAL RESEARCH

VOLUME VIII

Editor:

ANNA-TERESA TYMIENIECKA

JAPANESE PHENOMENOLOGY

Phenomenology as the Trans-cultural
Philosophical Approach

Edited by

YOSHIHIRO NITTA

University of Toyo

and

HIROTAKA TATEMATSU

Nanzan University

in cooperation with
The World Institute for Advanced Phenomenological
Research and Learning
Belmont, Mass.

D. REIDEL PUBLISHING COMPANY

DORDRECHT : HOLLAND / BOSTON : U.S.A.
LONDON : ENGLAND

Library of Congress Cataloging in Publication Data

Main entry under title:

Japanese phenomenology.

    (Analecta Husserliana : v. 8)
    Bibliography: p.
    Includes index.
    1. Phenomenology – Addresses, essays, lectures. 2. Philosophy, Japanese – 20th
century – Addresses, essays, lectures. 3. Husserl, Edmund, 1859–1938 – Addresses,
essays, lectures. 4. Nishida, Kitaro, 1870–1945 Addresses, essays, lectures I. Nitta,
Yoshihiro, 1929–   II. Tatematsu, Hirotaka. III. World Institute for Advanced
Phenomenological Research and Learning. IV. Series.
B3279.H94A129 vol. 8 [B5243.P48] 142'.7s   [142'.7'0952]   78–11758
ISBN-13:978-94-009-9870-4   e-ISBN-13:978-94-009-9868-1
DOI: 10.1007/978-94-009-9868-1

Published by D. Reidel Publishing Company
P.O. Box 17, Dordrecht, Holland

Sold and distributed in the U.S.A., Canada, and Mexico
by D. Reidel Publishing Company, Inc.
Lincoln Building, 160 Old Derby Street, Hingham,
Mass. 02043, U.S.A.

All Rights Reserved
Copyright © 1979 by D. Reidel Publishing Company, Dordrecht, Holland
Softcover reprint of the hardcover 1st edition 1979

No part of the material protected by this copyright notice may be reproduced or
utilized in any form or by any means, electronic or mechanical,
including photocopying, recording or by any informational storage and
retrieval system, without written permission from the copyright owner

TABLE OF CONTENTS

INTRODUCTORY ESSAY

PART I / PRESENT DAY
PHENOMENOLOGY IN JAPAN

PART II / PHENOMENOLOGY IN THE JAPANESE
INHERITANCE

PHENOMENOLOGY IN JAPAN:
A PRESENTATION

It is the privilege of the *Analecta* to present in this volume – for the first time to the West – essays by the most prominent Japanese scholars, and to record the existence and significance of Japan on the world map of phenomenological reflection.

Admittedly, many a philosopher in the West may have had some vague presentiment that the development of the phenomenological method and research in Europe and North America has not gone unnoticed in Japan. More likely than not, Husserl and his major disciples would figure in philosophical textbooks and in the intellectual surveys of the contemporary West available to the Japanese public. Moreover, from the little exposure we may have had to Japanese life and culture, we could perceive here and there some furtive echoes of our own discourse. But altogether we have been left in the dark as to the existence of a genuine and original phenomenological research and reflection in Japan.

In the present volume, including ten previously unpublished essays, Japanese scholars are given the opportunity to speak for themselves, hopefully, to shock the Western phenomenologists (and orientalists) into a first discovery of their Japanese colleagues.

This volume, consisting of two parts, opens with a study by Y. Nitta, of an unknown manuscript of Edmund Husserl, and with W. Mizuno's critical inquiry into the phenomenological method itself. Then follow the 'theory of space' by H. Kojima, an original treatment of phenomenological intuition by E. Shimomissē, and Sh. Takiura's penetrating essay on the 'reality of time'. The question of the 'life-world' is examined by H. Tatematsu, while J. Watanabe studies the nature of truth in Heidegger's *Being and Time*.

Part II, 'Phenomenology in the Japanese Inheritance', centers around Kitarō Nishida and the Kyoto School which is presented by T. Ogawa. It contains a first translation into English of a fragment from Nishida's writings made by a well known Nishida specialist D. Dilworth in collaboration with V. H. Viglielmo. This fragment from the period in which

Nishida stood in a dialogue with Husserl's thought is complemented by an essay by D. Dilworth showing Nishida's original reflection.

The "discovery" offered in the following pages is a complex proposition indeed. What appears here is not the image of "another phenomenology". And yet in this panorama of research and reflection belonging naturally to the common field of the phenomenological endeavor one may simultaneously discern a Japanese thread. This raises questions on the significance of the discovery itself.

First of all, it is not only the discovery that phenomenology is present and well-alive in Japan, but rather the discovery that there exists a trend of research and reflection which can be deemed, and recognized as both taking part in the common philosophical endeavor outlined by Husserl and as "Japanese" in the same way we speak about "French", "German", "Polish", "Italian" and "North American" phenomenologies. In Japan, with its own distinctive philosophical tradition, the pursuit of pheno-menological research might seem incongruous. We might surmise that a 'phenomenology in Japan' means that some of our Japanese colleagues have specialized in the inner workings of the rigorously defined method-ology, and developed the latter with the required accuracy. Yet, they would remain the experts in Japan of a philosophical method which originated within another tradition.

What the reader will find in this volume is quite different. Many Japanese philosophers have indeed obtained doctorates in Europe, and published works on Husserl, Heidegger, Max Scheler, and others, in Western Languages. But they have not confined themselves to the role of mere interpreters. Phenomenology became, on the contrary, a field of scholarship they could explore, evaluate and appropriate in their own terms: ultimately, as it seems, it became a genuine mode of Japanese philosophizing. The result is a *Japanese* phenomenology, that is to say: a reflection which is unmistakenly heir to Husserl, but reflects as much the Japanese intellectual legacy and the philosophical quest of contemporary Japan, from the Meiji era to World War II and the present.

Secondly, we discover that the Japanese phenomenology, so under-stood, is not a late-comer in the field. In fact it goes back to Husserl's appearance and teaching activity on the philosophical scene and we can see that it has developed a tradition of its own.

Central to that tradition is the figure of Kitarō Nishida (1870–1945), the most creative philosopher in twentieth century Japan. Familiar with the works of Bergson and Wiliam James, sharing Husserl's interest in logic and the foundations of mathematics, Nishida could feel at home with the phenomenological method and understand it in its significance for contemporary philosophy in the West. However, Nishida's world and his cultural roots were not Husserl's. From his experience of Zen meditation, Nishida criticized Husserl's identification of consciousness with "intentionality". Whereas the latter began with the distinction between the subject and the object, Nishida posited consciousness as prior to that distinction. The result was a philosophy of consciousness as an ontological investigation, and – as in Zen truth was, in opposition to Husserl, attained through losing one's subjective self. Thus already at that time phenomenological insights, albeit vicariously, entered Japanese philosophy through Nishida's controversy with Husserl.

Although other members of the Kyoto School did not share directly the concern with Husserlian philosophy, it is worth noting that, as early as 1911, a dissident phenomenological reflection was already present in Japan. From the bibliography of the selected translations of the major phenomenological writings into Japanese, which H. Tatematsu, one of the most distinguished translators of Husserl, prepared for our volume, we may see how thoroughly and intensely the interest in the phenomenological movement and its development was entertained in Japan from its early beginnings. One could validly argue that Nishida was not a phenomenologist in the strict sense; yet in his dialogue with phenomenology he may have uncovered issues which, sooner or later, will become crucial for phenomenology itself. For instance, the emphasis which he put upon action, as reflected here by D. Dilworth, is only now becoming a focus of our investigation (see *Analecta* Vol. VII).

Thirdly, while we assess the existence of a Japanese phenomenology, with a tradition of its own, we perceive also a host of questions which could not appear prior to the discovery of a non-Western phenomenology. Admittedly, the reader of the following essays could content himself with the reassuring view that Japanese philosophy proves the universal value of the Husserlian method. The conclusion would be that Husserl's concerns, his approach to the question of man's life-world etc., are significant for man as such, in spite of cultural differentiations. But once we have thus hailed the universal validity of the phenomenological method, we should let ourselves be confronted with the more disturbing interrogations inherent to this volume as to any East-West intellectual encounter.

For instance, one could be tempted to conclude that Nishida's attraction to Husserl's thought, and the subsequent phenomenological movement in Japan, reflects a hidden affinity between Husserl's project and the intellectual tradition of the East, especially Zen Buddhism. But why do we not find a similar encounter in China at the same time? It should then be clear that the existence of a Japanese phenomenology cannot be safely explained away by the magic invocation of an Eastern or Buddhist tradition akin to Husserl's intuition. The crisis of Japanese thought in the 1910's, and Nishida's personal temper and circumstances, are likely to have played as important a role as his Buddhist background in his attraction to phenomenology.

Moreover, did phenomenology provide Japanese philosophers only with another *language*, modern and intercultural, which allowed them to pursue their own tradition while overcoming intellectual insularity? Or was phenomenology a revolution in Japanese philosophy? If it was indeed a revolution, was Husserl important for the same reasons as he was in Europe? After all, maybe Nishida and others also found in Husserl insights which are not specifically phenomenological. Husserl may have become the privileged channel through which modern Japanese philosophers happened to have access to various, unrelated strata of the Western mind. However that may be, here is a phenomenology which we cannot understand simply by reading Husserl, Heidegger or Merleau-Ponty.

In a similar vein, one should not hastily conclude from this volume that phenomenology is the common ground for a philosophical understanding between the West and Japan, not to speak of "the East". On the contrary, it should be clear that a Japanese phenomenology does not offer a short-cut to the intercultural meetings of the minds. It simply opens up a common *Holzweg* into the forest of human experience and philosophical traditions. It will take time before we even find out where that *Holzweg* begins at all.

In conclusion, a "Japanese phenomenology" with its own relation to Husserl's inheritance, does not only constitute one more chapter in the phenomenological inventory. Its very existence may lead us to question the presuppositions of our own phenomenological discourse. At least it provokes us to consider whether the phenomenological approach might not be merely as Husserl himself claimed but, on the contrary, a *universal* avenue opened to the philosopher seeking to discover truth about man, his place in the cosmos and the significance of his existence.

The World Institute for Advanced Phenomenological
Research and Learning   ANNA-TERESA TYMIENIECKA

Department of East Asiatic Studies,
Harvard University   MICHEL MASSON

ACKNOWLEDGEMENTS

In bringing this long awaited selection of Japanese phenomenology to print, I would like to express my sincere thanks to Professors Yoshihiro Nitta of the Tokyo Psychiatric Institute and the University of Toyo and Hirotaka Tatematsu of the University of Nanzan for their collaboration in gathering the essays of the Japanese phenomenologists. Our thanks are also due to Professor David Dilworth for his dedicated collaboration in Kitarō Nishida's chapter. Our translator from the German, Dr. Barbara Haupt Mohr, has contributed greatly in making Japanese phenomenological thought available to the English-speaking scholarly public.

We would like also to express our gratitude to Professor Gen Itasaka and Mr. Alan Campbell of Harvard University for their most helpful collaboration in the preparation of this volume. And last but not least, we would like to express our gratitude to Professor Richard Stevens for his expert linguistic help and to Dr. Michel Masson for his recurrent invaluable editorial collaboration.

ANNA-TERESA TYMIENIECKA

xi

INTRODUCTORY ESSAY

YOSHIHIRO NITTA, HIROTAKA TATEMATSU,
AND EIICHI SHIMOMISSÉ*

PHENOMENOLOGY AND PHILOSOPHY IN JAPAN

INTRODUCTION

Phenomenology and phenomenological philosophy have been well received in the academic world and by intellectuals, since the introduction of Husserl's *Logische Untersuchungen*. Beginning with the Meiji Restoration, Japan quickly absorbed not only science and technology from the West, but also the economic system, the social structure, the legal and political organization, as well as art, literature, and philosophy. And it was thus natural that the introduction of phenomenology and phenomenological philosophy would be similarly appreciated.

Husserl's philosophical investigations have been introduced into Japan through several channels; Scheler's works are not only known in philosophical circles, but known also to scholars in sociology and jurisprudence; Heidegger has left his imprint and continues to exert a deep influence on Japanese philosophers. Sartre's existentialism was introduced in 1947; and Merleau-Ponty and Camus have been widely read. Japan is probably the only country where Heidegger's *Sein und Zeit* is available in five different translations; while the "complete works" of both Heidegger and Sartre have been published and a translation of Scheler's works is now underway.

It is also true that, since the first introduction of Edmund Husserl, phenomenology and phenomenological philosophy has continuously occupied a significant portion of the interest and investigation of academic philosophical circles, unlike other "philosophical trends" which came and went.

Nevertheless, it is our contention that the situation in regard to phenomenology and phenomenological philosophy in Japan was *far more complex*: it requires more attention and demands a more penetrating analysis. For the historical horizon of the traditional Japanese thought to which penomenology was introduced was quite unique. Furthermore, Japan during the Meiji 20-30s, i.e., at the turn of the century, revealed a similar spiritual-philosophical attitude receptive to a radical break from the past as well as an antipositivism and an antiintellectualism similar to that in Western Europe of the time. The return to a more primordial reality

Nitta | Tatematsu (eds.), Analecta Husserliana, Vol. VIII, 3-17. All Rights Reserved.
Copyright © 1978 by D. Reidel Publishing Company, Dordrecht, Holland.

was sought by both the man of letters and the philosopher. This quest introduces us to the development of phenomenology and phenomenological philosophy in Japan.

In order to comprehend phenomenological philosophy and place it in a proper perspective in the development of Japanese thought, it may be first of all necessary to make explicit what we mean by "phenomenology and phenomenological philosophy." It is a generally accepted fact that there is no such philosophical discipline or doctrine called "phenomenology"; Jeanson, for example, considers an attempt to provide an "objective definition" of it absurd.[1] Nevertheless, it is possible to characterize the fundamental features common to and underlying all phenomenological philosophies and inquiries. They may be summarized by the following six basic characteristics of their common approach: (1) a critical effort to make a complete break from the traditional philosophical approaches which are considered to be faulty "reductionisms"; (2) an anti-metaphysical, antispeculative attitude; (3) an antipositivism aiming at transcending mere empirical givenness; (4) an attempt at liberation from the dogmas of positive sciences and scientific explanations of reality; (5) an attempt to return to the primordial experience, to be unraveled as lived reality; (6) "bracketing" (*ausser-Kraft-setzen*) of the general thesis of the natural world, with all its biases, distortions, and concealments, in order that "phenomenological analysis" may disclose the primordial reality.

Accepting these working hypotheses as the basis for our interpretation, we shall be able to proceed, first, to explicate the historical background of Japanese philosophical thought prior to the introduction of phenomenology. Second, we shall be able in particular to construe Kitarō Nishida's philosophical investigations as phenomenological for the most part. Third, introductions and translations of phenomenology and phenomenological philosophies will be briefly described; and, fourth, some of the original contributions in actual pursuit of phenomenological inquiry in Japan will be discussed. Finally, a new direction of philosophizing as a dialogue between the East and the West will be pointed out as possible in the future phenomenological investigations in Japan.

I. PHILOSOPHICAL-INTELLECTUAL BACKGROUND OF
PHENOMENOLOGICAL THOUGHT IN JAPAN

What was the situation of the traditional religious and philosophical thinking in Japan prior to the introduction of phenomenology and phenomenological philosophy around 1910? As mentioned above, Japan at the turn of the century (1870–1910) represents a rather complex intellectual and cultural situation. On the one hand, the Meiji Restoration, which was supposed to "restore" the emperor's rule (thus return to the traditional past), began to abandon all the traditional values of the old culture and tried to reorient the way of life in order to absorb Western civilization under the slogans of "modernization" and "enlightenment." In a sense it was a social and spiritual revolution.[2] As Kosaka characterizes it, "young men of Meiji," who distinguish themselves from "old men of Tempo" and who held a new vision of the future of Japan, intended to make a clear break with the past by the Meiji 20s (1887–1897).[3]

Such a confrontation with the past cannot, however, be viewed in an oversimplified way. On the one hand, there is in fact a break from Japan's traditional past; on the other, there arose nevertheless a critical self-appraisal against blind Westernization which took two different directions: one went back to a nationalistic "Japanism"; the other, trying to trace back the original problems which Western civilization itself faced to their source, and confronting the problems which the Japanese civilization had been facing with them, developed a type of "philosophy of life" (*Lebensphilosophie*) particularly influenced by Nietzsche. In spite of these complex reactions to the preceding intellectual situations, a strong urge to go back to the *primordial* reality and to one's own explicit awareness of one's self in its primordiality is common to all the reactions and trends distinguished above and may be seen as *der Zeitgeist des Zeitalters.*

The Japanese intellectuals of the time went in various directions, but they were all antienlightenment, antirational, antipositivistic, and anti-reductionistic. Thus, Japanese "naturalism" and Sakae Osugi's "anarchism," for example, were primarily intended to unravel biases and distortions of current philosophies and to recover the concrete, full experience of man's own existence in the world. As a leading figure among them should be considered Kitarō Nishida, whose fundamental attitude of philosophizing is essentially animated by this same spirit.

The other important observable feature of the Japanese philosophical thinking may be found in its ontological apprehension of man's existence and the world. It may be traced back to three different origins – the Taoistic philosophy of Lao Tsu and Chuang Tzu, the *I Ching*, and Confucianism and Buddhism. None of them indeed affirmed "reality" *behind* the phenomenal world such as Being itself, or the absolute realm of Ideas, or the transcendent, personal God or Kantian thing-in-itself. Japanese philosophy, in fact, never accepted the ultimate reality or its principle as "existing" apart from and independent of the concrete, lived world of experience, although they are separable in our theoretical thinking.

I Ching states that in "I" there is *t'ai chi* ("the Great Ultimate"); it produced the two *i* ("Forms"); these two Forms produced the four *Hsiang* ("emblems"), and these four emblems produced the *pa kua* ("eight trigrams"). This Great Ultimate which is the *wu chi* ("the Nonultimate") is construed by Chu Hsi as the *li* ("Principle") immanent in all. According to the *Analects*, Confucius' disciples never heard him talk about "metaphysical questions." Gautama Buddha is said to have denied the substantiality of the "self" as well as of the "universes" which he has considered as illusory and ultimately *sunyata* ("void"). Lao Tsu spoke of *Tao* as the mother of "ten thousand things" and yet *Tao* itself is seen as being the immanent principle of all these entities distinguished by names (i.e., the 10,000 things) for their being, so that in turn Tao reveals itself as *wu* ("nothing").

It is also quite clear that each of these philosophies explicitly distinguishes itself as "wisdom," "enlightenment," or as the religious *praxis* not only from the mundane ontology of everydayness, but also from the pure *theoria* of Western philosophy. And yet an affinity with phenomenology is obvious. The presumed naiveté, distortion, and concealments present in the concepts of Western philosophical theories can only be radically unraveled and properly understood, when one's natural attitude or universal "common sense" belief is "bracketed" and what was taken for granted to be real ultimately becomes transparent as illusory and *sunyata*, i.e., "phenomenon." In Taoist terms, it was possible to acquire understanding only when the distinction by name, desire, and attachment to those "ten thousand things" were "dispossessed of their power and function," and when the self – the greatest attachment of all – was suspended to become decreased and nihil. Philosophizing and philosophical awakening is seen as an *open*, truly active approach to reality, whereas the absolved immersion in the everyday concerns is a *blinded* and shackled one.

This obvious similarity and affinity between the phenomenological approach and the East Asian way of philosophizing occurs in spite of the fact that the former, particularly in Husserl's case, was motivated mainly by pure theoretical interest and the self-justification of knowing itself, whereas the latter was led by practical religious concern about one's own existence. This is why Heidegger's existence philosophy finds such a strong appeal among Japanese philosophers and intellectuals today.

The East Asian way of philosophizing remained deeply rooted in the Japanese tradition despite the overwhelming influences of Western civilization. Therefore, it was not an accident, for example, that in 1897 Tetsujiro Inoue wrote an article entitled 'Gensho sunawachi jitsuzairon" ("Phenomenon as Realism") *prima facie* as a critique of Haeckel's naive realism *and* Kantian transcendental philosophy which, according to him, "hypothesizes" the "thing-in-itself" beyond and behind the "phenomenon." It was, however, Kitarō Nishida who has developed his thought in his native tradition and yet carried on a profound dialogue with phenomenology.

II. NISHIDA'S PHILOSOPHY AND PHENOMENOLOGY

Nishida's philosophy, which seems to make a bridge between the Western phenomenological philosophy and the Japanese thought, breathed and was alive in this tradition and at the same time has developed with a positive and critical confrontation with Western philosophy. In a similar attitude to that of his European contemporaries Nietzsche, Bergson, Dilthey, and Husserl, Nishida's own philosophy was developed. Although his reflection proceeded in a similar fashion to phenomenology,[4] going back to pure experience, and yet ultimately rejoining the East Asian conception of nothingness as the primordial reality.

In the above thesis, "Nishida's Philosophy and Phenomenology," the connective "and" is used rather ambiguously. On the one hand, it obviously refers to when and how Nishida encountered Husserl and other phenomenologists through their writings and how he reacted to, was influenced by, and became critical of Western phenomenology and phenomenological philosophy. On the other, it is intended by the use of this "and" to have Nishida's philosophy reveal itself as a *phenomenological philosophy*. It has been pointed out in particular that Nishida's "doctrine of pure experience" in his *A Study of Good* is a phenomenological inquiry of "psychological experience" in distinction from Nishida's later emphasis on "logic" in Hegel's sense.[5]

First, the *fundamental attitude of Nishida's philosophizing* may well be compared with that of Husserl and other phenomenologists. Second, like Husserl's, Nishida's philosophy undergoes an incessant radicalization of his *return to the more primordial reality and of its beholding and description*. At different stages of his philosophical development, it was called "pure experience" or "intellectual intuition," "self-conscious will" or sometime later absolute "reflection," "concrete, self-conscious universal," "locus" or "field" as nothingness, "intuition in action" in distinction from "expressive determination," and finally, the "world of historical reality." In a certain sense, Nishida's phenomenological inquiry is more radical than even that of the later Husserl.

It is in order, then, to discuss here briefly Nishida's relationship to Husserl. It is generally accepted that Nishida was the first Japanese philosopher who made reference to Husserl in his article entitled 'On the Theses of the Pure-Logic Schools in Epistemology' (1911). Nishida saw Husserl holding an almost identical position with Rickert's, although they belonged to different "trends" of thought.[6]

Although Husserl's *Die Philosophie as strenge Wissenschaft* was introduced by Kichinosuke Ito under the title "Husserl: Philosophy as Science" in 1915,[7] Nishida played the major role in introducing Husserl's phenomenology in 1910–1920. He gave a series of lectures called "Idealistic Philosophy of Today" in 1916 in which he sympathetically discussed Husserl's *Logische Untersuchungen*, introducing him as one of the significant representatives of the Brentano School. The object of Husserl's phenomenology is, according to Nishida, a concrete phenomenon of consciousness, and the realm of phenomenology is that of pure experience, which phenomenology intends to investigate by sustaining all possible attitudes. In this interpretation of Husserl, Nishida clearly anticipated the further development of his inquiry in the direction of transcendental philosophy.[8]

Although his article entitled 'Contemporary Philosophy' was published in the first issue of the first volume of *Tetsugaku-zasshi* in 1916, it was originally supposed to constitute Part 2 of the three parts of the aforementioned lecture series. However, it is interesting to note that this article contains a more detailed presentation of Husserl's phenomenology and makes reference to the title and content of *Ideen I*, and that the Husserlian notions of "noema" and "noesis" were mentioned and explained in it.[9]

Nishida's book entitled *Intuition and Reflection in Self-Consciousness* which was written between 1913 and 1917, includes some thirty-three references to Husserl along with Bergson, Richert, Cohen, Bolzano, Brentano, and Schapp, as well as to contemporary mathematicians. They indicate Nishida's penetrating understanding of Husserl particularly in regard to the impossibility of the reflective grasp of "the object-constituting act" itself as the concrete flowing experiencing present, long before Husserl himself wrestled with the same problem.[10] Nishida also criticizes Husserl's sharp distinction between the transcendent and the immanent by pointing out that not only the knowledge of an (external) object, but also the knowledge of mind (i.e., an immanent object and its relations), are often given "inadequately."[11] In order to emphasize the dynamic developing nature of the concrete, experienced reality, Husserl's intentional life (*intentionales Erlebnis*) is seen by Nishida as "self-conscious *will*" in a sense similar to Fichte's *Tathandlung*, in which meaning and fact are immediately unified.[12] Nishida also accepts Husserl's broad use of intuition extending beyond the empirical perception.[13] This "intuiting, self-conscious will" was further developed into his basic notion of "field" or "locus" understood as nothing.[14] He further follows Husserl's discovery of the correlation between a particular cognitive attitude and corresponding to it a particular region of "reality."[15] Nishida concludes in analogy to Husserl that the concrete, intuitively experienced reality, which is self-consciousness, by returning to its own primordial origin reveals itself as free, time-creating will. At this point the analogy breaks because the cognitive primacy in the consciousness/world constitution of Husserl, Nishida challenges by his conception of the will.[16]

A further pivotal point of confrontation between Husserl and Nishida occurs in the last four chapters of Nishida's book, *The System of the Self-Conscious Universals* (1929), where Husserl's fundamental pair of concepts of "noesis" and "noema" which characterize the intentional correlation of the intentional act and its object frequently appear.[17] However, instead of using "noesis," "noema," "intentionality," "intentional or conscious act" in Husserl's limited original sense, Nishida employed them rather freely.

Nishida considers in point of fact that Husserl's consciousness is primarily representative (*vorstellend*) and that the concepts "intentionality," "noesis," and "noema" are meant to make the basic structure and elementary parts of cognitive consciousness. As mentioned above, however, Nishida reproaches Husserl for having reified the self into a transcendental ego. Furthermore, he sees the essential structure or form of consciousness itself as self-consciousness rather than intentionality.[18]

As self-consciousness, the self "sees" itself within his own realm. Nishida calls the self in its function of seeing itself "noesis" and the self in the function of being seen (as its "object") "noema" (or "noematic" determination). Since the active self, as aboslute will, can never become an object, the self that apprehends itself transcends regressively to its ground. The concepts noesis and noema are used by Nishida to indicate the opposite poles of self-consciousness rather than those of an intentional structure of consciousness. Nishida calls Husserlian intention "weak will."[19]

It is Heidegger's hermeneutic phenomenology that appears to Nishida to be close to his standpoint of the self-determination of the intelligible self, for Heidegger takes the position of hermeneutic consciousness by moving away from the Husserlian position of immanent consciousness.[20] Since phenomenological attitude is, according to him, an intellectual self-consciousness of inner life, the transition from Husserl's phenomenology to Heidegger's fundamental ontology does in his view establish the transcendence of the self-consciousness of the acting self.[21]

To conclude, then, Nishida's critique of Husserl and Heidegger is intended to show the insufficiency of the radicalization of returning to the more concrete, more primordial reality from which our mundane, scientific as well as aesthetic and religious experience of the historical world emerges.

Nashida was the first to introduce phenomenology and phenomenological philosophy into Japanese academic circles and among intellectuals in 1910–1920, and although he possessed a keen insight into its strength and weakness, he encouraged some of his colleagues and students to study phenomenology and phenomenological philosophy at the University of Freiburg. His rather liberal use of some of the basic insights and concepts of phenomenology allowed him to elaborate effectively in an original way his own philosophy.

III. INTRODUCTION OF PHENOMENOLOGY IN JAPAN

In the previous chapter, Nishida's contributions to the introduction of Husserl's phenomenology were discussed. Nishida never went abroad. Many of his contemporary philosophers, however, went to Germany to study in Berlin, Heidelberg, Marburg, and Jena, where neo-Kantianism was the main stream of German philosophical circles. Upon their return, they assiduously introduced Windelband, Rickert, Lask, Münsterberg, Bauch, Cohen, Natorp, Cassirer, and even Mach, Riehl, and Vaihinger. Neo-Kantianism flooded the academic circles in Tokyo and Kyoto. Nishida's *Intuition and Reflection in Self-Consciousness* was at least partially his responsive and critical confrontation with that trend he encountered in Tokyo and Kyoto.[22]

Although Husserl's phenomenology was a reaction to and a critical confrontation with neo-Kantianism, it seems rather ironic that the neo-

Kantian transcendental way of philosophizing, too, paved the way for Husserl's smooth and wide acceptance in Japan.[23]

Many representative Japanese philosophers - Hajime Tanabe, Satomi Takahashi, Shyuzo Kuki, Tokuryu Yamanouchi, Goichi Miyake, Kiyoshi Miki, Risaku Mutai- studied at the University of Freiburg in the 1920s and published many translations and summaries of Husserl's writings, along with many articles introducing and commenting on Husserl's phenomenology. In 1934 Husserl's *Erneuerung-Ihr Problem und Ihre Methode* was published in German with its translation in the monthly journal, *Kaizo*; and Husserl's *Ideen I* was translated by Eiichi Kito in 1932 and a second translation by Kenzo Ikegami was published in 1934. The latter version was widely circulated in the postwar period.

Umaji Kaneko greatly contributed to the introduction and understanding of Husserl by his *A General View of European Thoughts* (1920) and *Introduction to Contemporary Philosophy* (1922). But the first book which specifically introduced phenomenology was Tokuryu Yamanouchi's *Philosophy of the Phenomenological School* (1926). His *Introduction to Phenomenology* (1928) was an extensive discussion of Brentano and Husserl with the most comprehensive bibliography on phenomenology of the time.

It should be mentioned that numerous articles for philosophical journals and books were written by Tetsuo Satake on Husserl's phenomenology and his efforts continued into the postwar period.

Kichiji Watanabe, who died young, should be remembered for his contributions in the field of aesthetics. In phenomenological aesthetics, Conrad, Hamann, Geiger, and Utitz were given a comprehensive treatment in his *Contemporary Trends of Aesthetics* (1927). K. Ohnishi's *Aesthetics of the Phenomenological School* (1938) was the most comprehensive book on aesthetics.

Satomi Takahashi's *Husserl's Phenomenology* (1931) was unique and very significant in that his treatment of Husserl's phenomenology not only covered his early period of *Logische Untersuchungen* and *Ideen I* but contains the development of his later philosophy, e.g., *Cartesianische Meditationen*. This was the only study in the prewar period devoted to Husserl's later period.

Taiji Okuma's "Psychotherapy and phenomenology" (1935) should not be forgotten for its influence on the field of science. Inumerable articles continued to be published even during World War II.

Heidegger's friendship with Kiyoshi Miki, and in particular, with Shuzo

Kuki must be mentioned. It was Kuki who introduced Heidegger's philosophy to Sartre and inspired him to study at the University of Freiburg later.[24] As early as 1927 – Heidegger's *Sein und Zeit* was published in 1927 – Miki published his article, "Fundamental Concepts of Hermeneutic Phenomenology" and introduced Heidegger's philosophy to Japan. Concise as it was, Kuki's *Philosophy of Existence* was the best introduction to Heidegger's philosophy before World War II. Terashima's translation was the first translation of Heidegger's *Sein und Zeit* into Japanese (in the prewar period) and was reprinted immediately after World War II.

During World War II, the study of Husserl's and Heidegger's phenomenology was mainly pursued at Tohoku University under the direction of Goichi Miyake and Tsuneo Hosoya; Husserl at Tokyo University by Kenzo Ikegami; Heidegger at Kyoto University by Keiji Nishitani; Scheler at Taihoku University by Akira Tanaka.

In the postwar period, a strong interest in Heidegger was immediately revived, and Sartre and Camus were introduced in response to the personal existential experience of Japanese intellectuals during and after World War II. As in France, however, *l'existentialisme* soon became a "fashionable trend of thought" in Japan.

Goichi Miyake's *Heidegger's Philosophy* took the form of a commentary and interpretation of Heidegger's philosophy and yet it was pursued phenomenologically: it was also quite unique in that this study was written with the intention of critically confronting himself with the prewar Japanese academic philosophies of Nishida, Tanabe, and Takahashi.

At present, the complete works of both Heidegger and Sartre have been published; and four different translations of Heidegger's *Sein und Zeit* are currently available. In addition there are now available translations of Husserl's major works: *Cartesianische Meditationen* by M. Yamamoto; *Die Philosophie als strenge Wissenschaft* by T. Satake; *Idee der Phänomenologie, Logische Untersuchungen*, and *Die Phänomenologie des inneren Zeitbewusstseins* by H. Tatematsu; *Die Krisis* by T. Hosoya; and *Erfahrung und Urteil* by H. Hasegawa. A translation of the complete works of Max Scheler is now under way, and the major works of Maurice Merleau-Ponty are presently available in Japanese.

An enormous number of articles on phenomenology and phenomenological philosophy have been written during this quarter of the century. They have been not only written for academic circles, but also have been read by high school and college students, as well as by other adult intellectuals in Japan.

IV. ORIGINAL CONTRIBUTIONS INSPIRED BY PHENOMENOLOGY

How Nishida was influenced by Husserl has been discussed above. Nishida's later interest in history and his notion of the "historical world" is inconceivable without Heidegger's influence.

Tanabe's two early books, *Recent Natural Sciences* (1915) and *An Introduction to Science* (1916), were written from an intuitionistic point of view, which was influenced and inspired by Husserl's phenomenological intuition.

S. Kuki's *Sur le temps*, S. Takahashi's "A Study of Time" in *Experience and Being* (1936) and his *The Standpoint of the Whole* (1932), T. Hosoya's *Introduction to Phenomenology of Knowledge* (1936) and *The Methodology of Philosophy* (1937), and R. Mutai's *Expression and Logic* (1940) and *A Study of Phenomenology* (1940) were written under Husserl's influence and inspiration. Particularly, in the last chapter of the above-mentioned work, Hosoya attempted to phenomenologically reconstrue what and how Nishida philosophizes by means of Heidegger's hermeneutic structure.

Heidegger's strong influences are obvious in Masaaki Kosaka's *The Historical World – A Phenomenological Essay* (1937) and in T. Yamanouchi's *The Phenomenal Mode of Being* (1941). The application of Heidegger's hermeneutic method resulted in many fruitful phenomenological investigations: S. Kuki's many concrete phenomenological analyses in *Man and Existence* (1938) and *The Structure of Iki* (1930), a hermeneutic analysis of the authentic, uniquely East-Asian mode of human existence; K. Miki's *A Study of Man in Pascal* and *Logic of Imagination* are also invaluable additions to the wealth of phenomenological inquiry in Japan; Tetsuro Watsuji's *The Climate* (1943) is a hermeneutic phenomenological inquiry into human existence in the *horizon of space* rather than in that of time in Heidegger's case; G. Miyake's *The Formation of Science and the World of Nature* (1940) is an application of phenomenological method to the history of Western philosophy designed to explicate the historical development of the correlation between philosophy as the formation of a system of *scientia* and nature as the revelation of the world. Watsuji's *Ethics as Anthropology* may also be mentioned.

In the postwar period, G. Miyake's *The Ontology of Man* (1966), *Philosophy of Morality* (1971), *A Study of Arts* (1975), and *A Study of Time* (1976) have resulted from his lifelong phenomenological investigations. Through his lectures and seminars dealing with those themes and problems, Miyake inspired many students to phenomenology and phenomenological inquiry at Tohoku University, Kyoto University, and Gakushuin University. This is quite obvious from the fact that most of today's young, currently active, phenomenologists in Japan are his disciples.

K. Nishitani published *What is Religion* (1961) in an attempt to have a significant dialogue between East and West in his existence-philosophical inquiry into religious experience as the central problem of philosophy. Koichi Tsujimura's *Discourse on Heidegger* (1971) is a profound and extensive interpretation of Heidegger and includes his critical confrontation with Heidegger from the standpoint of Eastern and Zen philosophy. Yoshinori Takeuchi's *Shinran and Today* (1974) is a penetrating existence-philosophical explication of Shinran's religious experience and reveals the perennial significance of Shinran's philosophy of religion.

In recent years, a new breed of young phenomenologists have started to make contributions.

Yoshihiro Nitta's *What is Phenomenology* (1968) is the first and most comprehensive interpretation of the later development of Husserl's phenomenology and pursues the contemporary significance of phenomenological philosophy itself. Gen Kida's *Phenomenology* (1970) deals comprehensively with Husserl, Heidegger, Sartre, and Merleau-Ponty from his own point of view. Some of the crucial and central problems in phenomenology have been approached by Shizuo Takiura's *Phenomenology of Imagination* (1971) on the basis of Husserl and Sartre, by Yosuke Yamazaki's *Developments of Phenomenology* (1974), by Hiroshi Ichikawa's *Mind and Body* (1975), and by S. Takiura's *Time* (1976).

Minoru Uchiyama's *Das Wertwidrige in der Ethik Max Schelers* (1966) is an attempt to interpret Max Scheler's phenomenology of value experience with a special emphasis on the problem of negative values. Eiichi Shimomissē's *Die Phänomenologie und das Problem der Grundlegung der Ethik* (1971) is an explication of the significance and task of phenomenological inquiry and its particular application to the problem of the foundation of ethics on the basis of an intentional analysis of emotional experience.

The four volume *Lecture Series on Phenomenology* is planned and will be edited by Gen Kida, Shizuo Takiura, Hirotaka Tatematsu, and Yoshihiro Nitta for publication in 1978, dealing with the formation, the movements, the problems of phenomenology, and its relationship to other sciences. To this, the psychopathologist Bin Kumura, the logician Shozo Omori, the cultural anthropologist Masao Yamaguchi, the psychologist Hiroshi Minami, and Junzo Kobata in the field of aesthetics will contribute articles from their special fields.

Masao Yamaguchi's *The Ambiguity of Culture* deals with the later thoughts of Husserl, and in the field of linguistics, and that of aesthetics much research in phenomenology and its applications appears in the journals. In the field of sociology, Schütz' *Phenomenology and the Social Sciences* has been recently translated by Shozo Fukaya.

V. THE FUTURE DIRECTION OF PHENOMENOLOGICAL RESEARCH IN JAPAN

First of all, an attempt should be made to reinterpret those European phenomenological philosophies from a new perspective within the context of the history of Western philosophy and to critically examine the central issues of the Western tradition (e.g., concerning "appearance," "world," "reason," and "difference," etc., as treated in Husserl's phenomenology) in their primordiality. For the problem of the basic structure of knowing which has arisen from Husserl's phenomenology and Heidegger's ontology must be made transparent phenomenologically from the more radically fundamental, ultimate ground of its own. It may be possible for the Japanese phenomenologist to more freely pursue this fundamental task, since he has not been immersed in the Western tradition, but has remained at a "distance" from it in order to be able to unravel what has been implicit and concealed through the development of Western philosophy.

Thus, a critical confrontation should be made by the Japanese phenomenologists with the intellectual and cultural tradition of the West, for example, to the universality of European reason and to the intellectual responsibility of mankind as conceived by Western philosophy. The question of the ultimate ground of knowledge in Western philosophy cannot be separated from the phenomenological clarification of the East Asian wisdom or awakening today. Not by means of the external comparison between the two civilizations, but through a radical return to the primordial ground of philosophical knowledge should the East and the West meet.

Further, methodological inquiry should be made into the possibility of the dialogue between East and West by following up new developments in the hermeneutic method as a possible direction in the future development of Western philosophy.

Finally, Japanese philosophy today has not yet succeeded philosophically to the point intended by Nishida. It seems that philosophers have confined themselves to commentaries or external criticisms of Nishida's philosophy. But an authentic succession and acceptance of cultural and philosophical heritage can only be accomplished by asking the fundamental questions of the inherited philosophical thought and method in their primordiality. In other words, the object must be to phenomenologically investigate what Nishida experienced and developed in his own philosophy. By means of this attempt, Nishida's philosophy and experience will become open and visible to the non-Japanese philosopher. Thus we shall open a possible way of inviting the non-East Asian philosopher to have a dialogue and pursue a "group research" with the Japanese philosopher.

Such a most fruitful exchange has not yet been quite accomplished in philosophy. The painstaking efforts of *Analecta Husserliana* will, it is hoped, contribute an important breakthrough in more meaningful and more inspiring communal inquiry in the field of phenomenology and phenomenological philosophy.

Tokyo
Nagoya
California State College, Dominguez Hills

NOTES

*This introductory essay is a cooperative venture of Yoshihiro Nitta, Hirotaka Tatematsu and Eiichi Shimomissē. Shimomissē is alone responsible for rendering the whole essay into English. Due to the lack of time, it has not been possible for Tatematsu and Nitta to review the final draft. Therefore, if there are any errors Shimomissē will accept the blame for them. The Introduction and Parts I and II were written by Shimomissē. Parts III, IV, and V were written by Tatematsu and Nitta. We are most grateful to Professor Anna-Terresa Tymieniecka, Editor of *Analecta Husserliana*, for her incessant encouragement, invaluable assistance, and compassionate understanding and patience to bring about this volume.
[1] F. Jeanson, *La Phénoménologie*, Paris, 1952, p. 22. Also cf. J. F. Lyotard, *La Phénoménologie*, Paris, 1954, p. 5 and P. Thévenaz "Qu'est-ce que la Phénoménologie?" in *L'homme et sa raison*, Neuchâtel, 1956, II, 2.

² The Meiji Restoration (1867) was a political revolution and is to be distinguished from this spiritual one.

³ M. Kosaka, *Japanese Thought in the Meiji Era*, Tokyo, 1958, p. 198ff.

⁴ I. Koyama, *Nishida Tetsugaku*, Nishida's own preface to this work, p. 4.

⁵ T. Shimomura, *Nishida Kitaro*, Tokyo, p. 210.

⁶ K. Nishida, 'Ninshikiron ni okeru jun ronriha no shucho ni tsuite,' *Bungei*, 2, nos. 8–9 (Tokyo, 1911). Cf. Nishida's complete works, I, 209, 222.

⁷ Husserl, 'Gaku toshiteno tetsugaku,' *Tetsugakuzashi*, 30, Nos. 343–46 (1915).

⁸ Nishida, 'Gendai niokeru risoshugi no tetsugaku,' in his complete works, XIV, 59, 64ff.

⁹ Nishida, 'Gendai no tetsugaku.' Cf. Nishida's complete works, I, 356–60.

¹⁰ Nishida, *Jikaku niokeru hansei to chokkan*, in his complete works, II, 149, 152, 154.

¹¹ Ibid., p. 172, 152ff.

¹² Ibid., p. 157.

¹³ Ibid., p. 156.

¹⁴ Cf. Nishida, *From the Acting to Seeing*, in complete works, vol. 4. There exists an interpretation that this intuitionism of Nishida is developed under Husserl's influence. Cf. M. Noda, 'East–West Synthesis in K. Nishida,' *East–West* (1961).

¹⁵ Nishida, *Jikaku niokeru hansei to chokkan*, in his complete works, II, 199.

¹⁶ Ibid., p. 316.

¹⁷ Nishida, *Ippansha no jikakuteki taikei*, in his complete works, vol. 5.

¹⁸ Ibid., p. 129, 147ff, 427ff.

¹⁹ Ibid., p. 129; also cf. Nishida's complete works, IV, 337ff.

²⁰ Ibid., p. 349ff.

²¹ Ibid., p. 463.

²² "Reflection" in the title of his book refers to neo-Kantianism, whereas "Intuition" suggests the influence of Bergson's philosophy. Nishida moved to Tokyo from Kanazawa in 1910.

²³ Even today an attempt has been successfully made to construe Husserl's phenomenology as nonintuitionistic, more Kantian transcendental constitutionism. Cf. G. Funke, *Zur transzendentalen Phänomenologie*, 1957.

²⁴ When Kuki went to Paris from Freiburg and hired a philosophy student at Sorbonne as his French tutor, it was Sartre who applied for the job. Kuki taught Sartre Heidegger's philosophy of existence. This episode was confirmed by Heidegger during Shimomissē's visit with Heidegger on October 28, 1966. Heidegger remembered his talks with Sartre about Kuki, when Sartre visited Heidegger in Freiburg after World War II.

PART I

PRESENT DAY PHENOMENOLOGY IN JAPAN

YOSHIHIRO NITTA

HUSSERL'S MANUSCRIPT 'A NOCTURNAL CONVERSATION'

*His Phenomenology of Intersubjectivity**

I

Phenomenological research currently exhibits two tendencies: on the one hand, a criticism of the metaphysical presuppositions hidden in the philosophical aspect of Husserl's transcendental phenomenology; on the other hand, the interest in the new efforts of phenomenological research to broaden its horizons into a phenomenology of history and society. The inquiry into the possibility of reciprocal relations between the transcendental and the historical has probably been the most important task of phenomenology from Husserl's time to the present. Thus H. Rombach's structural phenomenology, for instance, could have grown out of an awareness of this task.

If phenomenology, beyond reflective analysis of the essence of the constituting consciousness of the individual subjectivity, is to be suited to clarifying the essence of co-subjectivity, it must treat the phenomena of "the experience of Otherness" and of "the world" as factual themes. But, as is known, what has been published thus far of Husserl's theory of the constituting of the Other, that is, the analysis of the intersubjectivity of monads, which is developed above all in the fifth of the *Cartesian Meditations*,[1] has found little response thus far since it has not been able to fulfill the expectations of the new phenomenology of intersubjectivity that have been held in all quarters. Husserl's doctrine of "intuition," especially, has encountered resistance and rejection.

"Intuition" indicates the experience in which I, in constituting the Other as a subject different from me, posit my I as the constitutive zero point and proceed first from the organism of the Other as a given body, in order, then to extend, consecutively, my I-capacity and the structural connection of the rest of my inner experiences. Thus intuition is the experience of a sort of self-presentifying self-transposing, thus the experience of projecting the manner of inward intuition of my subjective life into the interior of the external appearance of the Other as an externally given. At the outset it is doubtful whether this theory of intuition opens up a possibility of overcoming the solipsistic interpretation, as Husserl himself attempted. Insofar as the experience of otherness, then, is viewed as a kind of positing

Nitta / Tatematsu (eds.), Analecta Husserliana, Vol. VIII, 21–36. *All Rights Reserved.*
Copyright © 1978 *by D. Reidel Publishing Company, Dordrecht, Holland.*

of consciousness, this inquiry into the Other proceeds from this: that the Other is regarded as an object. But thereby an inquiry into the factualness of the coexistence of I and the Other is overlooked. At the least, there is no clarification of how this question relates to the genetic analysis of the late Husserl. Further, the theory of the transcendental constitution of the Other scarcely explains why the immediate personal relation between me and the Other, to be found in the personalistic attitude of the second volume of Husserl's *Ideen*, left no trace in the *Cartesian Meditations*.[2] And finally one must ask whether it is not the case that the difference and the connection between the transcendental experience of Otherness and the experience of Otherness in the natural orientation displays a complicated, broken structure that is not found in any other experience. After the works that were published earlier, Husserl's theory of the Other in its entirety had to remain unclear for a long time; even its basic intention needed further clarification.

It was not until 1973 that I. Kern, in three volumes, posthumously made manuscripts on the phenomenology of intersubjectivity generally available.[3] Here are extensive studies of the problem of the Other, written between 1905 and 1935; out of these, what appeared suitable for a publication was selected. The editor added careful text-critical annotations and arranged the texts skillfully according to periods and subjects. As far as the content is concerned, the first part contains historically highly interesting materials, such as, for instance, a presentation of the representative classical doctrine of the problem of the Other, a critique by Husserl of Lipps, which became for Husserl himself an occasion for concerning himself with the problem of the Other, and, above all, the doctrine – essential to the analysis of the experience of Otherness – of the difference and identity of organism and body as constitutive conditions for the experience of Otherness. The first part includes, further, analyses of the complex function of kinaesthesis and organic movement, the comparative analysis of various kinds of presentifying consciousness and of intuition, etc. The second part contains manuscripts on the constitution of the community of monads as expressions of transcendental intersubjectivity, as well as text on the systematic constitution of the various stages of community as personal relationship. The third part is especially significant. Here Husserl recognizes the descriptions presented in the *Cartesian Meditations* as inadequate, and tries to improve them and at the same time to form a systematic theory of phenomenology. In these manuscripts, which date for the most part from the years 1932 to

1935, Husserl departs from his previous theory of the Other in order to take a new view of the basis for a possibility of the constitution of the Other from the point of view of genetic phenomenology. This analysis, by Husserl, of the experience of the Other is based on so-called static analysis, as will be shown. By contrast, in the theory developed in the thirties, he holds, without a doubt. to the genetic-temporal analysis. This is probably the reason why the editor also includes in this part some of the manuscripts in the so-called C-group.

Now, how does Husserl try to solve, in genetic analysis, the above-mentioned basic *aporie* of his theory of the Other? Or, in case no solution is found, what form does his basic intention take on now, and what direction does he give to the problem of the constitution of the Other, insofar as he grasped it then? How much could the analysis of the "living present," as the condition for the possibility of the experience of the Other, contribute to the clarification of the problem of the Other? On the basis of Manuscripts 29–38 in the third part, above all 32 and 33, 'A Nocturnal Conversation,' these problems will be discussed in the following. For that purpose it seems necessary at the outset to refer back to Husserl's preparatory analysis of the constitution of the Other and to set forth its problematic aspects as well as its aspects that are to be evaluated positively.

II

According to Husserl, the analysis of the experience of the Other must begin with a screening off, as a kind of methodological preliminary step. This preparatory method, also called "thematic *epoche*," consists of abstracting from what is unnecessary at the outset when one is thematizing something. The *epoche,* understood thus, is grasped in the *Cartesian Meditations* as a "reduction to the primordial sphere" or "primordial reduction." How is this "primordial reduction" distinct from the Cartesian reduction in the traditional sense?

In the Cartesian reduction, everything that is thought in the natural attitude is derived from phenomena of consciousness, therefore, from the noematic sense. If, in this process, what is meant by the Other in the natural sense is also derived from the phenomenon "the Other one," the phenomenon of the Other One is viewed, to be sure, as a phenomenon, but basically no longer as a constituting subject. The primordial reduction, on the other hand, is the necessary accomplishment for securing a sphere that will be useful as a starting point for the theory of intersubjectivity. Here only the sphere belonging to the I is used thematically, and thereby the effect of the Other on the sphere of my own experience is put aside. As long as – in the Cartesian reduction as in the primordial reduction – the reciprocal relation between the constituted object and the constituting achievement is used thematically, it is impossible to differentiate them. But because the primordial reduction does not try to be anything but a method for constituting intersubjectivity, the limitation it initially imposes is gradually relaxed, even eliminated, in the process of constitutive analysis, and thus the primordial sphere is transcended. Therefore it would be erroneous to assume that the phenomenological method regards only the individual subject as the theme of its analysis and that it proceeds straight from that base. Rather, the primordial sphere that is proper to the I is first of all thematized, in order, precisely in this way, to clarify the structure of the relation to the Other, already presupposed as an event, and also the structure of founded experience, namely the experience of the Other. Thereby primordial reduction is essentially distinguished from Cartesian reduction, both in its goal and in its method. Yet it seems that it was not yet clear to Husserl, when he began the *Cartesian Meditations*, that transcendental subjectivity as such is intersubjectivity.[4] It was just for this reason that he strove to make clear, by critical corrections of the *Cartesian Meditations*, that egological reduction is basically only a methodologically intended step toward intersubjectivity. Thus the analysis of the structure of the experience of the Other, which begins with this primordial reduction, is indeed the clarification of the "already" always existing relation to the Other; and yet it is not able to accomplish the explanation of the characters of the "Already" as such. In other words, the analysis cited is the structural analysis of the experience of the Other, but it cannot inquire into the condition for the possibility of this experience. In this way Husserl left genetic analysis with regard to the problem of the Other undone until shortly before the end of his work. It was only through analysis of the "living present" that he finally came to be able to assign to the problem of the Other its proper place.

The world that appears in the primordial sphere is corporeal nature, thus the phenomenal space made accessible by perception, or the surrounding world that allows my organism to come about as the "absolute here," the zero-point of orientation with its surroundings. According to Husserl, the Other appears "there" in this surrounding world as a body. The constitutive analysis of the experience of the Other begins when that experience is thematized and constitutes this body as that of the

Other, namely as an organism that is just as much an organism as my own. Husserl calls the experience of Otherness on this lowest level "analogizing apperception." But it is precisely here, on the level of the constitution of the organism of the Other, that the constitution of the Other encounters considerable difficulties.

III

The decisive difficulty lies in the question of how I can intuitively "translate" my consciousness from my organism, which is internally perceptible only to me, into a body as the body of the Other One. Before we pursue this question, however, the intention by which Husserl let himself be led in the constitution of the transcendental Other, must be clearly worked out, especially in view of the often exercised criticism that the experience of the Other in Husserl's sense is unsuitable for the grounding of the relation to the Other as a personal relation of "I and You." Whether Husserl's analysis of the constitution of the Other can ground a personal relation to the Other, is, to be sure, a meaningful question. But it is necessary to dinstinguish, from that question, the question of whether Husserl also intended to ground this relation, and this must be held in doubt. In Husserl's own words, the transcendental analysis of the Other was undertaken, instead, for the sake of the "constitution of the objective world." On the basis of this motivation, the transcendental analysis of the Other or of the experience of him must be the transcendental analysis of the transcendental Other or of the experience of him. In order to guarantee the constitution of the plurality of subjectivities necessary to the "constitution of the objective world," one must first ask how the Other could be constituted in his primordial world, which is basically like my own.

In the *Cartesian Meditations* the phenomenology of intersubjectivity is developed in five steps: 1. The constitution of the body as organism (analogizing apperception); 2. The experience of the pure alter ego (empathy); 3. The constitution of the objective world through identification of the primordial worlds; 4. The constitution of the Other as an object ('person'); 5. The extension of the objective meaning 'person' to myself (communalization). To solve the problem of how the objectivity of the objective world can exist, Husserl's phenomenology of intersubjectivity deals with the constitution of the community of subjects which is said to ground such objectivity, and precisely for that purpose it is essential for it to constitute the Other as subject also.

It should become clear from this goal-setting why Husserl concerns himself so relentlessly with the clarification of the "founding structure." If the subjective is what is immediate to me, then what is subjective for the Other is what is mediate to me. But sometimes that which in various ways appears as the immediate reveals itself as mediate, looked at from the point of view of the "founding structure." If that is the case, then the place where the most fundamental relation between immediate and mediate exists lies in the dimension of the basis of that structure of relation. It is here that one may first inquire into the relation itself between the mediate, that cannot be brought nearer to immediacy, and the immediate itself. The place where the mediacy of the Other as a subject that is not the subject of the I, can be established as mediacy, is, according to Husserl, the sphere of the immediately given, thus the sphere in which the body occurs. Husserl's goal intention, that it is only in the starting point of this sphere that it is possible to have a constitution of the Other aiming at the Otherness of the Other, must be seriously considered. Certainly there can be the experience of an immediate personal relation between me and another person. Husserl himself assumed that such a personal relation was grounded in the personalistic attitude, as is shown by the manuscripts of his studies of what could be called "various forms of community." Thus in Nos. 9ff. of the second part, for instance, various personal relations are dealt with, even though these are not grasped as radical personal relation in the sense proposed by dialogue philosophy. But the reason for the fact that an immediate personal relation of this sort is not applied to laying a foundation for the constitution of the transcendental Other, is not simply that Husserl carried out transcendental analysis, with a naturalistic attitude, on dead nature as a prototype, instead of understanding it from a naturalistic point of view. The central problem in Husserl's constitution of the transcendental Other is, instead, the undismissable necessity of having to constitute another subject that can never be my subject. This "never" is the first presupposition, since Husserl regards the "constitution of the objective world" as the real goal of the problematic of the Other.

Before we turn to this problematic, it is necessary to clarify the reason for that general criticism of Husserl which can be formulated as follows: "Why was living, natural, personal experience not analyzed phenomenologically although such experience, which is essentially different from experience of a physical thing, must necessarily be thought of concomitantly with the experience of something else?" Husserl himself is partly to blame for this criticism, since his descriptions in the *Cartesian*

Meditations lack the necessary clarity in some passages. One encounters some descriptions in which it seems that Husserl is carrying on phenomenological analysis of the constitution of the Other on the basis of the natural attitude. Kern traces this lack of clarity to an ambiguity, namely to the difference between the "First in the order for the foundation of philosophical reflection" – the realm of the respective monad – and the "First in the order of motivation of the natural attitude" – that is, the foundation of the motivation of the natural experience of something else.[5] Husserl failed to see the difference between the two motives probably because he thought that the Other could be attained as a transcendental Other without methodological objections by means of a transcendental analysis of the natural experience of something else. After he gradually became aware of this ambiguity, he held that the above described primordial reduction was the appropriate method for penetrating to the essence of intersubjectivity. Thereby he established again that his philosophical intention of a constitution of the transcendental Other could only be realized in the constitution of a transcendental monadology.

Let us return to the question of how much the constitution of the transcendental Other is aimed at the "constitution of the objective world" and why my subject does not suffice for that and why another subject has to be constituted. The fact that the essence of the subject has to be intersubjectivity as such, shows that the one world is necessarily accessible to several poles or points of view, and that therefore the one and the same world is not given except through the multiplicity and multipolarity of appearance. And it is precisely therein that appearance first evinces itself as appearance, as, correspondingly, also identity can be identity only through appearance. This insight follows necessarily from the unfolding of the original theme of phenomenology, insofar as phenomenology systematically analyses what exists into "appearance." Intentionality is the concept that discloses the manner of functioning of this system of appearance. To simplify greatly, phenomenology reduces what exists, and man's activity related to it, to the "phenomenon" of transcendental consciousness, in order in this way to be able to observe, thematically, what exists as a whole, as a system of appearances. Husserl's monadology undertakes only to hold fast to appearance precisely as appearance and to seek out its objectivity through the aspect of "appearance." Here one sees the effort to ground that "doctrine of appearance" and a philosophical position which modern philosophy had basically been striving toward.

IV

Even if Husserl intended the constitution of the transcendental Other, in the sense described, one must still ask whether, with this method, identity can really be constituted as identity. Does this method open up the horizon in which "appearance" is multipolarized? The substantiation of Husserl's theory of the constitution of the Other depends precisely on the answer to this question. In order to be able to establish appearance as appearance, the identical must be insured as given, throughout its various appearances. For that purpose it is not sufficient to establish that the various appearances are, for me as an individual subject, appearances of one and the same body. To be able to determine that the appearance given to me here and now is the appearance of something, I need another appearance of the same thing. By moving myself, my organism, from this place here to that place there, I can perceive another appearance. In the process, the appearance is given as appearance to my organism, and what makes its change of place possible is its kinaesthetic ability, its "I-can." My organism, as the point of reference of the appearance, is in itself the "zero appearance," and as the absolute center of my surrounding world it is the "zero point." Even when I move my organism from here to there, only the zero point moves. Husserl now inquires after the structure of consciousness in the instance that the I, while it stays at its place here, imagines that it is at a different place, there. In doing so, he tries to single out a kind of consciousness that duplicates and presentifies the I. What makes this consciousness possible is the doubleness of organism and body. The I is conscious of its organism both in feeling from within and as a corporeal object from without. In reality these two ways of consciousness coupled together form the double consciousness that the I has of its organism. The presentifying consciousness, which intends to double the I, consists precisely of the operation which, in objectivating memory or phantasy, separates and polarizes the two different ways of consciousness on the ground of the doubleness of organism and body, that is in a kind of experimental objectivating intention. I here can imagine that I am standing there or that I, standing there, am seeing myself standing here. Thereby I can assume that another appearance of a table, seen, namely, from another corner, would be given to me there than what is given to me here.

Husserl concentrates on the analysis of this presentifying consciousness, that is, the consciousness "if I were there," approximately in the years from 1915 to 1920. According to Husserl the possibility of the presentifying cited is, to be sure, "in reality refused," but nevertheless "not an empty possibility," but a motivated one. The doubling of the I and the here-there-difference from this time on form the framework of Husserl's conception of the Other. Thus he is convinced of the possibility of an analogizing apperception, that, corresponding to the doubling of the I, constitutes the organism of the Other out of its body which appears there, by a process in which the I extends its regard to its own organism to the body over there. "In order to gain the possibility of the experience, the external appearance of an other I, I clearly do not need real experience of such a one. It suffices for me to think of myself as corporeally moved-beyond, transposed-beyond, and to think of my corporeal appearance as transferred into an external appearance and at the same time in the original appearance, the self-appearance of my body, apperceiving it, therefore, as an organism with its sensitivities, etc."[6] Or, in another place, "Thus it lies in the essence of self-perception, the true apperception of my body, of my body as organism, of the unity of my I with the stream of consciousness . . . , that I can gain a possible idea of an Other before the actual experience of an other subject."[7]

The theory, set forth in the *Cartesian Meditations*, of analogizing apperception or of empathy, basically does not go beyond this framework. Here, to be sure, the passive function of "pairing" is introduced. But the introduction of such a genetic function into static analysis already signifies a confusion, and, beyond that, pairing, as Theunissen also says, because of its particular way of functioning, is not able to accomplish the union of body and organism.[8] Husserl assumes, further, that the identity of the primordial world of the I and of the primordial world of the Other can be grounded on the co-operation of presence and a-presence, specifically on the basis of the identity of the body that appears in my primordial sphere as the organism of the Other, with the body from which the Other constitutes his own organism. The analysis that makes up the core of the fifth chapter of the *Cartesian Meditations* repeats the thesis of the doubling of the I and thus continues to move within the same principle framework. Is the alter ego, with Husserl, therefore basically nothing but the disguised ego itself, as it is constituted in presentifying consciousness?

V

The Other, like me, must be an I. Because he is an I, he cannot be a product of constitution and he cannot be part of the sphere of my I or be dependent on it. The reason that the alter ego in my original sphere can be intended only in original mediacy, lies in the I-ness of the alter ego. But if the alter ego, because of this mediacy, were only apparently constituted, if it were namely a disguised ego, then it would lose its I-ness. The mediate givenness of the alter ego in my original sphere is grounded in the original coexistence of me and the Other. If, however, as Husserl intends, the alter ego were only mediately given, and only in the mode of presentifying, then the original coexistence of ego and alter ego could be established only by means of an equally original presentifying. If, further, the I is the absolute functioning presence, the "I function," the presentifying of the alter ego as a functioning I, if it is possible at all, can be achieved, but never in the manner of a presentifying constitution. Thereby one must ask whether the possibility of a non-objectifying, equally original presentifying could be found within the framework of Husserl's analysis.

With regard to this question, let us once more pursue Husserl's theory of intersubjectivity. The fact that my ego constitutes the alter ego as an ego, means at the same time that, vice versa, my ego is constituted by the ego of the Other. If the reciprocity of this constitutive function is granted, then it leads to the constitution of a plurality of subjectivities, thus to a community of monads. Husserl's first utterances about a coexistence that rest on such a "reciprocal constitution" go back approximately to the time when he wrote the *Cartesian Meditations*. To constitute one another reciprocally implies: always to have been with each other. Yet one may ask whether the reciprocal constitution, as it is conceived by Husserl, really can be characterized as equally original reciprocal constitution. Here we have to take into consideration an ambivalence in Husserl's concept of the transcendental community of monads. On the one hand it is maintained that the transcendental monad community itself is the totality of the absolute, for instance when it is said: "This inwardness of being-for-one-another as an intentional being-in-one-another is the 'metaphysical' original fact, it is an in-one-another of the Absolute."[9] The "We" is there-fore the absolute being-with-one-another, and the coexistence consisting of this being-in-one-another is "the being-for-one-another" which makes the assertion of an absolute 'Alone' senseless. When Husserl calls the totality of the transcendental monads the "Absolute," the plurality of transcendental subjectivities seems to be the final absolute fact. But on the other hand Husserl maintains that the plurality of subjectivities is constituted by the phenomenologizing Original-I as the "absolutely functioning I," and that therefore the community of monads arises originally from the pluralization of this single I. Accordingly, the community of monads appears at one time as a community of functioning I's, and at another time as the community constituted by the functioning I. Can this paradoxical state of affairs be further interpreted?

Here we must once more remember the goal of Husserl's phenomenology of transcendental intersubjectivity. It was to be found in the constitution of the "objective world." Since the constitution of the community out of the reciprocal constitution therefore means the transcendental constitution of the plurality of subjectivities that constitute the world, this reciprocal constitution does not necessarily imply an equally original reciprocal constitution. According to the intention of Husserl's analysis, I do not transform myself in the dimension of reciprocal constitution as an "absolutely functioning subject," but instead, as "one among the Others," into an Everyman. The fact that through the Other I become an Other over-against him, is called by Theunissen the "Othering." By becoming a member of the "We," through this "othering," I come to stand next to the Other, I become equal to him. If thereby, through my "othering," I am liberated from the "absolute Here," the zero point of appearance-space, then the "Here-There-relation" becomes irrelevant, since Here and There become interchangeable. On the ground of this decentralization and deperspectivizing, the Here-There-difference is nullified, as Theunissen says. If, then, the Subject, in a certain sense, is "everywhere," then its pluralization is accomplished and the communalization of subjects is made possible. To be sure, in this constitution of the community the Here-There-relation and -difference, which established the zero point of constitution, is not simply totally nullified but only, so to say, forgotten. Now, to be sure, Husserl does not formulate this constitution of the community of monads as distinctly as we have just attempted to do it. Rather, he vacillates in the determination of the character of his monadology. But if, for the constitution of the objectivity of the objective world, it is necessary to have the transcendental constitution of a plurality of subjectivities, then the constitution of the community of monads must mean the transcendental constitution of the transcendental condition for the constitution of the world. In this sense, before the constitution of the world, a multiplicity of pre-world subjects must always have been constituted. But these subjects do not correspond to the personal I which, as Husserl had said before, made itself worldly by means of the self-apperception of the transcendental I, but rather, as Theunissen indicates, they correspond to the I that was constituted by means of the Other-ing that is possible only inside the transcendental realm. Theunissen's interpretation, according to which the theory of transcendental intersubjectivity is "a transcendental grounding of the natural attitude," can probably also be accepted from the point of view of the philosophical grounding of the "doctrine of appearances," which was touched on above.

V I

The possibility of pre-worldly pluralization of the transcendental I is grounded in the temporalizing of the unique I. The manuscripts between 1932 and 1935 range around the theme of the absolutely functioning I and the problem of the pluralization of uniqueness.

Husserl's phenomenological reflection in these manuscripts searches for a key to primoridal time which, as enduring primordial aliveness, is carried out at the lowest level of subjectivity. At this level, "flowing" itself is thematized(?) as the mode of being of the I. When the "absolutely functioning" I is unveiled as a "continually flowing present," phenomenological analysis of intention goes into its last stage. When intuition reflects on the original flowing being, "the overcoming of naiveté"[10] comes about. "Flowing is primordial maturation, and transcendental analysis is unfolding of its implication."[11] The transcendental analysis meant here is nothing other than genetic analysis.

The absolutely functioning I is, as "enduring present," sheerly pre-temporal and absolutely unique; thus it allows no modification. As flowing present, however, this ego pluralizes itself precisely in its flowing. Thereby the temporalizing of this unique I results in its pluralizing. Husserl sees in the pluralizing of the I on the ground of temporalizing, its "monadizing" or self-alienation. "Self-alienation" means for Husserl actually the process in which the I, which is in itself unique, pluralizes itself, losing its uniqueness, and thus departs from absolute aliveness and removes itself into one place in time. Self-alienation, understood in this way, is called "monadizing." "The absolute ego constitutes in itself 'I myself and other I'."[12] According to this doctrine of the pluralizing of the unique I, the temporal self-constitution of the primordial I already includes in itself the I of the Other. That implies at the same time that the primordial I constitutes the I of the Other as Subject of constitution. Thereby the uniqueness of the primordial I excludes a uniqueness of the I of the Other, that is, the equal originality of the Other. In this way the pluralization of the unique I can be traced back to the pluralization of my I. Because flowing, in Husserl's sense, means streaming-out from the present, whereby the past represents a modification of the present, pluralization must take place in a direction toward the past. This relation to the past forms the basis for the thought that makes possible the plurality of subjectivities: namely, that the alter ego is a modification of the I. For, by means of the power with which "I move myself back into the Back-then," it becomes possible for me "to move over there."[13]

VII

When Husserl calls the totality of monads the Absolute or the Inter-penetration of the Absolute, as was mentioned above, this evinces the undeviating certainty with which he asserts the communality of the absolutely functioning I. Nevertheless he did not succeed in guaranteeing this assertion judgematically. Husserl was unable to accomplish this grounding because he neglected to inquire into the relation between the factualness of monads and the factualness of the functioning of the I.[14] Is there no longer a possibility of penetrating – by way of phenomenological analysis – to that equally primordial coexistence of monad-like subjectivities that Husserl himself had recognized? Using the time doctrine of the late Husserl, some systematic points can be made about this probelmatic.

(1) First of all, the "living present" as such possesses its uniqueness only in a plurality of I's. The plurality of I's, namely, results pre-reflectively, already before its temporalizing; therefore it is to be distinguished from the pluralization after temporalization. The latter becomes possible only by means of that pre-reflective plurality. Precisely this splitting and identity of the I before its temporalization now causes the "living presence" to come about. At the same time, precisely the pretemporality of the pluraliza-tion of the I, means that the monads are already primordially individuated.

(2) Consequently the structure of this primordial relationship can be analyzed, to be sure, in a reflection on a higher level. But since the "living presence" takes place only in such a reflection as this, its liveliness cannot be grasped again by this reflection. Thus primordial anonymity arises with reflection, but the latter cannot comprehend it. The pre-reflective functioning of the I is now the original self-consciousness insofar as this is pre-reflective consciousness. The pre-reflectiveness of self-conscious-

ness, which can never be grasped by positional reflection, and which first makes positional reflection possible, means at the same time that also the consciousness of the functioning of the alter ego is pre-constitutive and unthematic. The original mediacy of the alter ego, therefore, can also be considered radically clarified when it is proven how the I, in pre-reflective consciousness, becomes aware that the commanding of the alter ego functions together with it.

(3) Finally the flowing of time should not be regarded, as Husserl did, only in its direction toward the past, but should also be comprehended as that which is opened toward the future. For "in the destruction of the future by means of the unlimited domain of the past"[15] the transcendence of the Other is overpowered by my immanence. Husserl's protention also takes on, in accordance with the pre-eminence of retention, the character of anticipatory seizing-away; but this encroaching shuts out everything new and unexpected only in order to allow the I to come into relation to what is already presaged. Held rightly, it calls attention to the fact that in continually present transition not only the point of going-away but also that of arriving is to be found. For the Other is not a repetition of my past.[16]

The aporie in Husserl's analysis of the experience of Otherness arises from the fact that he proceeds from the thematic experience of Otherness when he interprets the experience of Otherness as a positional experience in which the Other comes to be constituted at all only by means of his objectivization. As Held says, Husserl's founding relation must be turned around.[17] In this way it can be revealed how the I, in consciousness of the present, is passively familiar with the anonymously co-functioning inter-subject already before the occurrence of the experience of Otherness. The Other is therefore not to be thematized as a wordly subject, but is always to be perceived, appresentatively-passively, as a function that remains unthematic, specifically as the other subject that I am not.

The phenomenology of the present emphasized the analysis of things themselves and, pressed by the thing itself, sought an appropriate method and thereby continually opened new horizons. Today phenomenology is faced with the task of freeing Husserl's theory of intersubjectivity from being wrapped up in itself. One of the reasons why modern philosophy has again and again gotten into the dead-end street of "philosophy of reflection" probably lies in the fact that it has interpreted self-consciousness according to the "model of reflection" and has absolutized this model. If self-consciousness can be freed from this model of reflection, then at the same time the experience of Otherness will be freed from the influence of this model. But it is necessary that reflection fail (run aground) first, in order to be able to open the way into the openness of the experience of the Other. Husserl's investigations into the origin of the experience of Otherness show intensively in their continually renewed starts, corrections and experiments, how hard he struggled to take this step.

University of Toyo

NOTES

*Translated from the German by Barbara Haupt Mohr.
[1] Cf. *Husserliana*, Vol. I. pp. 121 176.
[2] *Husserliana*, Vol. IV.
[3] 'On the Phenomenology of Intersubjectivity': *Husserliana*, Vols. XIII, XIV, XV.
[4] Cf. *Husserliana*, Vol. XV, Editor's Introduction, p. xxxiii.
[5] According to Kern, not only did this mistake lead to the confusion of the reader, "but thereby the whole thought sequence of the fifth Meditation proved to be ambivalent: Is it a question of the reflective-philosophical founding (grounding) of the transcendentally strange (other) and of the transcendental relation of one's own and other monads or of the constitutive analysis of the founding (motivation) of 'natural,' 'worldly' empathy?" See also pp. xix-xx.
[6] *Husserliana*, Vol. XIII, p. 253
[7] Ibid., p. 265.
[8] M. Theunissen, *The Other*, p. 62ff.
[9] *Husserliana*, Vol. XV, p. 366.
[10] Ibid., p. 585.
[11] Ibid., p. 584.
[12] Ibid., p. 640.
[13] Cf. M. Theunissen, see p. 149.
[14] In a more recent paper Landgrebe poses the question: "In how far can the latter (the primordial flow), in its originality, not be thought of as a diffuse flow which itself first constitutes particularly and individuation?" As Landgrebe continues, this question could not be answered by Husserl's transcendental reflection. According to him, the inquiry into the connection between I-activity as a teleological will that carries out the individuation of monads, and temporalization is the final and most difficult problem of Husserl's phenomenology. Husserl, he asserts, could not finally have considered the primordial I to be the factualness of monads, because the absolute factualness of the "I-think" is the limit of reflection, and only within this limit could reflection remain what it is. If the significance of this absolute factualness is to be grasped, he says, it is necessary to go from the method of reflection to interpretation. This indication on Landgrebe's part, of the limitation of Husserl's method simultaneously exposes Husserl's mistaken presupposition that the anonymous I-function that arises as "living present" can be totally illuminated. L. Landgrebe, 'Factualness and Individuation,' in *Being and Historicity*, p. 279.
[15] M. Theunissen, op cit., p. 151.

[16] K. Held remarks: "The thesis, stemming from dialogue philosophy, that the Other is the original future, can in this sense, but only in this one, become a constitutive part of a phenomenological transcendental philosophy" ("The Problem of Intersubjectivity and the Idea of a Phenomenological Transcendental Philosophy," in *Perspectives in Transcendental-phenomenological Research* (Phenomenologica 49), p. 59f.).

[17] "The reversal of the founding relation of thematicalness and unthematicalness in the experience of the other, begun by Husserl, is not only a methodological device for avoiding the destruction of the theory of intersubjectivity" (K. Held, see also p. 47).

KAZUHISA MIZUNO

THE PARADOX OF THE PHENOMENOLOGICAL METHOD

The phenomenological paradox,[1] which Husserl expounds in his *Crisis*, poses a methodological circle of transcendental subjectivity over against the world, and of incarnated subjectivity; on the one side of the circle we find the transcendental subjectivity as an absolute point of departure which can only be established by suspending the acceptance as being of the world; and on the other side of the circle the psychophysical subjectivity that presupposes the acceptance as being of the world. It is characteristic of the circle, that the constituting transcendental subjectivity is outside of the world, whereas the constituted subjectivity is in the world, and that—nevertheless—these two, as subjectivity are identical. In order to solve this paradox we must regard one of the two subjectivities as necessarily derived from the other. According to Husserl, to philosophize means to search for the act that is the absolute starting point and to derive all knowledge from that. For Husserl, transcendental consciousness is and remains that starting point. Therefore in order to resolve the paradox of the two subjectivities – the one outside us and the other in the world – we must follow the route of deriving incarnated subjectivity from transcendental subjectivity. The key to that is intersubjective reduction. For it is only through this reduction that the pre-given life-world can be philosophically grounded. After grounding this life-world, phenomenology turns to genetic investigation of the cultural, intellectual world. But grounding the life-world brings to light another paradox which is contained in intuition of an essence, one of the most radical conditions for establishing phenomenology. This paradoxical difficulty cannot be solved within the framework of Husserl's phenomenology. It is necessary to go beyond it. In order to do that, it is requisite that one takes into consideration Merleau-Ponty's development of the phenomenological movement. The attempt to solve this paradoxical difficulty offers a favorable opportunity to reflect on the philosophical situation in which phenomenology finds itself.

The following essay consists of three parts:

(1) What is the idea and method of Husserl's phenomenology?

(2) How can the above-described phenomenological paradox be solved with regard to idea and method? Finally

Nitta / Tatematsu (eds), Analecta Husserliana, Vol. VIII, 37–53. All Rights Reserved.
Copyright © 1978 by D. Reidel Publishing Company, Dordrecht, Holland.

(3) Observations on the paradoxical nature of the evidence, which remains a task for phenomenology even after the solution of the paradox to be dealt with in the second part.

I

The uniqueness of the modern way of thinking, according to Husserl, is the mathematical interpretation of the world,[2] which was accepted at the beginning of the modern era as a postulate, and whose method of idealization then became confused with real being itself. In thus confusing method with being, one forgets the immediately experienced life-world and also the sense-bestowing act of ideating itself, which transcends from the immediately experienced life-world into the objective world. This forgetting takes one to the unquestioned matter-of-course conviction of the objectivist attitude and at the same time to the hiddenness of the life-world's unquestionedness, by wrapping raw being in a "cloak of ideas." The objectivist attitude, for instance, takes oriented space as an incomplete index of geometrical space and measures internal time against an objective time scale.

In pointing this out, Husserl tries not to attack scientific research, but to warn philosophically that objectivist matter-of-course conviction derives from the forgetfulness of its own ground or origin. This warning is directed against the two objectivistically produced phases: against the founding of all of science on positivistic sciences and against the naturalizing of pure consciousness with measurable unities. Husserl confronts the first of these phases with the rehabilitation of the science of essences, and he confronts the second phase with the transcendental interpretation of consciousness. This confrontation of Husserl is done with the intention of basing all of science on the first Philosophy. In order to carry out his intention in the objectivist situation of the modern sciences, it is necessary, before transcendental egology, first of all to carry out the reduction of the sciences that goes back into the life-world by tearing off the objectivist "idea cloak," making explicit the forgetfulness hidden behind scientific unquestionedness, and by bringing the thought back into predicative experience as presupposition of objectivism.

Leibniz' *mathesis universalis* is not valid as a norm for the sciences: the exact system of modern formal logic, which derives from Leibniz' ideas does not qualify – owing to its philosophical naiveté – as a fundamental science for objective sciences.[3] Therefore the genetic presupposition of modern logic should be shown even more clearly. The guiding idea of phenomenology does not include the physicalistic idea, which comes up in the application of mathematical logic to the conception of all of being.[4]

Descartes had already put forth a new idea of the *mathesis universalis*. But after his emigration to Holland he was gripped by another thought: that the system of formal rationality requires a philosophical basis. What governed his meditation was "a previously unheard-of radicalism."[5] For Descartes, philosophical knowledge is the knowledge that has an absolute foundation. Husserl, to be sure, rejected Descartes' physicalistic realism, but held its transcendental direction in high regard. Like Descartes, he too sought an Archimedean Point.[6] Consequently it is of decisive and ultimate significance for Husserl's phenomenology to try to discover a beginning that has no presuppositions and to undertake, as a methodical beginning, an "overthrow,"[7] not only of habitual prejudices but also of all existing sciences. For all of knowledge, including the distinction between *doxa* and *epistémé,* fails to establish itself as truth as long as the idea of phenomenology is not realized as a foundation resting on the absolute beginning.

As is well known, the most important method for realizing the ultimate idea of phenomenology is to exclude the general thesis of the natural attitude. But phenomenological reduction has two steps, as mentioned above: an egological one and a life-world reduction. In his late works, Husserl fears a possible misunderstanding of the Cartesian way, which attempts to grasp transcendental consciousness through the *epoché* of the general thesis of the natural attitude;[8] and therefore he tries to make a detour.[9] The reduction in *Ideas I* has the following steps: the first step consists of excluding transcendent things and personal, bodily human existence; the second step, "an extension of the original reduction,"[10] that is carried out in the first step, consists of excluding value, the realm of practice, God, formal sciences of being, and all material sciences of essence (with the exception of transcendental phenomenology). On the other hand, in his work *Experience and Judgment,* Husserl indicates the following reduction steps: the first reduction is the one that goes back, from the objectivistic world view posited by the modern way of thinking to the life-world as the genetic presupposition of the objectivist world view; the second is the reduction to transcendental subjectivity, which possesses the life-world as its own constitutive component.[11] The reduction treated in the *Crisis* proceeds in this series: pre-given life-world, I-pole, and intersubjectivity.[12] The reduction in *Ideas I* is Cartesian. Descartes first rejected sensuous existence on the basis of a doubt as to the criterion for distinguishing between dream and reality, and, because of the possibility of the existence of the evil spirit *(malin génie)*, he did not consider intelligible essence to be a criterion of truth. Descartes established the evidence of the Cogito as the first principle only because of doubt as to the existence and essence of the external world. The evidence of the existence of the Cogito ranks higher than that of the essence of the external world, and therefore it is not formal, general science that forms the basis of knowledge, but instead egology. Husserl's work *Ideas I* lays a path to Cartesian egology, if one disregards the difference between "reduction" and "doubt." But *Experience and Judgment* takes the route of the inter-subjective reduction. *Cartesian Meditations* and *Crisis* also introduce intersubjective reduction. How can the contrast between the route of *Ideas I* and that of the other works be interpreted? That is to say, wherein lies the difference between the way of egological reduction from existence over the essence of the external world to the apodictic evidence of the Cogito, on the one hand, and the way of the life-world reduction from the scientific *epistéme* to the surrounding-world doxa? It does not mean, by any means, that the philosophical beginning is no longer to be found in the Cogito, but instead in the life-world. For, also in the *Crisis*, which deals with the life-world reduction, the life-world is nothing but "the mere transcendental phenomenon," namely a constitutive component of the transcendental subjectivity,[13] and the final residuum is nothing other than the transcendental consciousness and its correlative.[14] As long as the first task of a person who philosophizes consists of seeking the absolute beginning, his investigation cannot avoid proceeding from his own Ego itself, as Husserl says in *First Philosophy II*: "as a phenomenologist, I am necessarily a solipsist."[15] Phenomenology, if it is not a transcendental egology, does not, in his view, deserve the name of phenomenology as a radical science.

II

Husserl's theory of intersubjective reduction is understood in general as a doctrine whose purpose is to liberate oneself from the solipsism of egology. But such an understanding of Husserl's monadology, in which intersubjective reduction is accomplished, is superficial and inadequate. For the paradoxical connection of transcendental subjectivity to real subjectivity cannot be resolved with such external comprehension. In order to resolve it, Husserl created the constitution doctrine of inter-subjectivity which, without ignoring egology, leads to the overcoming of solipsism.[16] It is transcendental subjectivity, which constitutes my concrete Ego, the other Ego, and the intersubjective world. For it is none other than the phenomenologically absolute beginning. The transcendental Ego always operates in the following stages of the constitution doctrine of the world of monads: The first stage, at which the transcendental Ego is apperceived as a concrete Ego, as an original monad, is realized by means of a "thematic suspension" and a "self-apperception that makes it worldly."[17] Thus the transcendental Ego apperceives itself as a psychophysical subject that can move its own body in its environment. But one must not assume that a concrete Ego does not begin to exist until this apperception takes place. Constituting is not a creation of being. Constituting means, above all, to form a configuration around an objective core, that is "to form," "to give meaning," "to ensoul." The second stage, at which the original monad constitutes other monads (other concrete Egos), is realized by means of the "analogizing appresentation" of my concrete Ego.[18] This apperception means: to discern, as present, another psychic Ego in a pairing with the presently given in my primordial world as another corporeal body. In this sense another Ego is a constitutive modification of my Ego. But the other Ego, as a Doppelgänger of my Ego, comes to light temporally no later than the constitution of my concrete Ego. The third stage is a communalization of the original monad itself with other monads, which the former constitutes in itself.[19] With this communalization there is constituted a horizon which concrete Ego's possess in common with each other. The horizon is a common world which is fixed in the harmonious experience of my Ego. The intersubjective constituted world remains an already given world because its constitution is not a creation out of nothing. The common horizon is experienced, in everyday experience, as the life-world, since the normal concrete Ego does not by any means doubt the world's existential validity. So the life-world is protodoxic for the everydayness of the natural attitude. But the unquestioned obviousness of this immediately experienced world first derives its philosophical foundation from non-everyday reflection, i.e., from the phenomenological attitude. For the understanding of being of the natural attitude cannot escape naiveté. It is only by bracketing this naiveté that phenomenological reflection reaches the beginning of the transcendental Ego and returns, via the constitutive route, to the original immediacy. The movement of the first step is the phenomenological reduction of egology, and the movement of the latter is the development of monadology, in which the transcendental Ego always remains a constituting subject.

In order to view the concretizing of the Ego systematically, the inter-subjective world-constitution emerges as only one side of the spatializing of the transcendental Ego. The transcendental Ego has, as a still more fundamental constitutive stage, its self-temporalizing. Husserl says: "The current 'I' executes a performance in which it constitutes a variational mode of itself as existing (in the mode of past). Starting from this we trace how the current I, flowingly-constantly present, constitutes itself in self-temporalization as enduring through all 'its' pasts. In this same way the current 'I' already having duration in the primordial sphere of duration constitutes in itself an Other as Other. The so-to-say self-temporalization by means of moving-away-from-original-presence (by means of recollection) has its analogy in my moving-away-from-otherness (empathy as a moving-away-from-original-presence on a higher level – that of my original presence – into a merely presentified original presence)."[20] Thus the constitution of the life-world communality in monadology is a moving-away-from-original-presence on a higher level, namely a self-spatializing of the transcendental Ego as an analogy of its self-temporalizing. From a systematic point of view, this self-spatializing is based on the self-constitution of the inner time-consciousness, which is carried out in egology.

The transcendental consciousness in the *Phenomenology of the Inner Time-Consciousness* is described as divided into three layers: in the first layer the stream of consciousness, i.e., experience; in the second, appearance, which contains the time object and sense, i.e., the object in the Now; and in the third, simply the object.[21] If one considers the noetic-noematic connection in *Ideas I*, the time object belongs to noesis and sense belongs to the noema. An act that animates the objective, throughout these three levels, is called intentionality. Intentionality operates as a "horizontal intentionality" and a "vertical intentionality."[22] Horizontal intentionality is, like external perception, an action which, outside of the stream of consciousness, posits the transcendent as the identical, while on the other hand vertical intentionality, like internal perception, is an action that constitutes the thickness of time around the Now in the midst of the stream of consciousness. The Now is by no means as pure as a point in a straight line. Such a "fine Now" is, to be sure, an original consciousness, that is, an original impression, but only an ideal limit, namely something intellectually abstracted.[23] The real Now, a "coarse Now," has in itself, on the other hand, a temporal of retention and protention.[24] It is just because of this temporal of "beforehand" and "afterward" that consciousness is always a self-consciousness,[25] even though at present not-positing, and is thus distinguished from the thing. The asserted past and future is a posited phase of time, i.e., presentified by means of memory and expectation. The absolute claim of consciousness consists in its presentness, while the presentified modifications of consciousness do not have an absolute claim, but only a relative one.

The present Ego concretizes itself into one having duration by means of the self-constitution of the transcendental time consciousness. This Ego having duration is not yet a corporeally concretized Ego, since it remains inside the pure consciousness. The corporeal concretizing of the Ego is not realized until the development of monadology.

The complete concretizing of the Ego promotes its self-spatialization in monadology, which in Husserl's phenomenology presupposes a self-temperalization of the Ego in egology. Here it emerges that the phenomenology of the late Husserl neither banned transcendental subjectivity from its area of research nor made the life-world its philosophical beginning. The life-world always remains a primordial one in the inter-subjectively constituted regions; this doxic originality of the life-world is not apprehended in the transcendental attitude, but only in the natural one. The phenomenology of the life-world presupposes the suspension of its acceptance as being. For Husserl it is the egologically pure act of consciousness that is transcendentally without question. On the other hand it is only in the doctrine of constitution of transcendental consciousness that the naive unquestioned obviousness of the world is established. Thus the life-world is philosophically justified only when it is intersubjectively constituted. Similarly to the way the existence and the essence of the external world, which were rejected by Cartesian hyperbolic doubt, were finally taken back at the end of his metaphysics, so with Husserl the being of the life-world is taken back by having it established by intersubjective constitution.

Consequently the intersubjective development of phenomenology does not by any means signify a change in phenomenological motivation. It is always the consistent motivation of absolute grounding that develops phenomenology into a doctrine of constitution of the life-world. Therefore one cannot uphold the interpretation that Husserl, who in 1911 rejected the naturalization of pure consciousness and defined philosophy as a strict science, entirely abandoned the route of radical egological reduction – which was taken two years later in *Ideas I* – in 1928 when he left the university of Freiburg. The idea of the First Philosophy retains its sense, for him, for his whole life. Husserl says the following in a manuscript that he wrote in 1935:[26] the present state of philosophy emphasizes a departure that no longer regards philosophy as a strict science but tries to deal with it on a level of art or religion. In the present philosophical situation, in which the dream of strict science is at an end, it is necessary first of all to have a historical sense for this negative state of affairs and to regard it from a historical standpoint. In order to overcome the crisis of the sciences that has arisen from objectivism, phenomenology needs the reduction that goes back to the life-world before it begins with egological reduction. Phenomenological reflection on the modern sciences requires above all a way of thinking that tries to tear away the intellectual veil of objectivism and then to take away the immediate unquestioned obviousness of the life-world. In this sense Husserl carried out first the life-world reduction and then the egological reduction, in his later phenomenology, by clearly and strictly distinguishing the objective a priori from the life-world a priori.

To be sure, the life-world is established as the given by the phenomenological development from egology to monadology. But the phenomenological movement does not end with this grounding. For the distinguishing thing about Husserl's phenomenology lies in the fact that the constitutive backward movement toward the immediacy of the life-world does not come to an end, like a circle, in its beginning, but instead in the pre-given being. For Husserl transcendental consciousness is a thoroughly suitable philosophical beginning. But consciousness cannot produce a real being, but only ideal objectivities. Even if the constituting movement of transcendental consciousness goes back to immediacy, this immediacy is the being that is pre-given before the life-world reduction. The constituting movement is not a return to its beginning of the transcendental Ego, but a taking-back of real being. After the revival of life-world immediacy, phenomenology investigates the genesis of the culturally formed world. Thus phenomenology proceeds from constitutive "grounding" to genetic "founding." Genetic phenomenology considers its task to be to examine how both the objective sciences and also cultural phenomena arise from prepredicative experience. But it is founded on a life-world phenomenology, namely "a pure essence doctrine of the life-world"[27] as the primordial layer of founding. But with the exception of the analyses of *Ideas II*, Husserl did not always adequately carry out the descriptive analysis of the life-world. It was many of Husserl's followers who carried out the many-sided and original analyses of the life-world. *Formal and*

Transcendental Logic and *Experience and Judgment* give many important indications as to the genetic investigation of the objective a priori. Nevertheless, the whole development of genetic phenomenology cannot be considered therein complete, since we must reflect on what the evidence of the life-world a priori is and how the life-world a priori grounds the cultural or objective a prioris.

Husserl points to the grounding relation in two points: the grounding relation of the original doxa to its doxic modifications and to value positing and volitional positing. In the first connection the original doxa grounds the present apprehensions and the presentified apprehensions.[28] In the presentified modifications are contained uncertain presumption, doubt, negation, assumption, memory and expectation. In the second connection, fact-formulating positing grounds value-formulating positing, or emotion positing, and volitional positing.[29] For value is always a value of the thing. Doxology corresponds to fact formulating; axiology corresponds to value formulating, and theory of action corresponds to volitional formulating.

III

Grasping of facts and axiological interest, however, are not separate in everyday experience but penetrate one another in habitualities.[30] In the investigation of prepredicative experience, which, for example, grounds a region of linguistic expression, prepredicative experience must be intellectually analyzed, in methodical, thematic abstraction just as in intersubjective reduction.[31] In order to concentrate on the area of linguistic expressions thematically, interest in praxis and value must be abstracted from all prepredicative experience and at the same time it is necessary to prescind from inspective perception and its explicating modifications, which consist of the internal horizon of the inspective perception, the modifications of an object and consist of its external horizon, the relation of the same object to the others.[32] Without in this way removing from the entire interest in the life-world an inductive foresight peculiar to inspective perception, the linguistic expressions could not be genetically grounded.

With genetic considerations, however, not all regions may be handled like those of the linguistic expressions. Genetic examination of logical articulated knowledge is entirely different from the examination of individual cultural and axiological subjects in such areas as art, religion, history, etc. For the former is the examination of a system of the formal, objective a priori, while the latter is the examination of a knowledge of the so-called life-world, subjective a priori. The evidence of the formal a priori is dependent and basically distinct from the self-sufficient evidence of the life-worldly,[33] although both are ascertained with the free variations peculiar to intuition of an essence.[34]

At the same time one must not overlook still another distinction of concrete material essence: a distinction between transcendental consciousness and the life-world. For the essence structure of transcendental consciousness includes a constitutive function, and the essence structure of the life-world includes an interpenetration of fact and value.

To be sure, phenomenology presupposes intuition of an essence insofar as it is one of the material sciences of essence. But intuition of an essence for its part needs a constitutive analysis of fulfilled intentional experience, which phenomenology carries out. As a result there arises a circular proof: that intuition of an essence is a necessary pre-condition for phenomenology and also is explicated only by phenomenology. We already tried in the second section, to solve the phenomenological paradox. Now we confront another paradox, the paradox of evidence.

In my view, Husserl's description of transcendental consciousness analyses three essence structures: intentionality, horizontality and stratification connections. Consciousness is first of all not a state of having contents but a kind of intentional relation. Therefore it is always an act directed to Something. Second, this act of consciousness does not intend a merely explicit object but always directs itself to an object in its two potentialities, namely, to an object in its spatial and temporal horizon. Third, the act of consciousness forms a stratification connection of grounding. In its basic strata there are the being-doxic, presentified apprehension of facts of memory and expectation and the not-being-doxic apprehension of phantasy. Upon such doxic apprehension, value apprehension and positing of will are built up.

Phenomenologically psychological and phenomenologically transcendental consciousness possess one and the same essence structure as long as they, so differently oriented, both belong to an eidetic science. Life-world concrete consciousness and transcendental consciousness therefore also possess one and the same essence structure. Husserl explains the essence structure of psychophysical subjectivity by means of the idea of a kinesthesia as the ability to move the body. According to his explanation, the whole experience of perception and action can be apprehended only by means of the description of consciousness, not as the intentionality of pure consciousness, but as that of incarnated consciousness. In this sense, the essence of transcendental consciousness and that of immediate experience are basically different, since the former consists of its constituting function and the latter consists of the mutual interpenetration of fact and value. The phenomenology of the life-world therefore drives the constitutive structure of transcendental consciousness into the background and emphasizes, with corporeal motricity, the opaqueness of experience. Phenomenology now goes from Husserl's analytical grasp of the grounding relations to the total grasp of such experience, as it habitually embraces past and future, necessity and freedom, fact and value, etc. Therefore we cannot limit our observation just to Husserl's phenomenology. The concretization of the transcendental Ego, its self-temporalization and self-spatialization are to contain, already implicitly, an idea of incarnation. Merleau-Ponty detects something that Husserl did not explicitly express, and grasps Husserl's three structures of consciousness as that of the body: intentionality as the corporeal motricity (*motricité du corps*), the horizontality of consciousness as that of the body and the grounding stratification relation as the "integration" relation of the physical, the vital and the psychic order of human beings. Subdivision appears only in the case of "disintegration" of the superordinate order, e.g., in the case of schizophrenia, because integration means to gather up and take place the subordinate into the superordinate instead of erecting an analytical synthesis of the fundamental elements, which would be nothing but an intellectual abstraction. Corporeal motricity does not react to the atomic impression of the object, but only to its total being, which depends on the degree of integration of the motricity of the body. The conditions of the body express either a sense as integration of subjectivity or a counter-sense as its disintegration. Counter-sense is, after all, also a modification of sense. Corporeal motricity is a primordial sense-bestowing act. The appearance of an object is an expression of the subjective gestic effect on the same object. Perception is not mere taking up of the impression, but already an expression of the object.[35] The world's encounter with the Ego itself is an accession (*avènement*) of the world by means of its expression.[36] Between Ego and world a kind of encroachment (*empiétement*)[37] takes place around the axis of the body.

In just this way an encroachment between the Ego and the Other takes place. The ego is, for itself, the beginning and, at the same time, mediated by the Other.

Il est donc bien vrai que les 'mondes privés' communiquent, que chacun d'eux se donne à son titulaire comme variante d'un monde commun. La Communication fait de nous les témoins d'un seul monde, comme la synergie de nos yeux les suspend à une chose unique. Mais dans un cas comme dans l'autre, la certitude, tout irrésistible qu'elle soit, rests absolument obscure; nous pouvons la vivre, nous ne pouvons ni la percer, ni la formuler, ni l'ériger en thèse. Tout essai d'élucidation nous ramène aux dilemmes.[38]

A mutually interpenetrating relation of the Ego and the world, the Ego and the Other, the past and the future, the rational and the irrational, etc., a mediating relation, where one thing is the "figure" while the other is the "ground," and vice versa, and, finally, an ambiguous relation, in which the prototype vanishes at once in analytical thinking, which divides it into atomic elements – such a relation forms the essence structure of immediate experiences of the life-world. Is it not already a kind of dialectical relation? The manner of thinking that proceeds from such a relation does not seek an Archimedian Point, but instead sees its beginning in relation itself. Description of the life-world essence structure transforms unequivocal derivation from an absolutely still point into an ambiguous "back and forth" (va-et-vient) between several points.

The evidence of such an obscure relation can therefore no longer be an adequately fulfilled experience. Round about the adequately experienced core there is always in reality an indistinct fog. "Voir, c'est par principe voir plus qu'on ne voit, c'est accéder à un être de latence. L'invisible est le relief et la profondeur du visible, et pas plus que lui le visible ne comporte de positivité pure."[39] In this sense the evidence of the life-world essence-structure does not so much have a character of intuition as one of reflection, of "reflection on what is unreflected" (réflexion-sur-un-irréflechi).[40] This reflection means the continual guiding of thought into the right course in order not to deviate from the "impenetrability" (opacité) of the immediate, and it even means the firm resolution to accentuate the encroachment (empiétement) in the life-world.

The tradition of European thought sees it as a criterion of truth, to illuminate the difference of an object clearly and distinctly. It built a temple of reason by despising thoughts about the child and the body and condemning anomalous areas to the realm of heresy.

C'était par exemple une évidence, pour l'homme formé au savoir objectif de l'Occident, que la magie ou le mythe n'ont pas de vérité intrinsèque, que les effets magiques et la vie mythique et rituelle doivent être expliqués par des causes 'objectives', et rapportés pour le reste aux illusions de la Subjectiv.té.[41]

It is right on the horizon of unreason that now, according to life-world reflection, the nature of reason can be explained. A normal person's perception can be understood clearly only by taking into account the loss of his normalcy. Thus a mythologic space, e.g. a schizophrenic space, cannot be explained with regard to geometric, homogeneous space, but rather, it is to be understood as an oriented space, which also has a gate that opens to objectivity.

Merleau-Ponty also includes Husserl's transcendental subjectivity in the thought of the "flying over" (*survol*), by undertaking to turn the coordinate axis of evidence around. With Husserl, the doctrine of evidence is brought to the most extreme limit of reason. The phenomenological development from transcendental subjectivity to intersubjective self-constitution signifies a convergence at this most extreme limit. In the phenomenology of the life-world the doctrine of evidence unavoidably encounters an aporia. Therefore Merleau-Ponty finds his starting point of thought in the origin of aporia, by changing his attitude to evidence and, according to his opinion, deciphers what Husserl did not clearly express.[42]

Merleau-Ponty's procedure of solving the paradox of evidence paradoxically will deviate from the radical route to the First Philosophy, which Husserl pursued all through his life. With the philosophical beginning of the life-world ambiguity, the unambiguous reorganizing of science out of the absolute absence of presuppositions will be impossible. Just because the life-world is primordial, the doctrine of constitutive grounding through transcendental subjectivity may not be dismissed from view so easily. For the acceptance as being of the life-world remains of no interest, at least to the phenomenological examination of the life-world, whether the life-world is a constitutive component of transcendental subjectivity or not. But Merleau-Ponty's phenomenology lacks a philosophical explanation for the life-world's acceptance as being. Merleau-Ponty did not give to the attitude of a phenomenologist itself enough philosophical thought.

According to Husserl the phenomenologist is a "disinterested observer." Merleau-Ponty would say that it is not permitted to man to take the position of such an observer. But with just this statement in the name of reiterating reflection the phenomenology of phenomenology would never be established. According to Husserl even "the disinterested observer" takes an interest, namely an abiding interest in the idea of the First Philosophy.[43] The disinterested, observing interest, as an interest in a practically-speaking valueless metaphysical appearance, is hard to contradict. Rather, the phenomenologist must determine where in the whole of phenomenology the interest in First Philosophy should locate the ideas, undeniable for life-world phenomenology, of incarnation and encroachment. Then we must now read from Merleau-Ponty's text what he does not clearly express. According to Merleau-Ponty it is not a philosophically true reflection, and there is constantly the danger that this may be forgotten, to explain the progress of sensuous experience with the Cartesian *inspectio mentis,* scientifically, in objective reflection. True reflection is a more basic conversion than scientific reflection, namely objective idealization. It is, rather, an inverted reflection, a kind of "super-reflection" (*surréflexion*).[44] For super-reflection also takes account of the landscape that it changed by reflection and tries, with the existing positive language, to go beyond its sense and to express the silent contact between perception and the perceived, our pre-logical contact with the raw world. In doing so it is not possible to avoid the non-agreement of ordinary language with its positive sense. It is precisely the non-agreeing expression that presents to us, in reverse, the raw world, which does not fit into the positive sense system. The world reaches us first with the creative expression. "La réflexion s'enfonce dans le monde au lieu de le dominer."[45] Reflection seeks to find the secret of our connection with the world not in the subject, but in the world itself.

Transcendental subjectivity, to be sure, never produces being, but constitutes the world out of the ground without presuppositions, as the philosophical beginning. It is not a subject that "makes" the world, but a subject that "sees" the world as a whole. But the disinterested observer, who in this way transforms the world into the predicate of the transcendental first person, is similar to the creator, except that the former undertakes the contemplation of the world and the latter undertakes the creation of the world. Reason, with its light, from the position of contemplation, illuminates the world in all its parts. On the other hand, the way of thinking that wants to investigate the truth that is hidden from reason and even by reason, proceeds from the encroachment of man and the world. To be sure, it cannot prove the mutually interpenetrating connection, but it can experience through discovery that transcendental subjectivity, which surveys the world as a whole, is check-mated. Such a discovery is less a grounding in evidence than an acceptance of the gift from the absolutely Other. The gift is a stumbling block to the interest in the absolutely subjective point of view. Its acceptance is not an intentional conversion. Thus reason, with its volitional self-negation, cannot carry out absolute criticism on itself. It finds itself beaten after the absolute critique of reason is completely carried out. While the change of attitude of phenomenological reduction is therefore a subjectively active conversion, because it has been chosen by the philosopher, the change of attitude that experiences the self-collapse of the interest in absolute subjectivity is an absolutely passive conversion. It is precisely in this way that the phenomenology of phenomenology would reach its own ground, where the gate to the world stands open.

Kobe

NOTES

*Translated from the German by Barbara Haupt Mohr.
1 *Husserliana*, Vol. VI, pp. 182 190; Vol. I pp. 129 130.
2 *Husserliana*, Vol. VI, pp. 50 52: *Erfahrung und Urteil*, pp. 39–44 (Klassen Verlag, 1954).
3 *Husserliana*, Vol. VI, pp. 143 144.
4 *Husserliana*, Vol. VI, p. 269.
5 *Husserliana*, Vol. VI, p. 77.
6 Descartes, *Meditationes de prima philosophia*, meditatio II. *Husserliana*, Vol. VIII, p. 69.
7 *Husserliana*, Vol. VIII, p. 165.
8 *Husserliana*, Vol. VI, pp. 157 158.
9 *Husserliana*, Vol. VI, p. 175.
10 *Husserliana*, Vol. III, p. 143.
11 *Erfahrung und Urteil*, p. 49.
12 *Husserliana*, Vol. VI, p. 175.
13 *Husserliana*, Vol. VI, p. 177.
14 *Husserliana*, Vol. VI, p. 154.
15 *Husserliana*, Vol. VIII, p. 174.
16 *Husserliana*, Vol. VI, p. 186.
17 *Husserliana*, Vol. I, pp. 124 138.
18 *Husserliana*, Vol. I, pp. 138 149.
19 *Husserliana*, Vol. I, pp. 149 163.
20 *Husserliana*, Vol. VI, p. 189.
21 *Husserliana*, Vol. X, p. 26; ibid., p. 76.
22 *Husserliana*, Vol. X, p. 31.
23 *Husserliana*, Vol. X, p. 40.
24 *Husserliana*, Vol. X, pp 27 34; ibid., p. 40.
25 *Husserliana*, Vol. X, pp 116 120.
26 *Husserliana*, Vol. VI, pp. 508 513.
27 *Husserliana*, Vol. VI, p. 144.

28 *Husserliana*, Vol. III, pp. 249–256.
29 *Husserliana*, Vol. III, pp. 283 291.
30 *Erfahrung und Urteil*. p. 52.
31 *Erfahrung und Urteil*. pp. 56 58.
32 *Erfahrung und Urteil*. pp. 69 71; pp. 114–116; pp. 171 173.
33 *Erfahrung und Urteil*, pp. 432 443.
34 *Erfahrung und Urteil*, pp. 410 420.
35 Merleau-Ponty: *Résumés de cours*, p. 14 (Gallimard, 1968).
36 Merleau-Ponty: *La prose du monde*, pp. 69 117 (p. 113., p. 116) (Gallimard, 1969).
37 Merleau-Ponty: *Résumés de cours*, p. 165. Merleau-Ponty: *L'oeil et l'esprit*, p. 46, p. 80,
p. 90 (Gallimard, 1964).
38 Merleau-Ponty: *Le visible et l'invisible*, p. 27 (Gallimard, 1964).
39 Merleau-Ponty: *Signes*, p. 29 (Gallimard, 1960).
40 Merleau-Ponty: *Phénoménologie de la perception*, p. 76 (Gallimard, 1945).
41 Merleau-Ponty: *Le visible et l'invisible*, p. 43.
42 Merleau-Ponty: *Signes*, p. 202 203.
43 *Husserliana*. Vol. VIII, pp. 195 197.
44 Merleau-Ponty: *Le visible et l'invisible*, p. 61.
45 Merleau-Ponty: *Le visible et l'invisible*, p. 61.

BIBLIOGRAPHY

Bossert, Philip J.: 'A Common Misunderstanding concerning Husserl's Crisis Text,' 1974, *Philosophy and Phenomenological Research*. Vol. 35.
Brand, Gerd: 'The Material Apriori and the Foundation for its Analysis in Husserl' 1969, *Analecta Husserliana* II.
Brand, Gerd: *Die Lebenswelt* (Berlin) 1971.
Carr, David: 'The "Fifth Meditation" and Husserl's Cartesianism,' 1973, *Philosophy and Phenomenological Research*. Vol. 34.
Eley, Lother: *Die Krisis des Apriori in der Transzendentalen Phänomenologie E. Husserls* (The Hague) 1962 (*Phaenomenologica* 10).
Funke, Gerhard: *Phänomenologie-Metaphysik oder Method*? (Bonn) 1966.
Gadamer, Hans-Georg: *Die Phänomenologische Bewegung*.
Gadamer, Hans-Georg: *Die Wissenschaft von der Lebenswelt*, (Kleine Schriften III, 1972).
Landgrebe, Ludwig: *Der Weg der Phänomenologie* (Gütersloh) 1963.
Landgrebe, Ludwig: *Phänomenologie und Geschichte* (Gütersloh) 1968.
Levin, David Michael: 'Induction and Husserl's Theory of Eidetic Variation,' 1968, *Philosophy and Phenomenological Research*, Vol. 29.
Levin, David Michael: *Reason and Evidence in Husserl's Phenomenology* (Evanston) 1970.
Mohanty, Jitendranath: '"Life-world" and "A priori" in Husserl's Later Thought,' *Analecta Husserliana* III.
Monasterio, Xavier O.: 'Paradoxes et mythes de la phénoménologie,' 1969, *Revue de Métaphysique et de Morale*, Vol. 74.
Patočka, Jean: 'La doctrine husserlienne de l'intuition eidétique et ses critiques récents,' 1965 *Revue internationale de Philosophie*, Vol. 71–72.

Ricoeur, Paul: *Husserl – An Analysis of His Phenomenology* (Evanston) 1967.

Ricoeur, Paul: *Le conflit des interprétations* (Paris) 1969.

Schütz, Alfred: 'Type and Eidos in Husserl's Late Philosophy,' 1959 *Philosophy and Phenomenological Research,* Vol. 20 (*Phaenomenologica* 22).

Sokolowski, Robert: *The Formation of Husserl's Concept of Constitution* (The Hague) 1970 (*Phaenomenologica* 33).

Tymieniecka, Anna-Teresa: *Phenomenology and Science in Contemporary European Thought* (New York, Toronto) 1962.

HIROSHI KOJIMA

THE POTENTIAL PLURALITY OF THE TRANSCENDENTAL EGO OF HUSSERL AND ITS RELEVANCE TO THE THEORY OF SPACE

So far as we know from the text of Husserl's *Cartesianische Meditationen,* the motive for the construction of the theory of "Others" was mainly to repudiate the charge of "solipsism" against him.[1] He calls this criticism "as it might seem, a very important objection."[2] Indeed, one might think that here lies one of the Achille's tendons of Husserl's phenomenology. According to his theory, the world and the nature are as a whole constituted as the noematic meaning by the intentional act (noesis) of the transcendental consciousness. They are the immanent transcendence of the transcendental consciousness of ego. Since this transcendental ego is taken for my ego, we might say, the whole world is here nothing but *my* world. That is to say, the whole world is only given as a perspective view to me. The wholeness of the world remains only as the unlimited horizontalness of the view, which develops panoramically before my ego.

Is such a world really able to contain an ego other than mine? That is to say, does my intentionality constitute even other consciousnesses than mine? As all the objects of the world are, according to Husserl, constituted only by my intentionality, nothing will remain which is not constituted by me. If the world-constitution theory of Husserl is not solipsism, the consciousnesses of other egos must necessarily be constituted by my consciousness. That is the problem which Husserl confronted at that time.

But we cannot help wondering whether this was not a pseudoproblem, though Husserl himself does not seem to have thought so. This doubt has its deepest ground in our interpretation of the relation of Husserl's transcendentality to the natural attitude of man. Even if the phenomenological reduction excludes the naïve setting of the objective world as a whole, the unlimited horizontalness of the transcendental consciousness still remains. This horizontalness will be a priori incompatible with solipsism,

Nitta/Tatematsu (eds.), Analecta Husserliana, Vol. VIII, 55–61. *All Rights Reserved.*
Copyright © 1978 by D. Reidel Publishing Company, Dordrecht, Holland.

because it will suggest the potential plurality of the transcendental consciousness; this is the point of our argument.

I

According to Theunissen, Husserl's effort to develop the theory of "Others" is guided throughout by the aim to establish constitutively the objective world.[3] But in our opinion the theoretical establishment of the objective world should not be seen as the aim of the theory of "Others." Rather, they are the same thing. Because the objective world also means for Husserl "the single world for everyone," which envelops me and others equally originally. The theory of the objective world and that of "Others" should not be separated. Both are two sides of the transcendental founding of the natural attitude of man.[4]

In my natural attitude, I am always with others in the objective, single world for everyone. I am one of the egos who are scattered here and there throughout the world-space. I am not an absolute, solitary ego, but an averaged one, which is already mediated and penetrated by others.

Once we adopt this point of view, Husserl's egological theory of world-constitution reveals a new aspect to us. For if the ego of the natural attitude of man is already penetrated by the existence of other egos, even the phenomenological reduction would be unable to change him to a mere solitary ego, just as it was unable to change the objective world to a mere perspective world. Rather the ego of egology will probably anticipate the potential plurality of the transcendental ego, just as the perspective view of the reduced world anticipates other perspectives as its horizontal background.

It is true that Husserl himself was not sufficiently aware of the fact that the transcendental ego might be a priori plural. So he has chosen as his procedure for reaching other transcendental egos a detour through the reduction to the primordial sphere and the introjection of myself into other bodies (empathy). Husserl thinks, indeed, that by excluding all alien elements from the domain of my constituting ego, he can establish the primordial sphere of myself, namely my monad. But, in fact, the a priori plurality of the transcendental ego survives this reduction. Other egos are not found only after the introjection of myself into other bodies. Already within the primordial sphere of my ego, anonymous other egos carry on their functions. If so, from the beginning there was no problem of the overcoming of solipsism, or the constitution of other consciousnesses through my consciousness. Other consciousnesses are not to be constituted by my consciousness; rather, they are equally original with mine. Their intentionality are not noemata of my intentionality, but they are co-noeses with mine.

In support of cur argument, we will take up Husserl's concept of "appresentation." He says, "an appresentation occurs even in external experience, since the strictly seen front of a physical thing always and necessarily appresents a rear aspect. . . ."[5] Appresentation is a kind of co-presentation, or an analogical apperception. In my primordial sphere things appear always with their rear aspects, which I cannot see from here in my perspective. The presentation of things accompanies always and necessarily a kind of co-presentation, namely an analogical apperception of the unseen sides of things. Every perception contains something more than mere sensual reception. Husserl has attributed this function to the neotic composition of the transcendental consciousness of my ego.

But, is it truly my intentionality, which appresents the backs of things? I grasp the world only from here in my perspective. I can never see things from any other place than here at this moment. How can I appresent their rear aspects? Do I imagine it according to my remembrance? Then I shall see the image of the (ever seen) back side confronting me, just doubled above the present presentation of the front side, but never as the image of the back side in this very moment. So the back side could not be my image.

Husserl also calls the appresentation an empty intention, which is not yet filled with sensual hyle. So he considers it possible, by turning things, to change my appresentation into presentation, that is, to fill the empty intention. But we must say that such an argument is rather scandalous for him. For in order to turn a thing I must know in advance that it has another side! So the back of a thing cannot be defined by my empty intention, which should be filled through its turning. Rather my intention to the back is empty, or preliminary, because it is not originally my intention.

In this moment, I can constitute the back side neither from my image nor from my empty intention. From the standpoint of Husserl, namely that of my single transcendental consciousness, the appresentation reveals itself, so to say, as an enigma. But from the standpoint of plural transcendental consciousness, it is not at all an enigma. The consciousness of others are already performing their function before being constituted by my consciousness through the introjection of myself. Namely, the intention of an anonymous other ego, which aims at the thing from another side in this very moment, defines the back as itself. The back is nothing but the front which is confronted by another ego. The appresented existence of the back is never founded without the intentionality of an anonymous other consciousness from behind the thing.

But how is my intentionality from here related to the intentionality of others from beyond? My intentionality is a priori mediated by the intentionalities of others. With other words, I anticipate always and necessarily the co-operating other egos, while I am intending the front side of a thing. They are anonymous, but nonetheless actual. So the unity of presentation and appresentation of a thing, which is often emphasized by Husserl, means that this thing is constituted not by me alone, but together by me and others at the same time. I realize the act of other intentionalities in my own intentionality, not as a part of it, but as their equally original cooperations. This is what I mean by the phrase: the potential plurality of the transcendental ego. Of course it is a type of intersubjectivity, but I would like to call it *Interintentionality*. It is an immediate inter-subjectivity without the mediation of bodies.

Here other transcendental consciousnesses are not constituted by my transcendental consciousness. Rather both are participating in the same *single horizon* around the constituted thing. While I acknowledge the unity of presentation and appresentation of a thing, I find myself as a transcendental ego set in this single universal horizon, which is beyond my perspective, knowing that all the perspectives of other transcendental egos also participate in this horizon. So one might say that this horizon is what gives the transcendental founding to the so-called objective world-space for everyone. It is aptly to be called the transcendental scheme of inter-intentionality.

II

On the contrary, Husserl's intersubjectivity is made possible by the intro-jection of myself into other bodies. And according to his theory an objective world for everyone is founded by intersubjectivity in this sense. But, in fact, his intersubjectivity already anticipates in advance an objective world for everyone. Let us examine his argument. He says, "As reflexively related to itself, my animate bodily organism (in my primordial sphere) has the central 'Here' as its mode of givenness; every other body, and accordingly the 'other's' body, has the mode 'There.' . . . By free modification of my kinesthesias, particularly those of locomotion, I can

change my position in such a manner that I convert any There into a Here –
that is to say, I could occupy any spatial locus with my organism."[6]

If Husserl really stands on a central Here, never will he describe the
situation in such a way. For then no There could ever be changed into my
central Here. My central Here is absolute and so never changeable.[7] Only
every There might come nearer and be absorbed into the central Here, or
come apart from Here and go farther away. My body as a central Here will
never walk about, but only things before me come nearer to me, or go
farther away. One must consider where the central Here of Husserl "walks
about." The central Here "walking about" is not the center any more,
because it is already so situated on a certain unmovable plane (in a space)
relative to its also unmovable There. And what is this plane or space other
than an objective world, which is beyond the difference between the
central Here and every There?

Husserl's mistake in describing the situation is critical, because it is not
only a matter of describing, but also involves the fundamental
construction of his intersubjectivity. He says, "since the other body there
enters into a pairing association with my body here and, being given
perceptually, becomes the core of an appresentation, the core of my
experience of a coexisting ego, that ego . . . must be appresented *as an ego
now coexisting in the mode There,* 'such as I should be if I were there'."[8]
So, we might say, the essence of his intersubjectivity-theory is not a
coupling association or appresentative introjection, but the possibility of
removing the center (Here) from here to there. The coupling is only the
factual motive of this removal. So he says, "not only the systems of
appearance that pertain to my current perceiving 'from here', but other
quite determinate systems, corresponding to the change of position that
puts me 'there', belong constitutively to each physical thing."[9]

But, as we have seen, the absolute central Here as such will never be
removed. If it could be removed to There, it will no longer be absolute and
the space on which Here moves will have an indefinite center of
coordinates, which could be displaced here and there arbitrarily. Even the
imaginary removal of the "as if I were there" already anticipates this kind
of space, which is a kind of objective space, quite independent of the
position of the center. That is to say, this space potentially has the center of
coordinates equally upon every There.

So it will be clear that Husserl's intersubjectivity-theory anticipates in
advance an objective world-space, where potentially the plurality of
perspectives dominates. And the space in which I know that everything is

seen not only from here out, but also from there out (at the same time), is nothing but the space where the presentation is always accompanied by the appresentation. That is to say, Husserl's intersubjectivity-theory anticipates the existence of other intentionalities, or consciousnesses, before it thematizes the body of others. So, we might say, Husserl's procedure is only to fix an indefinite anonymous intentionality of others factually upon the definite body.[10] In this respect, inter-intentionality is far in advance of Husserl's intersubjectivity.

I have said that the inter-intentionality of transcendental consciousness is possible in so far as it has as a guide the transcendental scheme of the spatial horizon. This scheme is not a content of my consciousness. Rather, my consciousness is a participant in this scheme of space, in which other consciousnesses also participate. Modern philosophy has begun with the discovery of this scheme, as infinite space, but it has failed to grasp its essence, because consciousness has rigidly been bound to my ego and space has been driven into mere objectivity. The phenomenological reduction of the objective world, as the core method of Husserl, is, in this respect, nothing but the liberation of space from mere objectivity. Space as the scheme of inter-intentionality is neither mere subjectivity nor mere objectivity. It is the place where subjectivity and objectivity meet. Everything appears in my perspective with its definite horizontal composition, but this very fact indicates immediately the existence of intentionalities other than mine.[11]

Moreover, this relation between the space-scheme and inter-intentionality seems to suggest to us, in a kind of analogy, the relation between reason (*Vernunft*) as a transcendental scheme of thought and *co*-intentionality as the transcendental apperception, which is however no longer an ego.

Yokohama

NOTES

[1] Edmund Husserl, *Cartesianische Meditationen*, 2nd ed., The Hague, Martinus Nijhoff, 1963, S. 121ff.; *Cartesian Mediations*, trans. by Dorion Cairns, The Hague, Martinus Nijhoff, 1960, pp. 89ff.
[2] Ibid.
[3] Theunissen, *Der Andere*, S. 51ff.
[4] Ibid., S. 94ff.
[5] Husserl, op cit. S. 139; Cairn's translation (p. 109).

6 Ibid., S. 145ff.; Cairn's translation (p. 116).

7 The phenomenological description of the mode of being of the absolute Here in the property primordial monad I have given in a treatise, which has been published in Germany: Hiroshi Kojima, 'Zur philosophischen Erschliessung der religiösen Dimension', in *Philosophisches Jahrbuch*, 85Jg., 1978, S. 56ff.

8 Husserl, op. cit. S. 148; Cairn's translation (p. 119).

9 Ibid., S. 146; Cairn's translation (pp. 116–117).

10 Cf. Sartre, *L'être et le Néant*, pp. 336ff.

11 In this treatise I have completely refrained from the problem of time. This must be necessarily treated in relation with the investigation of monadology.

EIICHI SHIMOMISSĒ

PHILOSOPHY AND PHENOMENOLOGICAL INTUITION

In this paper, it is intended to explore and re-establish a way of philosophizing as the "revelation of the world," following the tradition of phenomenological research in terms of problems, methods, and the historical horizon. It will be pursued through searching its anticipating mode of philosophical inquiry in Max Scheler's philosophical investigations.

Max Scheler, unlike Edmund Husserl, never made phenomenology and phenomenological "method" (*Einstellung*) as such the central "theme" of his philosophical inquiries, nor is it correct to characterize Scheler's entire philosophical investigations as phenomenological (even if it were possible to do so).[1] Rather than being concerned about the self-justification of philosophy as *mathesis universalis*,[2] or developing a philosophical system, or being "preoccupied with Being", hidden in everydayness and forgotten by the Western philosophical tradition, Scheler's central attention was directed toward *doing philosophy phenomenologically*. The relationship of Scheler's philosophical contributions to "phenomenology" as stated in his own words is that his philosophy was developed "on the broadest basis of the phenomenological experience."[3] Indeed, his "doing philosophy phenomenologically" was so extensive and so involved that, by reflecting upon it, we gain insight into our own questions, methods and perspectives. It is hoped, therefore, that the reason Max Scheler's philosophizing was chosen for this investigation rather than that of Husserl, Heidegger, Sartre or Merleau-Ponty can be justified by the completion of this inquiry itself. However, it must be emphasized that the question raised here is by no means concerned with the "objective definition" of phenomenology, nor with the "formal characteristics" of the phenomenological approach. Rather, our question is: what does phenomenology and the phenomenological method mean to us and to our philosophizing? It is indeed properly answered by Maurice Merleau-Ponty: *"c'est en nous-mêmes que nous trouverons l'unité de la phénoménologie et son vrai sens."*[4]

Three quarters of a century have passed since the publication of

Nitta / Tatematsu (eds.), Analecta Husserliana, Vol. VIII, 63–67. All Rights Reserved.
Copyright © 1978 by D. Reidel Publishing Company, Dordrecht, Holland.

Edmund Husserl's *Logische Untersuchungen*, and the so-called "phenomenological movements," both the old and the new, *seem* to already belong to the past, despite the fact that new volumes of *Husserliana* continue to appear and much of the new research based upon them has been published. Even in the United States many translations of phenomenological research have been published, and renewed interest in phenomenology seems to have arisen not only in philosophy, but also in empirical sciences. And yet phenomenology as a way of doing philosophy *appears* as if it were a *"finished"* (*abgeschlossener*) philosophical thought, which might be studied as a historical phenomenon!

Against such apparent contention we must point out that the authentic nature of phenomenology as a way of philosophizing still remains *"open"* (*offen*) and phenomenological inquiries will continue to constitute a unity of the philosophical trend. Phenomenological philosophy has not yet accomplished the task of terminating philosophy's imitation of mathematics and natural sciences, nor has it completely succeeded in showing that we ought to "go back" to the primary and concrete experience of reality as it reveals itself as the world. Instead of "openness," mistrust of experience and what Scheler calls *Welthass* (hatred of the world) still hinder us from experiencing reality as "phenomenon." For the most part, philosophy has remained a kind of indirect, vicarious and yet possibly "exact" knowledge of reality today.

As a starting point, a general, non-thematic conception of phenomenology will be taken and the genuine essence of phenomenology as philosophizing will reveal itself through the process of this inquiry. Phenomenology can only be approached phenomenologically.

Instead of phenomenology as a philosophical *doctrine* or *system*, it has been preferable to talk about the Munich phenomenological circle or the Göttingen circle, a phenomenological movement or a phenomenological school. This purports *eine gemeinsame Gesinnung* (a common intention) in philosophizing, a new philosophical ideal commonly shared among some philosophers, or a new philosophical *Grundeinstellung*, i.e., a new basic philosophic "method" or "style" in approach. Edmund Husserl and his co-editors of *Jahrbuch für Philosophie und Phänomenologische Forschung* saw in phenomenology a new, immediate approach to fact itself. The well-known preface to its first volume states: "It is not an academic system that binds the editors and that ought to be presupposed by all the future editors. Our unity is furnished by the common conviction that only by means of returning to the original sources of intuition and to the insights into essences derived from them can the great traditions of philosophy be intuitively clarified. Then the problems of philosophy can be newly set forth on an intuitive basis and can be solved once and for all."[5]

The so-called leading motto: "Return to fact itself!" which well summarizes this common conviction neither specifies the facticity of this very "fact" itself, i.e., what constitutes the "phenomenomenality" of the "phenomenon" of phenomenology, whether it be an "a priori self-given-ness" of any kind, or an "intuited essence," or a "transcendentally reduced meaning," nor does it elicit from itself why we ought to return to this "fact" itself.

If instead of attacking the problem of what constitutes the facticity of the phenomenological "fact" we direct our attention to "returning," it is quite clear that the new common conviction of philosophizing expressed in this motto assumed that Western philosophy had been *somewhere other* than where it should have been. What is the *locus* in which the preceding Western philosophy pursued and some contemporary philosophy still pursues its inquiry into problems by means of the traditional use of basic concepts? What is it from which we ought to "return" in philosophy?

First of all, needless to say, it is all the forms of "reductionism" in philosophy[6] which were popular particularly at the end of the 19th century. In the face of threats of the overwhelming "advancement" of the sciences and an enormous amount of scientific and technological findings, philosophy at the turn of the last century took refuge in an attempt to dissolve philosophy into "something else." It was prevailing then, for example, to "reduce" transcendental philosophy (i.e., ontology and epistemology) to the investigation of the foundations of the sciences and mathematics by Neo-Kantians on the one hand and to replace epistemology and logic by psychology on the other. In philosophical reductionism, "definition," "analysis" and "explanation" are extensively employed to replace "that which is to be reduced" by "that to which it is reduced," e.g., definiendum by definiens, analysandum by analysans, and explanandum by explanans, etc., whereby the identity of the former and the latter, being unjustifiable within the system itself, is simply presupposed,[7] i.e., the latter is normally so "constructed" to *symbolize* the former.

Any form of reductionism, however, involves abstraction; either by divorcing a certain element, one-sided aspect or an epiphenomenon from the concrete fact to stand for the latter, or by symbolizing the concrete fact by something else, by defining this by that, analyzing this into that, explaining this by means of that, or inferring this from that. And it is normally the case that such an abstraction fails to recognize (*verkennen*), cover up (*verbergen*) or deceive (*täuschen*) us in regard to the nature and meaning of that concrete fact. This is why phenomenology categorically rejects "analysis," "explanation," any type of "logical inference," and, thus, all forms of "scientific"[8] endeavor as reductionistic and vicarious in consequence. The construction and constitution in terms of a theory, whether it be scientistic philosophy or a form of idealism, also involves symbols and symbolization; Scheler calls the phenomenological approach *de-symbolizing* and *a-symbolical*.[9]

Now we stand before the most crucial question in phenomenology: what is, then, to be described? Now we come back to that to which we ought to "return," the "fact" itself, the "phenomenon" that phenomenology describes. At least from what has been mentioned above, it is quite obvious that the "fact" itself, the "phenomenon" to be described in pheno- menology, is *not* the referent of the sciences, i.e., scientific "datum," entity, e.g., atom, quantum, stimulus or "id," "superego," etc., or scientific "law."[10] Rather, it is the concrete *world of our lived experience* that any science or theory (and further any philosophical doctrine or system hitherto developed in Western history) ultimately must be related to: even what Aristotle calls *epistēmē* (philosophy) and *technē* (art and special sciences) must start with and refer back to the reality of our actual experience, so that it may be explained scientifically, *in terms of* its cause (by inference), its analyzed moments, and its law (by theory).

This concrete experience of reality with its fullness and its constituent entities are those which we, in our everyday life, incessantly deal with, sensorily perceiving and imagining, inferring and planning, acting and reacting, and so on.

As Heidegger pointed out, in our mundane life we are normally and for the most part totally *absorbed* in our day-to-day "business" and the "immediate needs" of living. It is only the absence of something "being ready to hand" (whether its awareness be implicit, or explicit) that interrupts our experiencing (including our activities) which generally flows with habitual smoothness. We take care of our "*pragmata*" one after another, including taking a break and enjoying recreation. In short, we are *engaged*.[11]

Precisely because of this pragmatic, concernful and executive *immersion* in this experience of reality of everydayness, we are in principle (more fundamentally than by means of sciences due to the "*closeness*" of our everyday immersion) *blinded* to this concrete reality as it is. The mundane, absorbed way of concern and execution of those everyday "pragmata" does nothing other than *remove* (*verstellen*) and *conceal* (*verbergen*) this full, concrete reality as such *from us*. This basic attitude of our pragmatic interest in reality is inseparably one with our ontic belief in the "disguised" appearance (*das verstellte Erscheinen*) of this reality and they reinforce each other to enable our absorbed engagement in it. As long as we *live in it,* we can in no way raise even the slightest doubt about it as a whole, but being accustomed to it, continue to believe that what we are experiencing is the sole reality.[12] Our universal belief in this reality thus experienced and understood may be called the mundane ontology of every-dayness.[13]

To philosophize is the questioning search as Heidegger correctly characterized it. Whether it is by means of wonder (*thaumázein*), or it is through the Cartesian doubt, it has been considered and known in the philosophical tradition that philosophical thinking is supposed to liberate the questioning searcher from that universal doxa *and* enable him to seek, "behold," indicate, and even (sometimes) "describe" genuine reality in its fullness and concreteness, no matter how indirect, indefinite, incomplete, or rather (historically) restricted it might be in its givenness to us.

Through the history of Western philosophy in particular, various attempts have been made to develop and formulate philosophical thought and knowledge of the way toward "experiencing" that genuine reality. This reality is to be understood in terms of some basic concepts, which are to be philosophically purified and critically questioned as to their mundane ontological use. They are, for example, "one" and "many," "idea" or "form" and "matter," "being" and "nothing," "potency" and "actuality," "means" and "end," "cause" and "effect," "necessity" and "accident," "time" and "space," "to be" and "ought" (or "fact" and "value"), "appearance" and "reality," "drive" and "will," "doxa" or "opinion" and "knowledge," "sense," "understanding" and "reason," to name a few.

The more radical the questioning search becomes, the more indeterminate and open should philosophical thinking become. For not only through critical examination of the mundane ontology of every-dayness, but also by probing confrontation with the preceding philosophies, philosophizing can liberate itself from all possible, both hidden and explicit, assumptions and biases in order to be able to return to and behold the concrete, lived world of reality in its primordiality rather than a substitute (such as a theory, a language, or any system or model "representing" reality). In Scheler's words, we should have an open attitude of "the free, compassionate (*liebevoll*) and active (*tätige*) *devotion* to the world."[14]

It may be significant to note that, as time proceeds historically, those fundamental notions and philosophical conceptions previously examined and carefully "liberated" usually *deteriorate back* to the mundane, accustomed use of language and thinking; they serve again uncritically to develop and shape the mundane ontology of everydayness. More seriously, due to their fundamental nature, those concepts and the ways of thinking *with ostensible apparentness and conceptual depth* inhere in and thus constitute the philosophical tradition itself.

Phenomenology as a new, most radical style of philosophizing has found itself in the midst of the situation not only heavily burdened by the long and respectable tradition of Western philosophy since Ancient Greece, but also buried in the staunchly believed mundane ontology of everydayness ever more backed-up with many unexamined implications of the overwhelmingly successful contemporary mathematics and natural sciences.

The philosophical situation at the turn of the century was far more complicated than described above due to the self-understanding of Western philosophy since Decartes regarding how to philosophize. Contemporary philosophy on the one hand imitated mathematics and the natural sciences and, at the same time, made its own task to provide the foundation for mathematics and other sciences (*die Wissenschaftslehre*). On the other hand, precisely because of this attempt *and* this task, contemporary philosophy had to pursue its own self-justification of philosophizing as *the* primary science (*die Selbstbegründung der Wissenschaftlichkeit der Philosophie als der Wissenschaft*) by means of *self-reflection*. Following Descartes this self-reflecting took the form of apodeictic evidence of the immediacy of the self reflection upon itself in "cogito, ergo sum." Thus, philosophy was considered "as the study of the ultimate grounds of *knowledge* as a study of the knowable as absolutely certain" (*als die Lehre von den letzten Gründen des Wissens als Lehre vom Wissenkönnen als gewissem*).[15]

The turn to subjectivity and probing of the way of knowing in terms of self-reflection in contemporary philosophy resulted from the destruction of confidence and mistrust in the most fundamental principle of Western metaphysics, i.e., the "identity of knowing (experience) and being (reality)."[16]

Thus, contemporary philosophy had to start with the examination of knowing – experiencing – as a "tool" or "medium" to grasp reality[17] and pursue its inquiry by means of self-reflection of human reason. The endeavors of this philosophy of reflection to restore the confidence in experience failed and continued to be obstructed from the concrete world of lived experience. The philosophy of reflection culminated in Kant's transcendental philosophy.

Various endeavors of the German Idealists Fichte, Schelling and Hegel were nothing but attempts to overcome the predicaments of this philosophy of reflection mentioned above and to restore the Parmenidean principle through the "philosophy of identity." Thus, metaphysics was reinstated by the German Idealists.

Husserl misconstrued the essays of German Idealism as arbitrarily speculative constructions of a philosophical system totally divorced from concrete, actual reality. Sharing with contemporary logical positivists the strong conviction that speculative thinking is not the way of philosophizing, Husserl considered that one of the tasks of phenomenology was to overcome such a speculative metaphysics. Quite different from the logical empiricism and philosophy of language today, phenomenology is the way of philosophizing that radically explores and strips off the distortion, the covering-up, and deception which hides the concrete, full reality and hinders it from revealing itself as it is in its primordiality.

Husserl's critical confrontation with mathematics and the natural sciences through his investigation into the foundation of logic lead him to believe that phenomenology should be *Wissenschaftslehre* as *mathesis universalis*. Further, this was understood by Husserl to be the complete overhaul of epistemology as the *ontologia generalis* and, as a result, the critique of human reason. In this manner, Husserl was still living heavily in the continuing tradition of Western philosophy. In spite of the revolutionary motto and appeal of his "phenomenology" and "phenomenological research," Husserl was, at least at the beginning, *blindly*, later more critically involved in the tradition of the "philosophy of reflection." He even criticized Kant's careful avoidance of the intuitive "representation" of self-reflection. For Kant was apparently aware that it is impossible to "grasp" the hidden, functioning self by means of the objectifying, representing self-reflection. Rather, Kant took the way of the hypothetical, inferential transcendental method for his philosophizing. It was not an accident that Husserl read and was influenced by Kant's *Critique of Pure Reason* in 1905–1907. It was also quite natural for Husserl to see continuity and affinity between his investigations and the philosophy of Kant (as well as that of Descartes).

It is, however, one of our fundamental contentions that phenomenology and phenomenological method has *in principle* nothing to do with transcendental philosophy, its turn to subjectivity and self-reflection. This turn to subjectivity and self-reflection was merely Husserl's way of pursuing phenomenology as the way of philosophy. Against the early and middle period of Husserl's investigation, Scheler's philosophical inquiry was, *from the beginning*, motivated by a radical mistrust and deeply rooted dissatisfaction with traditional philosophy.

As pointed out before, the development of Western philosophy since Descartes was grounded upon the principle that the way of being, i.e., the essence of the known (*das Wesen des Festegestellten*), is determined by the way of knowing (*die Methode der Feststellung*)[18] and furthermore the philosophical pursuit consists ultimately in grounding a *special kind of "knowing"* guaranteeing that it indeed knows what it claims to know. This self-reflection upon self, i.e., the regressive, representing grasp of subjectivity by itself, however, involves itself in a self-referential paradox. The actual, primordial self is not the transcendental ego which is already "fixed" by reflection as the noetic pole intentionally related to its noema. Rather it is the primordial, self-temporalizing, ever-flowing functioning self as the *lebendige Gegenwart* (actual, living present). This paradox involves the fruitless attempt to capture the ongoing, functioning, spontaneous self regressively through a representing, objectifying act of reflection. The actual, streaming primordial functioning self continues to escape from the reflecting sight no matter how many times the self-reflection may be reiterated. For this attempt itself hides or conceals the very object of its quest, i.e., the primordial spontaneous, functioning self. It is well known that, through detailed, painstaking and reiterated "phenomenological analyses," Husserl's life-long efforts to *make* everything (including the *alter ego* as well as that primordial functioning self) *transparent* by means of self-reflection proved itself as unfeasible finally at the ultimate stage of the philosophy of reflection. Husserl even called such a "primordial fact" as "self-temporalizing anonymous self," "*nunc stans*" or *das absolute Faktum*. This was also pointed out by Sartre, Merleau-Ponty and Thévenaz.[19]

Together with Nietzsche, Bergson and Dilthey, Scheler's critical confrontation with the traditional Western philosophy was far *more radical* than Husserl's; Scheler *explicitly* attempted to *uncover* all concealed "idols," "dogmas," and "biases" deeply rooted in the mundane as well as the philosophical ontology of the preceding philosophies which constitute the implicit and hidden foundation for the contemporary sciences and philosophy. To Scheler, reflection as the representing, objectifying intentional act is by no means phenomenological intuition. Scheler was aware of the paradox in self-reflection when he spoke of the impossibility of grasping and seeing the "person" as the concrete essential ontic unity of various intentional acts.[20] Also he was able to phenomenologically reveal that a self-deception (*Selbsttäuschung*) may be involved in Descartes' so-called indubitable evidence of immediate self-reflection.[21] Thus, Scheler's phenomenological investigation began precisely where he clearly and explicitly broke off from the traditional "philosophy of reflection." Whereas Husserl, being influenced by the *Critique of Pure Reason*, developed his phenomenological inquiry in the direction of transcendental philosophy, Scheler took the opposite direction and tried to uncover the presuppositions and biases hidden in the basis of Kant's philosophy in order to criticize the latter.

To Scheler, phenomenology is a new way of philosophizing in order to liberate man and the world from all possible mistrust, concealment and separation. It is an attempt to recover the right and validity of "experiencing" as the open, harmonious unison between man and his world in the primordiality of the full, concrete reality as it reveals itself.

Scheler inherited from Husserl the following fundamental conceptions: the "phenomenological reduction" as the necessary procedure to philosophize phenomenologically, the "phenomenological experience" as a priori, pure intuition, the "pure fact," or what Scheler calls "essence," and the "intentionality" of our experiencing as the inseparable co-relation of the "act" and its "object."

Husserl talked about various phenomenological reductions or "epoches," which, for example, may be enumerated as the "eidetic," the "transcendental," the "intersubjective," the "universal-transcendental," and the "phenomenological-psychological reduction," etc. These various epoches were developed by Husserl to deal with specific philosophical problems phenomenologically, but there is the fundamental nature of the phenomenological reduction common to all: it is, in short, a philosophical procedure to enable the change and shift of one's attitude (*Einstellung*) from one to the other in such a way that by means of the epoche, i.e., in the new attitude, the previously *implicit* and *overlooked, covered-up,* and *misconstrued* becomes the *explicit* and *visible, uncovered,* and *adequately construed.* In general, Scheler speaks of the phenomenological reduction as a shift from the natural attitude to the phenomenological attitude.

As long as we are totally immersed in mundane intercourse (*Verkehr und Vernehmen*) with the "world," there exists neither necessity, nor chance of our being able to philosophize, as described previously. The involvement of our transcending ourselves to and projecting ourselves in the world *and* the pragmatic value and import of this manner of being in the world is so strong and deep that even if doubt might arise, it would be necessarily a particular doubt about our concern about a particular object as an implement, which C. S. Peirce calls a "living doubt." This doubt never, by its nature, rises above the horizon of our particular concerns (e.g., solving a disturbing problem such as too much tolerance of the automobile's brakes) and is capable of "moving around" solely within that domain.

Quite different is the situation in the following case: a dream experience can never be understood as a "dream" until some other experience such as the buzz of the alarm clock intercepts, undermines and destroys our "general thesis" of that dream experience as being *real.* Indeed, sometimes it happens that we do doubt spontaneously whether we might be "dreaming" in that very dream experience itself and yet such "feigned doubt" in isolation could usually not be strong enough to "mediate" our dreaming as "dream experience" rather than "actual experience." In the "dreaming" itself, its being a "dream" is implicit, concealed, unnoticed, and misconstrued as being "real." When we become awake, then the mode of that previous experience reveals itself explicitly uncovered, visible and adequately construed as un-real, that is as a "dream."

As is obvious from this metaphor, the phenomenological reduction or epoche *artificially* (motivated by our innate urge to philosophize) performs a violent change in our attitude, an awakening from the dogmatic slumber in the mundane ontology of everydayness to the questioning search for truth. Thus, we become liberated from the naive, universal belief in the reality of the "natural world" in order to be able to "bracket" (*ausser-Kraft-setzen*) the operative power of latency, concealment, misunderstanding, and deception. This Scheler sometimes calls "adduction" rather than "reduction."[22]

What thus discloses itself in the phenomenological attitude is immediately (i.e., without the "filter" of operative latency, distortion, concealment and misunderstanding) "beheld" by the pure intuition (*die phänomenologische Schau oder Erfahrung*) and is called the "pure" or "phenomenological fact" or the "a priori essence" in Scheler's term. It must be emphasized here again that this pure intuition does not have to be self-reflection of self upon itself.

What is blinded in the natural attitude becomes or should become "visible" in the phenomenological attitude for various reasons. It may be due to the total immersion of ourselves in everyday concerns through the distortion by abstraction, and practical or even some theoretical interest, because of the concealment by language, symbols and theories, or even other philosophical conceptions traditionally inherited.

However, it is Eugen Fink in his famous article titled "L'analyse intentionnelle et le problème de la pensée spéculative"[23] who initiated the contention that the problem of reality being implicit, overlooked, distorted, misconstrued and covered-up in the natural attitude reappears in what should be given as "transparent" in the phenomenological attitude due to the *limitation* of the phenomenological intuition *and* the *speculative thinking* which supplements it. This interpretation of Husserl's phenomenology and his "intentional analyses" has been now widely accepted and many efforts have been made to substantiate it by the miscarriage of Husserl's continuous painstaking attempt in his later period to make the total structure of the intentional correlations "transparent" by means of the self-reflection.[24] Fink points out that in Husserl's phenomenology, for example, the phenomenality of the "phenomenon" as the "entity for us" (i.e., *Gegenstand*) and the facticity of the pure fact given in the phenomenological intuition *cannot* be *intuitively disclosed*, but is to be *speculatively conceived*, and that the necessity of Husserl's exactingly detailed intentional analyses year after year indicates the lack of the intuitive proof and the need for speculative thinking.[25]

Such a criticism of Husserl's phenomenology and of the phenomenological philosophy in general that speculative thinking is inevitable even in the phenomenological philosophizing is based upon the following two assumptions: the one is Husserl's heavy dependence on traditional Western philosophy as mentioned above; the other assumption, which was also pointed out in this paper before, is Husserl's as well as Fink's and others' identification of phenomenological intuition with self-reflection. There is no doubt that both Husserl and Scheler inherited and employed many traditionally stained, variously conceived notions of philosophy which, in fact, could possibly lead Husserl as well as Scheler away from the primary task of phenomenological inquiry. In case of Scheler, such concepts as "essence" and "person" are very misleading, although they facilitated Scheler's ability to deal with certain problems. However, it has been made clear above that these two assumptions are more unique to Husserl's approach and partially resulted from his naiveté, and that Scheler is quite free from such accusations.

The crucial problem which is now left to us is the clarification of phenomenological intuition. As repeatedly emphasized before, first of all, phenomenological intuition should be clearly distinguished from most of the "self-reflection" on which Husserl's "intentional analyses" depended.

Although our explication does not concern itself with the ambiguous use of phenomenological insight, it may be necessary to call attention to the ambiguity of Scheler's use of "phenomenological intuition" and "experience." On the one hand, it is employed methodologically, i.e., the phenomenological intuition constitutes the *locus* in which the so-called phenomenological facts – the pure essences – are self-given, namely disclosed intuitively. On the other, to Scheler, the phenomenological experience means the "free, compassionate, active *devotion* to the world" as a new, open way of the world's revealing itself in its full concreteness as it actually is. Scheler's ambiguous use of phenomenological insight, intuition or experience seems deliberate in that philosophizing as phenomenological experience consists ultimately in man's open, fully confided, perspicuous intercourse with the world itself.

As obvious from the fact that phenomenological reduction or epoche is a continuous process of the laborious, non-natural endeavor and never completes itself due to the finitude and ambiguity of our body (*Leib*), phenomenological insight or experience, although it is called intuition, is *not* an instant, mystical beholding of reality, unless it is idealized. Even though Scheler wrote that the phenomenological fact or a priori essence is the experienced or intuited, which is completely and immediately given in the experiencing or intuiting act, the complete and immediate self-given-ness of the pure fact shows the complete independence and non-inter-mediation of the senses and the way of empirical experience. This allows the essence to give itself directly in its intending act.

Phenomenological experience or insight as intuition is "immediate" in the above-mentioned sense, and yet the *process* of attaining the unity of the intended (*Gemeintes*) and intuited (*Erfülltes*) discloses itself as "mediated" as a long, toilsome *modus operandi* going through many steps of *negation* in Hegel's sense of "*Vermittelung*." Scheler speaks as if it were possible to have a phenomenological intuition of an individual object (the empirically perceived redness of a rose) as its essence (the a priori color "red of this very rose") *in isolation* in our experience. Although theoretically this may be possible, in actuality it requires a total change of one's attitude in the general thesis of one's own experiencing. For *as a process,* the phenomenological reduction and phenomenological experience are one and the same. In this sense, the previously given example of the metaphor of the "dream" experience does not faithfully portray its concrete nature and its experience at all in the *process* of phenomenological intuition.

Rather from a different point of view, Stephen Strasser contends in his short article titled "Intuition und Dialektik" that the principle of intuition should be complemented by the principle of dialectic and that "the grasp of an object *uno intuitu* is not a human experience. Human intuition has the character of an endless process rather than that of a simple, perfect, ultimate 'visio.'"[26] However, the question is *not* whether the principle of intuition is to be complemented by that of dialectic. Rather, phenomenological intuition indeed is a *process* and it is identical with the phenomenological reduction which *reveals in itself the structure of dialectical mediation by means of negation.* It is not because of speculative thinking that the phenomenological description forms a series of negative, distinguishing statements, e.g., "perception is not memory," "perception is not hallucination," "perception is not anticipation," etc., in order to *disentangle (entwirren)* the essential core from the unessential elements to bring it to an immediate phenomenon, but precisely that structure of the negative *Vermittelung* is reflected in the phenomenological description of the process in phenomenological intuition.

The person who reads and follows the phenomenological description is guided by this process of "negative mediation" step by step, is freed gradually from the distortion, confusion, concealment and finally, in the ideal case, arrives at the intuitive grasp of the identity of the intended and the intuited. This can be formally stated as a tautologous statement, that is, "perception is perception." This is why Wundt argued against Husserl that phenomenological description is "tautologous" in its final analysis. [27]

It is quite interesting and appears philosophically significant to point out that Husserl's and others' phenomenology inevitably needs speculative thinking, but this may miss the entire point of the phenomenological investigation, which is to let the world gradually open up itself as it actually is, by bracketing step by step the operative latency, distortion, confusion, miscomprehension and covering-up. Might the contention of the functioning of the speculative thinking in Husserl's intentional analyses and the failure to see the primordial, authentic|nature of phenomenological intuition or experience itself be an indication for one's being still trapped in the traditional Western philosophy and operating with speculative thinking by oneself in a hidden and unnoticed manner? Phenomenological inquiry must be pursued further as long as the world as the horizon in which entities (objects) appear and the totality of entities in it will not be completely revealed with its full concreteness, as it actually is. This task, as long as it is pursued by man, will, however, never find its end.

California State University
Dominguez Hills

NOTES

[1] Scheler wrote 'Phänomenologie und Erkenntnistheorie' and 'Lehre von den Drei Tatsachen' around 1910 15 which were published posthumously in his G.W. Bd. 1, Franke, Bern, 1957.
[2] This fundamental conception has preoccupied the philosopher's mind since Descartes. Husserl himself was caught by this idea as were his contemporary Neo-Kantians.
[3] Max Scheler, *Der Formalismus in der Ethik und die materiale Wertethik*, G.W., Bd. 2, Franke, Bern, 1954, p. 29.
[4] Maurice Merleau-Ponty, *La phénoménologie de la perception*, Gallimard, Paris, 1945, p. II.
[5] Edmund Husserl, Vorwort zum Bd. I, Teil I, *Jahrbuch für Philosophie und Phänomenologische Forschung*, p. VI.
[6] The "reduction" of "reductionism" should not be confused with the so-called "phenomenological reduction"; the former purports a "μετάβασις εἰς ἄλλô γένος," whereas the latter is a veritable philosophical procedure in which the hitherto covered-up and unseen "authentic" reality reveals itself as it is. In other words, the latter is the unravelling of the former's wrong "μετάβασις." Cf. Husserl, *Ideen zu einer reinen Phänomenologie und phänomenologischen Philosophie*, Erstes Buch, *Husserliana*, Bd. III, Den Haag, 1950, p. 145. M. Scheler, Nachlass, Bd. I, G.W. I, p. 380.
[7] The pragmatic principle of simplicity and strength of "theory" is normally applied for justification.
[8] Unlike the older usage of the German *Wissenschaft*, this notion of "science" and "scientific" excludes "philosophy and philosophical inquiry." This is why Scheler could say *wissenschaftliche Philosophie* is *Unding*, cf. Scheler, *Vom Ewigen im Menschen*, G. W., Bd. 5, p. 77.
[9] Scheler, *Der Formalismus*, p. 69 for example. *Entsymbolisierung* (de-symbolizing) does not necessarily destroy, annihilate, or remove the symbols, but it is a form of phenomenological reduction in the sense of bracketing.
[10] For example, a physicalistic "description" of "phenomenon-red" does not require any

actual visual experience of "red" to recognize it and is exactly that which is perfectly intelligible to a *blind* person. Cf. Rougier, *Traité de la connaissance*, p. 298.

[11] To the philosopher, it is indeed trivial and redundant to mention here that this description of our absorbed activities in everydayness is not possible in the mundane attitude, but only in a critical, philosophical attitude which takes an (interested) distance from it. The following descriptions will make this evident.

[12] Cf. Plato's well-known 'Allegory of the Cave' in Bk. VII of *The Republic*. Also cf. Eugen Fink, '*Was will die Phänomenologie Edmund Husserls?*', *Tatwelt*, Bd. X, 1934.

[13] Husserl calls this belief the "universal Urdoxa" in the natural world.

[14] M. Scheler, *Schriften zur Soziologie und Weltanschauungslehre*, G.W., Bd. 6, p. 65.

[15] Wilhelm Teichner, *Rekonstruktion oder Reproduktion des Grundes*, Bouvier, Bonn, 1976, p. 1.

[16] " τὸ γὰρ αὐτὸ ιοεῖυ ἐστίυ τε καὶ εἶναι." Diels-Kranz, *Die Fragmente des Vorsokratiker*, Bd. I, Weidmann, Dublin/Zürich, 1966, Parmenides B. Fragmente I-19-3, p. 231.

[17] G. W. F. Hegel, *Die Phänomenologie des Geistes*, Glockner Ausgabe, p. 67. It is interesting to note that this magnum opus of Hegel in the first edition bore the title 'Wissenschaft der Erfahrun des Bewusstseins,' which the Blockner's edition lacks. Cf. M. Heidegger, *Holzwege*, Vittorio Klostermann, Frankfurt a. M., 1952, p. 105.

[18] M. Scheler, 'Lehre von den Drei Tatsachen,' G.W., Bd. 1, 1975, p. 445.

[19] Cf. J.-P. Sartre, 'La transcendence de l'ego,' Recherches Philosophiques VI, 1936–37. M. Merleau-Ponty, *La phénoménologie de la perception*, Gallimard, Paris, 1945, Avant-propos. P. Thévenaz, 'Réflexion et Conscience de soi,' *L'homme et sa raison*, Éditions de la Baconnière, Neuchâtel, 1956.

[20] M. Scheler, *Der Formalismus*, Pp. 92, 97, 393ff.

[21] M. Scheler, 'Die Idole der Selbsterkenntnis,' G.W., Bd. 3, 1955, p. 226ff. Scheler, 'Idealismus-Realismus,' Philosophischer Anzeiger II, 1927–28, Bonn, p. 258ff.

[22] M. Scheler, 'Lehre von der Drei Tatsachen,' G.W. Bd. I, p. 444.

[23] This article was published in *Problèmes actuels de la phénoménologie*, Desclée de Brouwer, Bruxelles, 1952.

[24] E. Fink, 'Die Spätphilosophie Husserls in der Freiburger Zeit,' in *Edmund Husserl 1859–1959*, Martinus Nijhoft, Den Haag, 1959. L. Landgrebe, *Der Weg der Phänomenologie*, Gütersloh, 1963. K. Held, *Lebendige Gegenwart*, Martinus Nijhoff, Den Haag, 1966.

[25] E. Fink, 'L'analyse intentionelle et le problème de la pensée spéculative,' p. 74ff.

[26] *Edmund Husserl 1859–1959*, Martinus Nijhoff, Den Haag, 1959, p. 151.

[27] W. Wundt, *Kleine Schriften*, Bd. I, Leipzig, 1910, p. 613ff. Edmund Husserl, *Ideen*, Bk. I, *Husserliana*, Bd. III, Martinus Nijhoff, Den Haag, 1950, p. 355. M. Scheler, *Nachlass* I, G.W., Bd. I, p. 391.

SHIZUO TAKIURA

IS TIME REAL?

In our ordinary language, there are two ways of speaking of time. We say in some cases that each position in time is "earlier" than some positions and "later" than others, and in other cases that each position is either "past," "present," or "future." In the article entitled "The Unreality of Time" in *Mind* (1908), J. E. McTaggart distinguished these two ways, and for the sake of brevity spoke of the series of positions running from the past through the present to the future as the A series, and the series of positions which runs from earlier to later as the B series. And, regarding the A series as more fundamental to time than the B series, he asserted that time is unreal. For, according to him, the A series is not a series of the real. This view is also put forward in his posthumous *Nature of Existence*, II (1927). On the contrary, P. Bieri, regarding the B series as the more fundamental, asserts in his recent book *Zeit und Zeiterfahrung* (1972) that time *is* real. In his opinion, the A series involves the B series as its "principle of construction." But I think that these two arguments are not quite contrary, because the B series should also be considered unreal. I will state my reasons in this paper. As for myself, I believe that it is only motion or change of things that is real, and it may be possible to draw this conclusion *mutatis mutandis* from Husserl's theory of the constitution of time.

To begin with, let us review briefly McTaggart's argument for the unreality of time. First, he argues that time involves change, and to express the notion of change, we must refer to the A series, because the B series is "permanent." Take an event, the death of Queen Anne for example. That it is a death, that it has certain causes, that it has certain effects – every "characteristic" of this sort never changes. If her death is later than the death of William III, this relation can never change. McTaggart therefore said in the above-mentioned article, "If N is ever earlier than O and later than M, it will always be, and has always been, earlier than O and later than M." According to him, any event which has a position in the B series "will always be, and has always been, an event," and consequently "cannot begin or cease to be an event."

No doubt, we might be able to object against the view that it is not an

Nitta / Tatematsu (eds.), Analecta Husserliana, Vol. VIII, 79–88. All Rights Reserved.
Copyright © 1978 by D. Reidel Publishing Company, Dordrecht, Holland.

event, but a thing that can change. The death of Anne happened, and did not change. Thus B. Russell proposed the definition of change as the difference, with regard to truth or falsehood, between a proposition concerning an entity and a time T and a proposition concerning the same entity and another time T'.[1] So there is a change, in Russell's view, if the proposition "at the time T my poker is hot" is true, and the proposition "at the time T' my poker is hot" is false. Nevertheless, McTaggart asserted that it was erroneous to say that there was any change in the poker, as far as T and T' are positions in the B series. For, in this case, the poker has a quality such that it is hot on a particular Monday, and a quality such that it is not hot at any other time, and these qualities are true of it at any time – the time when it is hot and the time when it is cold.[2]

But, according to McTaggart, we can speak of change in one respect, and only in one respect. That is to say, the hotness of the poker "began by being a future event. It became an event in the nearer future. At last it was present. Then it became past, and will always remain so."[3] Thus the only change we can get is from future to present, and from present to past. If the A and B series are equally essential to time – since "earlier" and "later" are temporal determinations as well – the one must be more fundamental than the other. And if the B series is also temporal, it is because the B series can arise only when the A series, which gives change and direction, is combined with a third series, which does not determine the direction but only gives an order like that of the alphabetical letters or the natural numbers. This he called the C series. Since every change must be in a particular direction, the time which involves change can not be given by the C series alone. It may be noteworthy that so-called absolute time is for McTaggart nothing but the time of B-relations. For this reason, he said that no change could be looked for in the numerically different moments of absolute time. The B series is, as it were, time spatialized by the mediation of the C series.

McTaggart's second thesis was that even if without the A series there would be no change or time, time could not be true of reality. For, "the existence of any A series involves a contradiction." Obviously "past," "present" and "future" must be incompatible determinations, since this is essential to the meaning of the terms, and yet they are all in fact predicable of each event. For example, if M is *past*, we can rightly say that it has been *present* and *future*. If it is *future*, it will be *present* and *past*. And if it is *present*, it has been *future* and will be *past*. This is McTaggart's so-called contradiction. To the objection that there is no contradiction in the fact than an event has all temporal determinations "successively," he would reply that such an explanation involves a vicious circle. For it presupposes the A series in order to account for the A series. When we say for example that, if M is past, it has been present, we are appealing indeed to two time-series, which are both A series, the one in which M is situated as past, and the other in which the same M is regarded as present. And according to him this strange fact is possible because the A series consists, perhaps, of the relations of events and moments to something not itself in the time-series. For this reason also, McTaggart asserted the unreality of time.

Now I believe that we need not invoke McTaggart's view to affirm that the time which consists in the A series is unreal. For it is evident that the past and the future at least don't exist now. Certainly, we often roughly say that the present sinks down or passes away into the past. But this can not mean that the moment which is present should pass away into the time-point which has in itself the property of pastness, since the past is called past only because it has already ceased to exist. When we speak of the present sinking into past, we think in reality of the fact that events or things which exist now will soon disappear. This must also be the case with the future. Although it is usually said that the present advances indefinitely towards the future, there can be no moment that exists at present and yet has the property of futurity. Is it not absurd indeed to affirm now the existence of future time, in spite of the absence of the future event?

Perhaps it must be kept in mind that time is not any sort of real thing, whether it is conceived in terms of the A series or not. Time is not even a stream, since a stream is nothing but a mode of being of things. Only either we or things can flow in time; it is contradictory to say that time itself flows in time. Certainly, every motion or change takes place in time and is described as a function of time (and space). But no motion or change occurs because there is time in the world. For time cannot act upon material bodies to make them move or change. Even if there remained time in the world, a body which has lost all reasons to subsist would have to cease to exist.

Shall we say then, "Time is the indefinite milieu, analogous to space, where events, marking a date, succeed one after another; but which is given in itself as a whole and indivisibly to thought?"[4] When motion is said to occur "in time," most of us probably think of such a milieu. But the problem is how to grasp the mode of its being. If it were a sort of real framework of the universe, there could be neither past nor future in it, since there can be no parts of a real framework that have already ceased to exist or have not yet come to be. In such a time, as Bergson once pointed out, the past and future would be juxtaposed on the same plane as the present, and consequently in fact be present. What is time in a world which has neither past nor future?

It is well known that Newton said in his *Principia Mathematica* that absolute time, "of itself, and from its own nature, flows equably without relation to anything external." This passage shows that Newtonian absolute time was a sort of real framework of the universe, for, as Koyré interpreted it, it has "its own nature, independent of everything external, that is, independent of the existence or nonexistence of the world."[5] This is why the length of "intermundium" was often discussed in those days. But it appears absurd to suppose the existence of such a time. For we cannot think of a state of affairs in which there is only time, while the world does not exist. When I measure a temporal process, for example, I can never do so by comparing it with absolute time itself, but only with some other process, the rotation of the earth, for example. It is only because one regards time as a sort of thing, that time is conceived as a framework of the world. Thus we must rather say, following Augustine, that time exists because it tends to nonexistence (*tendit non esse*).[6]

However, as mentioned above, Bieri asserts the reality of time on the basis of the logical priority of the B series to the A series. No doubt, between the A and B series there is a semantic relation, which is asymmetrical and seems to show the semantic priority of the A series. It is scarcely possible to describe our ordinary experience of time in terms of the B series alone. When we posit a question "when" concerning an event, we could not content ourselves with information concerning B-relations. If it is said to be "earlier" than some other event, we should also like to know whether this event is earlier or not than our present. It seems at first glance, therefore, that we can reduce most B-relations to A-determinations and that the A-determinations are more fundamental for our intersubjective understanding about time. But semantic priority does not always imply logical priority. So Bieri calls attention to the following passage from McTaggart: "The term P is earlier than the term Q, if it is ever past while Q is present, or present while Q is future."[7] This passage clearly shows that there can be distinctions of "more past" and "more future" within the past and future themselves, and these distinctions can never be described in terms of the A series, since it leads to a vicious circle. In other words, our concept of time involves in addition to the three unanalysable concepts of past, present, and future the notion of a structural order, and it was because of the latter that we had to introduce the concepts of "more past" and "more future." Thus, Bieri's conclusion is that the B series is the "principal of construction" of the A series. For this reason, the propositions about the A series imply *prima facie* those about the B series and not *vice versa*. Of course, it is "successively" the case that one and the same event is future, present and past. But this change of determinations of events along with the A series can be understood only when the state of affairs in question is considered as a whole as ordered along with B-relations. The A series itself must indeed stand in a relation, in order to be able to form a series, and this relation is a B-relation. And according to Bieri, the logical dependence of the A series upon the B series suggests that the B series is the real one, since semantic priority is nothing but the priority on the level of our subjective or intersubjective understanding.

Now, from this point of view, Bieri gives a sharp criticism of the transcendental theory of time, including Husserl's phenomenology of internal time consciousness. As is well known, the point of his phenomenology consists in the method which is called "transcendental reduction," that is to say, in the discovery of things as phenomena, as being given to consciousness, and at the same time in the discovery of transcendental subjectivity. In the theory of time also, Husserl devoted his whole effort to the reduction of time to certain modes of operating subjectivity, which is itself however not "in time," since it must not be the constituted, but the constituting. So Bieri postulates that if one can show that the experience of time, to which time itself is going to be reduced, can be understood only by presupposing a time-structure, one has proved the reality of time.

It is widely known that in Husserl's theory of time the most important acts of subjectivity are "now-consciousness" of primal impressions, "retention," and "protention." In his view, it is because of these subjective functions that sensory data are ordered preconsciously in a form of a temporal flux. For example, retention is an act by which the elapsed sound, for example, is still retained in our consciousness and we become aware of it as "what has just been (*das eben Gewesene*), or as the "just past" (*das soeben Vergangene*); it is a privileged act in which "the immanent temporal unity of the sound and also the unity of the flux of consciousness itself are constituted."[8] Therefore, it must be an indispensable act to the constitution of time. For, according to Husserl, "it pertains to the essence of the intuition of time that in every point of its duration it is consciousness of what has just been."[9]

However, Bieri points out first of all that Husserl almost thought of the

A series in speaking of time. His vocabulary, including "now-conscious-ness," "retention," and "protention," shows that his notion of time consisted of the A-determinations. This can be considered as a necessary result from his method characterized by "self-reflection" of experience. And Bieri asks if Husserl really performed "the exclusion (*Ausschaltung*) of objective time," in other words, if he did not succeed in the transcendental reduction of time. Certainly, we can realize with Husserl that retention is an act by which the temporal unity of immanent object and even of consciousness itself is constituted. But Bieri rather doubts that retention is an act by which immanent objects are subject to retentional modification. For retention, taken in its literal sense, is an act by which we can "take hold" of the past in our consciousness as "the just-having-been," and not an act which in itself makes immanent objects pass away. The modification itself does not occur because we have an act called retention, but it is presupposed in retention; retention can operate only by presupposing the modification of immanent objects in time. That is to say, retention itself can operate only "in time." And time in this sense is, in Bieri's opinion, nothing but the time of B-relations, for otherwise we cannot help falling into a vicious circle. Since it is obviously the same with "protention," he concludes that even the program of describing the experience of time cannot do without real time, in which the phenomenon must stand.[10]

We have considered Bieri's sharp and interesting criticism of Husserl's theory of time. But since the point of phenomenology, as mentioned above, consists in the method of transcendental reduction, we must ask ourselves if in Bieri's criticism the problem of reduction was fully taken into consideration. For, properly speaking, it seems very probable that the time of the B series, which was supposed real by him, should also be excluded by reduction, and that his criticism does not fit for Husserl's proper intention. What indeed is the residuum after the reduction? Is it really the relation of "earlier" and "later"?

Now, what most naturally suggests the notion of time to us is the phenomenon of change or motion of things. Any change or motion does not occur in a moment. And between its beginning and its end, we usually constitute some interval, which we call time. So our question here is how to consider this interval. Is it real? But before proceeding to this problem, we must attempt the reduction of this interval itself.

In our ordinary life, change is described in terms of space and time; time appears always entangled with space. So, in order to consider time as pure as possible, we must choose an example in which space is reduced to the minimum. Such an example may be offered by a melody, which was a privileged object of observation for Husserl, too. A melody is not as simple, however, as it seems at first glance. To perceive in it the essence of time, we should simplify it as far as we can, and exclude pitch, timbre, rhythm, and so on. What shall we get as a residuum? Perhaps sounds scarcely differing from each other, linked by pair, like the ticktock of a clock.

Let us suppose that we are listening to the ticktock of a pendulum; we shall realize that our attention follows those sounds by a kind of internal gesture. We recognize this rhythm, because we notch it, and make a series of movements within ourselves which accord with the beats of the pendulum. It must be the same with a melody, and for this reason we can be deeply affected with it or be surprised when unexpected sounds are heard in it.

What is carried out then, strictly speaking, in this inner movement? It goes without saying that, in the couple tick-tock, we have neither one sound without the other, nor two sounds together. We perceive the passage from "tick" (X) to "tock" (Y), and this is the only reality that is given immediately to our consciousness. This passage consists neither in the superposition of Y on X, nor in the souvenir of X associated with Y. The fact is that, when Y arrives, we have consciousness of X as "just past" (*soeben vergangen*). We have already seen that this consciousness was called "retention" by Husserl. So, as G. Berger rightly pointed out, retention gives us an actuality in the form of "just past," while memory is an intention toward the past, i.e. the absent.[11] This is the reason why Husserl counted retention as a sort of perception. "Only in primary remembrance (= retention)," Husserl said, "do we *see* what is past; only in it is the past constituted, i.e., not in a representative but in *a presentative way*."[12] Retention gives us then a change as a real process of passage, and at the same time we have the "dense" present.[13] Therefore, after phenomenological reduction, we have not time but the present which is changing.

In a passage cited above Huserl said that we see in retention "what is past." It seems, therefore, as if retention involved the consciousness of the past. But we must keep in mind that retention was not for Husserl an "apprehending act," but as it were a preconscious act. He said in Appendix IX of *The Phenomenology of Internal Time-Consciousness*, "Retention itself is not an 'act,' but a momentary consciousness of the phase which has expired and, at the same time, a foundation for the retentional consciousness of the next phase." In respect of retention, there can be no question of a clear consciousness of the past. Then, it is also evident that retention cannot involve even the consciousness of the "earlier," as far as a positional consciousness is concerned. The "exclusion of objective time" must necessarily imply the exclusion of time as the B series. And this will be confirmed by reference to Husserl's concept of the "living present" (*lebendige Gegenwart*). For K. Held tells us by means of Husserl's manuscripts that the reduction to the "living present" was the most radical one and Husserl's greatest concern in his last years.[14]

What Husserl called the "living present" is an original and primordial present of the ego which is operating at present. But it must be remembered that this primordial present was not for Husserl a position in time, a temporal phase among others, and that "I am" or "I operate" in the most original sense of the term was distinguished from my present in the sense of the actual now-phase. This present was the persisting and constant "there" (*Da*) of my presenting act, and therefore an anonymous, prereflexive, and pretemporal present, from which the constitution of time should be started. On the other hand, the constancy of this present cannot be considered as the immobility of "transtemporal" (*überzeitlich*) substance, for example. It was called "living" because the ego itself of this present always emerges as co-streaming (*mitströmend*) in each formation of its life. It is why this present was often called by Husserl the "standing-streaming" (*stehend-strömend*) present. Such a present could be obtained only by the insatiable deepening of reduction through parenthesizing every sort of past and future. Therefore, we can still assert that even from Husserl's point of view the residuum after the reduction should not be time, but the dense, changing, and pretemporal present. In Husserl's view also, time must be seized as the result of self-temporalization (*sich zeitigen*) of the transcendental ego which has a mode of being of "living present," in a word, as the constituted.

Again, I must add that it is also the case with time as the B series. No doubt, compared with the A series, the B series may be characterized as being more abstract or intersubjectively more elaborated. For this reason, it seems as if we could speak of "earlier" or "later" as objective relations which events themselves might have among each other, and we should not need the existence of any subjectivity to be able to speak of such a relation. But it is evident that an event can be neither "earlier" not "later" without being compared with some other events. If every kind of comparison pre-supposes some act, such a relation must be the constituted. And indeed, the primordial changing present does not even involve a distinction of "earlier" and "later." For this distinction presupposes an actual and positional act of consciousness, while the present is almost preconscious.

Does one yet object that the present can change only because of real time? Certainly, as Bieri pointed out, it is not because we have made it change that the present changes. Even the acts like retention and protention cannot help operating on the ground of the real change. Moreover, we cannot reverse the direction of time or arbitrarily replace the order of temporal sequence, while, in the case of space, we can make "here" and "there" emerge everywhere by changing our position. Even if we can arbitrarily regard any moment as "past" or "future" by defining some suitable moment as "now," we can never change the temporal order as we please. Thus we are inclined to suppose the existence of real time. But we must know that the reality of change does not imply the reality of time. As stated above, it is not because there is time in the world that things change. If there are some processes of change which are irreversible, it is not because time has made them irreversible. Time does not operate upon reality: we merely describe change in terms of time, which is irreversible in the linguistic sense of the term, no matter whether the real process of change is irreversible or not. It may not be useless to notice here that there have been many misunderstandings in this respect. For example, sometimes the direction of time has been confounded with the causal relation of events, and sometimes the second principle of thermodynamics has been regarded as the demonstration of the irreversibility of time itself. But it is evident that the causal relation of events does not imply the temporal one, for cause and effect must always be simultaneous, and there can be no relation of "earlier" and "later" between them. For this reason, the event as cause can produce some effect, and at the same time receive its reactions. Likewise, the second principle of thermodynamics can in itself tell us nothing about time. For even in order to discuss the problem of the quantity of entropy in some process, the irreversibility of time must already be presupposed. If the process is said to be "reversible" or "irreversible," it is only on the basis of the unique direction of time.

In this sense, we could say that time pertains only to our language for expressing change, and is a human meaning of change. As Berger said, we have perhaps constituted time so that it might be a revolt against change or death. For change means first of all the disappearance of what is present, and it is that which annihilates rather than realizes. Our "nostalgia of sufficiency"[15] probably has made us constitute time to enable us to grasp change as a whole. What we call the interval between the beginning of motion and its end is nothing but such a change conceived as a whole.

Tôhoku University, Sendai

NOTES

[1] B. Russell, *The Principles of Mathematics*, I, Cambridge, 1903, § 442.
[2] J. M. E. McTaggart, *Nature of Existence*, II, Cambridge, 1968, § 315.
[3] J. E. McTaggart, 'The Unreality of Time,' *Mind*, 1908.
[4] A. Lalande, *Vocabulaire technique et critique de la philosophie*, P.U.F., 1956.
[5] A. Koyré. *Newtonian Studies*, Chapman & Hall, 1965, p. 103.
[6] Cf. L. Wittgenstein, *Tractatus Logico-philosophicus*, 6.3611.
[7] J. M. E. McTaggart, *Nature of Existence*, II, § 610.
[8] E. Husserl, *The Phenomenology of Internal Time-Consciousness*, translated by J. Churchill, M. Nijhoff, 1964, §39.
[9] Ibid., §12.
[10] P. Bieri, *Zeit und Zeiterfahrung*, Suhrkamp Verlag, 1972, p. 199.
[11] G. Berger, *Phénoménologie du temps et prospective*, P.U.F., 1964, p. 130.
[12] E. Husserl, op. cit., §17.
[13] Cf. 'L'épaisseur du présent préobjectif,' M. Merleau-Ponty, *Phénoménologie de la perception*, Gallimard, 1945, p. 495.
[14] K. Held, *Lebendige Gegenwart*, Nijhoff, 1966.
[15] G. Berger, op. cit., p. 135.

WATARU KURODA

PHENOMENOLOGY AND GRAMMAR

A Consideration of the Relation Between
Husserl's Logical Investigations
and Wittgenstein's Later Philosophy

I

Edmund Husserl's research in the problems concerning phenomenological
reduction is a beautiful exemplification of philosophical strictness. In fact,
we cannot begin the activity of philosophizing unless we go beyond the
naiveté of the natural standpoint in accordance with which we believe
unconditionally in the existence of the world, and live completely
absorbed in the matters therein. We have to suspend this belief and, with
a full determination, to think over everything again on our own
responsibility. At the same time, without our internal tension and effort
to maintain this determination, we cannot pursue our philosophical
consideration with success. However, it is quite another matter whether we
accept Husserl's conviction that the field of experience and cognition
opened by phenomenological reduction must be that of transcendental
consciousness, and solve all the problems therein. The assumption that all
the existence and significance must be reduced to the constitutive acts of
consciousness may be, conversely, an encumbrance to the thorough
pursuit of that philosophical task. I do not believe that we can solve all the
problems of philosophy by one and the same method. We should rather
think of the procedure of reduction as an open method unrestricted by the
standpoint of transcendental subjectivism. I want to justify and develop
this simple and perhaps "naive" belief of mine as far as possible. This is the
motivation of my offering this thesis here.

For an illustration of this general remark, we take up the problem of
linguistic meaning as one of the fundamental issues of philosophy. That is,
we inquire into the meaning of "meaning." This is an activity of thinking
to be called *nonnatural*. In our ordinary linguistic consciousness, the
meaning of a word is inseparably connected with things and processes of
the world which the words denote, classify, or describe. If we concern
ourselves with the problem of meaning as a matter of philosophy, we must

Nitta/Tatematsu (eds.), Analecta Husserliana, Vol. VIII, 89—107. *All Rights Reserved.*
Copyright © 1978 *by D. Reidel Publishing Company, Dordrecht, Holland.*

suspend for a while our belief in the existence of these objects, and put in parentheses all the knowledge of empirical science on linguistic or semiotic processes. I should like to call this methodological procedure "semantical reduction." All the "naive" interpretations of meaning which took no account of the necessity of this reduction actually collapsed, though I can give no detailed explanation concerning these circumstances here. Consider the denotation theory which regards the real object which a word points to as the meaning itself, or the ideational theory which identifies idea, image, or another immanent object with meaning, or the behaviouristic theory which defines "meaning" in terms of some pattern of stimulus-response. These all represent a standpoint that considers the linguistic meaning in connection with the things, qualities and relations in the world, or identifies it with them. None of them goes over the limitations of the natural standpoint, but each of them presents a naturalistic theory which arises from adhering to one aspect of ordinary linguistic consciousness.

In contrast to this, Saussure's *Course in General Linguistics* offers an example of semantical reduction, giving us some valuable suggestions. According to his theory, the meaning of a linguistic symbol is determined by its difference or opposition to other symbols belonging to the same linguistic system, and no symbol bears significance independently. This important insight was preceded by the methodological step of denying and eliminating natural relations between the signs and things signified completely. However, I am afraid even this insight of Saussure's, if not understood accurately, will only lead us to another form of naturalism. If we think of difference and contrast among symbols as present independently from the description of linguistic rules, we immediately fall into an error of reifying the linguistic system. We shall be subjected to the illusion that the whole of the linguistic system must be present, in some way or other, in the actual situation of linguistic utterance. Can it not be said that philosophical theories of meaning inclined to psychologism are obsessed with this illusion? As a matter of fact, we use words meaningfully, and yet cannot read the whole system of linguistic rules through a sentence or a speech act. Therefore, one comes to believe in the existence of some kind of grammatical mechanism in the realm of mind. I suppose the main line of their thinking can be traced like this. The fundamental conception remains unchanged whether that system is regarded as the mechanism of association between ideas, or as behavioural disposition built up by the repetition of stimulus and response, or, so to speak, as innate ideas. In short, that is nothing but an idle duplication of grammar. On the whole, we cannot talk of any other linguistic rules than those described by language, and there exists no linguistic system except those constituted by the systematic description of linguistic rules. This simple and fundamental truth has been overlooked.

How about the case with Husserl? Can his consideration of the problems of language and meaning have overcome the naiveté of the natural standpoint? If we ask for an answer to this in Husserl's works, we must first examine his *Logical Investigations*. In fact, we can find descriptions corresponding to what I call semantical reduction in the opening chapter of the first investigation titled "Expression and Meaning," in Volume 2 of this work. There Husserl tries to distinguish *Ausdruck* ("expression") from *Anzeichen* ("indication") and to reduce our language to a system of genuine *expressive* signs, with no communicative function. This process excludes all the elements of language necessary for the communication with others, and consequently there remain only isolated mental life and intentional acts of consciousness which constitute the meaning of expression therein.[1] In this way Husserl's semantical reduction was from the start directed toward the realm of consciousness. In fact, most of the ideas presented in the process of this reduction was incorporated into the description of phenomenological reduction in "Ideas, Book I." Furthermore, we can recognize a distinct correspondence between this part of the *Logical Investigations* and the consideration of intersubjectivity in the *Cartesian Meditations*, especially that process of reducing transcendental experience into the proper sphere of monadic consciousness. The process of reduction in the *Logical Investigations* can be regarded as the first sketch of phenomenological reduction, and as Derrida's *The Voice and the Phenomenon* admirably proved, Husserl's semantical investigations at this early stage fundamentally conditioned the subsequent development of phenomenology.

Tracing the process of reduction in the *Logical Investigations*, I venture to sum up how the process went on and what result came out. Though detailed discussion cannot be expected here, I want to be exhaustive about the important points for criticism. This examination will illuminate the point where Husserl's approach to transcendental phenomenology parts from the path which we must follow.

II

As stated previously, what dominates the whole process of reduction in the *Logical Investigations* is the contrast between indication and expression. First, what sort of sign is *indication*? We shall take up some examples

Husserl presented; "a flag is the indication of a nation," "a brand is that of a slave," and "a fossil vertebrae is that of the existence of prediluvian animals." "In these we discover as a common circumstance the fact that certain objects or states of affairs *of whose reality someone has actual knowledge* indicate to him *the reality of certain other objects or states of affairs* in the sense that *his belief in the reality of the one is experienced* (though not at all evidently) *as motivating a belief or surmise in the reality of the other.*"[2] In short, what combines an indication and the thing it points to is the relation of empirical motivation, and is essentially different from such an ideal and necessary connection as we find between a proposition and its logical consequence.

On the other hand, what kind of sign is expression? According to Husserl, each instance or part of speech, as also each sign that is essentially of the same sort, will be counted as an expression, a meaningful sign. Then what qualifies linguistic signs and their equals as expressions? An answer to this question will be given for the first time in the stage where our language is reduced to the pure state of expression; for the present, we shall change the order of description and take a short cut to it. There Husserl says that meaningful expression is not a mere verbal sound, but a sign enlivened by the act of meaning (*Meinen*) or meaning-intention (*Bedeutungsintention*) which relates this sign to a certain object, and the act of meaning-fulfillment (*Bedeutungserfüllung*) which intuitively realizes that relation to the object. In pure act of expression performed in "isolated mental life," the verbal sound is experienced as one with the meaning-intention, and in case the intended object comes into sight intuitively, the act of meaning-intention is naturally made one with that of meaning-fulfillment. Consequently expression and its object are united by internal and intentional connection, which is quite different from external and empirical relation in the case of indication signs.[3]

In short, the difference between expression and indication is based on the difference in referring function of the sign, and it is not a substantial, but a functional distinction. However, from my viewpoint, Husserl often betrays the principle of this functional classification, and that at very crucial points. This is his distortion of method caused by a certain pressure from theoretical assumptions. We shall confirm the point at each stage of semantical reduction in the *Logical Investigations*.

We can consider Husserl's semantical reduction by dividing it into three parts. The process of restricting the extension of "expression" to the sphere of linguistic sign corresponds to the first stage of reduction. Here the signs of gesture and facial expression are separated from expression in a genuine sense and excluded from the field of phenomenological semantics. According to Husserl, facial expressions and various gestures which accompany our speech are only of the nature of involuntary utterance (*Äusserung*), and they are not phenomenally one with the experiences made manifest in them, as is the case with speech.[4] However, are finger language of a deaf-and-dumb person and talking eyes of lovers only involuntary utterances, and can we say that they express nothing? It seems that the functional distinction between indication and expression cannot be a justification for separating gesture and facial expression from speech or part of it. At least Husserl must have tried, at the first stage of reduction, a detailed investigation into the extension and function of signs that are *essentially of the same sort* as linguistic expressions.[5] The importance of the problem Husserl evaded was shown in Merleau-Ponty's discussion of language, and Wittgenstein's criticism of phenomenological semantics also laid a focus there, as we shall see below.

What corresponds to the second stage of reduction consists in removing the indicating function of speech, which is closely interwoven with its expressing function in the social process of communication. In short, that is the process of separating the intimating function (*Kundgebung*) which tells us the mental states of a speaker, from the meaning-expression, and restricting the object of semantical consideration to the expressive function of soliloquy performed within the limit of lonely mental life. There are many points to be doubted in this reductive procedure. Does the intimation of experience by words only serve to play the part of indication? When we engage in verbal communication with one another, do we not actually distinguish the case in which we have a definite knowledge of the partner's experience by means of his words, from one in which we only guess it? There are not a few cases in which the words expressing experiences remain only opaque, indicative signs for the speaker himself. This is not a rare case even in soliloquy. However, such objections will never move Husserl. His reduction is from the outset not founded on the observation and description of actual linguistic life, but is predestined by the *essential difference* as follows:

The hearer perceives the speaker as manifesting certain inner experiences, and to that extent he also perceives these experiences themselves; he has not an 'inner' but an 'outer' percept of them. Here we have the big difference between the real grasp of what is in adequate intuition, and the putative grasp of what is on a basis of inadequate, though intuitive, presentation. In the former case we have to do with an experienced, in the latter case with a presumed being, to which no truth corresponds at all.[6]

Semantical distinction between indication and expression, or between intimation and an act of expression derives from more fundamental distinction, which is both ontological and epistemological, between transcendence and immanence. Criticism of phenomenological semantics must aim ultimately at this presumption.

As the third process of reduction Husserl excludes the physical existence of the linguistic sign from his consideration,[7] but this may well be said to belong to the second process as a part of it. According to Husserl, the sign we understand as an indication must be perceived by us as really existent, but in soliloquy we are content with the imagined words. The very expressive function of linguistic expression is not at all damaged by the nonexistence of physical signs. However, this discussion is obviously futile. Physical existence and nonexistence of signs have no concern with the functional distinction between indication and expression. For instance, the imagination of the Stars and Stripes reminds us of Mr. Lincoln or Mr. Kennedy and the verbal image of "President of the United States" can do the same. What can be observed in both cases are sign-relations empirically motivated, and the origin of this relationship lies in the psychological facts which may be called by the classical term of "association of ideas." Indication never changes into expression, simply because it does not exist as a physical sign. The second instance above will show that the functional distinction between expression and indication does not coincide with the boundary between linguistic sign and nonlinguistic sign.

So far we have examined the reductive procedure in the *Logical Investigations*. In the process of this reduction, the fundamental characteristics of phenomenological semantics have become conspicuously clear. First of all, we can point out the epistemological orientation of this semantics or, in other words, its subordination to epistemology. This orientation is made manifest by the titles given to the second volume of the German edition and to the second part of the same volume, and it may be said that this needs no mention here.[8] However that feature is far more significant and has to do with the core of Husserl's semantics. Husserl describes meaning-action in general in accordance with this fundamental orientation. His defining the meaning of expression as *its relation to an object* will directly show this. The aim of semantical investigation was from the beginning fixed at the expression of *objectifying act*, i.e., of presentation and judgment. If we compare the first investigation, 'Expression and Meaning,' with the sixth assigned to the phenomenological elucidation of knowledge, it seems evident that the fundamental scheme of his semantics was transferred unchanged to an epistemological dimension. In the case of Husserl, the cognition of an object and the intuitive completion of the meaning-intention were made one and the same. Therefore, the act of identification confirming the agreement between the object just meant and the object given to us in intuitive fullness, is regarded as "cognition," and such fundamental concepts of epistemology as "truth" or "evidence" are also defined from this viewpoint. The definitions are as follows. "Truth" is, as the objective correlate of identifying act, a *state of affairs*; and as the aim of identification it is *the full agreement of what is meant with what is given as such*. "Evidence" is just the ideal of this adequation.[9]

In short for Husserl, the concept of meaning is generally subordinate to that of truth. Meaning is the possibility of fulfillment and insofar as the object is not given to intuition, it cannot but remain in a negative status of the privation of truth. However, when Husserl introduced the concepts of the meaning-intention and meaning-fulfillment for the first time, he called our attention to the fact that the former is an autonomous action, not dependent on the latter. "A *name*, e.g., names its object whatever the circumstances, in so far as it *means* that object."[10] This autonomy of meaning-intention in fact plays an important part in the *Logical Investigations*. For instance, Husserl thinks of *pure grammar* as the basic discipline of logic, and makes the pure theory of meaning-forms precede the doctrine of the validity of meaning. This important proposal is naturally based on the presupposition that meaning-intention is autonomous and independent from meaning fulfillment. However, the idea of pure grammar is not discordant with the initial orientation of Husserl's semantics. As proof of this, the name of "pure grammar" is changed into "pure logical grammar" in the second edition. Husserl acknowledged himself that even this idea of "pure grammar" is limited by the epistemological aim of objective validity, and does not comprise "the entire a priori of general grammar." That is, the autonomy of meaning-intention is no other than provisional and is to be absorbed into the process of cognition in due time. Summarizing this according to Derrida's concise formulation, the meaning precedes the truth only as its anticipation.[11]

Thus, in the *Logical Investigations*, the meaning-action is to be analyzed as a process of verification which consists in confirming the accordance of thinking and intuition. It may be said that this is a semantics with a distinct mark of verificationism. Furthermore, this verificationism is determined from the beginning by the presuppositions of immanentism which regards the verification by inner perception as the ideal of cognition. Needless to say, from Husserl's standpoint, the final evidence cannot be secured by the outer perception including the existential presupposition beyond what is immediately given to our present consciousness, and the cognition based on inner perception definitely takes precedence concerning the matter of evidence. The whole process of semantical reduction is led by this ideal of cognition. The language which displays pure expressiveness within the enclosure of "isolated mental life" is nothing but one of phenomenological cognition. Apparently the final aim of *the Logical Investigations* is laid on the proof of completeness of phenomenological cognition. In the final, sixth investigation, his semantical and epistemological consideration developed mainly with expressions of inner experience as its subject matter. Phenomenological investigation into cognition is completed with the theory of phenomenological cognition.

In this way, semantics is subordinate to the conception of cognition in general and the latter, to the ideal of immanent, phenomenological cognition. In this teleological correlation, the phenomenological description of language becomes fused with the description of phenomenological language. I do not think this is a situation to be positively appreciated. What difficulties does this unification bring into phenomenological semantics, and how much is Husserl to pay for it? I shall consider these problems in the last part of my article. For the present, a method of indirect illumination will be employed in order to show the points of importance more clearly. In the next part, I will examine Wittgenstein's consideration of the problem of meaning, contrasting it with the semantics of the *Logical Investigations*. This is not at all an arbitrary comparison. There lies quite a close and important relation between the two theories of meaning. Wittgenstein's criticism of "private language," which formed the central theme of his philosophical activity in the period of *Philosophical Investigations*, can be comprehended thoroughly only when we take this relation into account. So far as I know, this relationship has remained unnoticed. I want to present a part of what I have investigated recently.[12]

III

If we try to trace Wittgenstein's contact with the *Logical Investigations* as a historical fact, we must go back to the time when *Tractatus Logico-Philo-*

sophicus was written. Some of the basic terms in this book, for instance, such words as "state of affairs" (*Sachverhalt*), "situation" (*Sachlage*), or "logical grammar" (*logische Grammatik*) indicate the link between this book and the *Logical Investigations*. In his posthumous work, *Philosophical Remarks* written in about 1930 when he was trying to go beyond his thoughts in the *Tractatus*, the word "phenomenology" came to be one of his important terms, which has already become widely known. This phenomenology was to illuminate the internal or essential relations among sensible qualia pertaining to such fields as sounds or colors, and Wittgenstein called this discipline also by the name of "grammar." For instance, the phenomenological description of color-space shows *a priori* meaning-rules which determine the compatibilities and incompatibilities of terms belonging to the system of color concepts.[13] This idea of Wittgenstein is undoubtedly based on Husserl's idea of "pure grammar."[14]

However, correspondence of vocabulary does not count for much. It seems to me that the grounding of Wittgenstein's later thoughts was built up by his confrontation with Husserl's semantics. Another posthumous work, "Philosophical Grammar" (1931–1934), especially Part I, can be regarded as a record of this critical examination. The title *Philosophical Grammar* may have been chosen to indicate its relation with the *Logical Investigations*. This is an expression which Husserl used when he explained the idea of "pure grammar."[15] Wittgenstein inherited Husserl's idea of philosophical grammar, according to which one must cut off all the empirical facts concerning linguistic processes and concentrate on the study of pure forms of meaning or meaning-rules. In addition, from the grammar he pursued, the supposition of intentional acts constituting the meaning of linguistic expression had to be thoroughly excluded. The grammar had to be liberated from the epistemological presuppositions and limitations of phenomenological semantics, and had to be expanded to all the spheres of linguistic meaning.

Why was Wittgenstein interested to such an extent in Husserl's *Logical Investigations*? There seems to be no answer but the following one. Between the semantical conceptions developed in the *Tractatus* and the intentional analysis of meaning in the *Logical Investigations*, there can be found a basic agreement. Therefore, for Wittgenstein, to criticize and go beyond the semantics in the *Logical Investigations* means at the same time a thorough departure from the domain of thought circumscribed by the *Tractatus*. To confirm this point, we shall give a glimpse of the so called picture theory of meaning in the *Tractatus*.

In the language of the *Tractatus*, every proposition is a truth-function of elementary propositions, and the sense of a proposition is a function of the sense of elementary propositions. An elementary proposition itself is considered to consist of the combination of simple signs whose only proper function is naming a particular object. That is, it is regarded as a picture representing a certain state of affairs by the disposition of simple signs. To understand the sense of an elementary proposition is to know what is the state of affairs or the configuration of simple objects the proposition represents. In other words, it is to know in what case a proposition accords with reality and is true. This is an instance of verificationism which subordinates the problem of meaning to that of truth, and is consistent with the standpoint of the *Logical Investigations*. According to the *Tractatus*, if we have no possibility of verifying our proposition by collating or comparing it with extralinguistic reality, there will remain no room for taking the sense of a proposition into account. As is the case with Husserl, meaning precedes truth only in so far as the former anticipates the latter.

Of course, Wittgenstein in the *Tractatus* makes no mention of the mental action which enlivens the linguistic signs. It seems that all his semantic consideration is developed along the line of demarcation between physical reality and propositional sign which is also nothing but a part of physical reality. However, I think it necessary to pay special attention to the following point. When he talks of the structure and function of a picture in general, he makes a definite discrimination between two relations, that is, between the relation among various elements constituting the picture and the relation between the elements of the picture and those of reality itself. The picture has a certain structure determined by the disposition of its elements. It is not until pictorial relationships between the elements of the picture and those elements or things outside the picture are established that the picture becomes a representation of a particular fact.[16] Then what is it that correlates the elements of the picture with another object? Of course, it is not the picture itself but, as Miss Anscombe says, it is *we* ourselves.[17]

Here is a paradoxical state of affairs. The *Tractatus* view of meaning which defines the sense of a proposition in terms of a pictorial relationship between two facts must be supplemented with the supposition of a kind of mental action that makes the proposition reach a certain fact, by correlating names with objects. Even though we bring in another proposition or picture as an intermediate, we cannot but refer to an act of interpretation which relates this both to the original reality and to the proposition. If this is a conclusion of the picture theory of meaning, the semantics of the *Tractatus* will not be alien to that of the *Logical Investigations*. Rather we can say the former is fundamentally in accord with the latter. In fact, Wittgenstein himself recognized this. In *Philosophical Remarks*, the first work after the *Tractatus*, he presents the following proposition. "If you exclude the elements of intention from language, its whole function then collapses."[18] Here lies a starting point of his speculation on meaning in the transition period. His examination was focused on the tacit assumption of intentional act or *Bild-Auffassung* which supported the *Tractatus* view of language from behind. Only after the element of intention had been eliminated from Wittgenstein's view on language, could the conception that the meaning of a linguistic expression is its use come to be fixed as the leading idea of his philosophy. And it was about this time that he began to write *Philosophical Grammar*.[19]

Then how did Wittgenstein in *Philosophical Grammar* succeed in excluding the moment of intentionality from language? I will try to summarize Wittgenstein's discussion on the two most important problems.

The first is his investigation into ostensive definition.[20] According to Wittgenstein, ostensive definition is not, as is understood generally, a rule which combines linguistic signs with the objects outside language. When we teach a child a word by pointing to an object which actually exists within our visual field and by saying "this is a rose" or "this is red," what we are actually doing is explaining the meaning of "rose" or "red" by the sign of gesture. In other words, we replace the sign of gesture-language with a sign of word-language, and show the rule of translation between them. In short, ostensive definition is like ordinary verbal definition, namely, nothing other than a rule for replacing signs, and can be interpreted as the presentation of a rule of grammar. Accordingly, to teach words by the process of ostensive definition is a work which remains within the boundary of language. When interpreting it, one need neither allude to transcendent objects nor suppose the act of consciousness which fills up the gulf between words and objects.

This interpretation of ostensive definition is apparently too simplified and includes several overstatements. Wittgenstein could not reach consummation here and had to delay elaboration of this interpretation until a later period. Anyway, in this interpretation of ostensive definition, two important and far-reaching ideas are included. The first involves getting rid of the distinct division between word and gesture, and considering both of them equally in the status of linguistic sign. This, as I have stated above, forms just a radical contrast to Husserl's theory which excludes gesture signs from the sphere of linguistic expression, in accordance with his fundamental intention of dividing two kinds of signs, that is, expression and indication. Wittgenstein must have been conscious of this opposition. The second is a sort of operationism which defines "meaning" in terms of "explanation of meaning," and rejects the reification of meaning definitely. This view had been established already in the *Tractatus* so far as signs of logic and mathematics are concerned. However the language depicting the real world was left outside the interpretation at that stage. In *Philosophical Grammar*, Wittgenstein grasps both ostensive and verbal definition as rules of grammar, so the dualistic interpretation of meaning, which contrasts the rules combining language and reality with those rules of combination among signs remaining inside language, has been overcome at this stage of his thought. As the conclusion of his meditation on ostensive definition, Wittgenstein remarked: "The connection between language and reality is made by definitions of words, and these belong to grammar, so that language remains self-contained and autonomous."[21] The assertion that language is autonomous is the fundamental proposition of *Philosophical Grammar*, and we can conclude that all the inquiry of Wittgenstein in the transient period was converged on this proposition. This autonomy signifies more than anything else the autonomy against intentional act of consciousness. The supposition of meaning-act which enlivens the linguistic signs from behind must be excluded from all the field of semantical investigation: "It is grammatical rules that determine meaning (constitute it) and so they themselves are not answerable to any meaning and to that extent are arbitrary."[22]

Here we shall turn the focus of our inquiry to another problem. Wittgenstein in *Philosophical Grammar*, having dealt with the problem of ostensive definition, directed his attention to the analysis of such experiences, which are concerned with the state of affairs not yet realized, as expecting, wanting, wishing, hoping, intending, looking for, etc.[23] I suppose this discussion was also held with some parts of the *Logical Investigations* as the object of criticism. We can confirm this without difficulty if we contrast it with the description of meaning-fulfillment in the sixth investigation of Husserl's work.[24] The acts of consciousness Wittgensteins referred to have a common characteristic of intending towards a certain goal, which belongs to "a narrower concept of intention" in Husserl's case. According to Husserl, the contrast between intention and fulfillment in these experiences is essentially similar to that we have already seen between meaning-intention and meaning-fulfillment. Meaning-intention is in fact an instance of intentional act in this narrower sense. Consequently, it was quite natural for Wittgenstein that he applied the method, which proved to be effective concerning the problem of ostensive definition, to the analysis of the experience of this kind, and tried to reaffirm the proposition that language is self-contained and autonomous.

We shall trace the outline of Wittgenstein's discussion, taking up the case of expectation as an example. For instance, when someone expects Mr. N. to visit him, he will show various sorts of behavior of expectation. However we cannot be sure that his expectation is just that of Mr. N.'s visit, in case we are observing only his expectant behavior, in other words, the pictures of expectation. Therefore, we are inclined to suppose some act of consciousness which connects his present expectation with Mr. N.'s visit in the near future. The fundamental construction of the problem is similar to the case with ostensive definition.

However, what will happen if he says there, "I expect Mr. N. will come in"? If these words are added to his behavior of expectation, there will be no doubt that his expectation is that of Mr. N.'s visit. The expression of expectation by words first defines his behavior as the expectation of Mr. N.'s visit. A certain expectation and an object which satisfies it are combined, so to speak, grammatically and logically by the verbal expression of expectation, and their relation is certainly an internal one and different from the empirical and hypothetical relation, for instance, between hunger and a meal which appeases it. However, one need not think of this internal connection as brought about by some intentional act of unification. Expectation and its object are immediately combined by the first person expression of expectation. "It is in language that expectation and its fulfilment make contact."[25] We do well to think that the expression of expectation is the act of expectation itself.[26]

There seems to be included in this discussion a very important and essential criticism of phenomenological semantics. Husserl, who subordinates the investigation of meaning to an epistemological purpose and tries to see an ideal state of expression in the words of phenomenological description, reduces the function of enunciation into that of judgment. In short, sentences are considered to be expressions of judgment which consists of subsuming each experience under the concept of expectation, wish, hope, etc., and naming them. The experiences of expectation or wish are objects of judgment, and only play the part of fulfilling the meaning-intention of judgment by their actual presence.[27] Against this, if we admit, after Wittgenstein, the case that the expression of expectation by words is not the result of reflective judgment but the act of expectation itself, we can no longer adhere to the conviction that the function of language is to be restricted to that of description (expression of judgment). The verificationism which combined the *Tractatus* with the *Logical Investigations* must be abandoned. Actually, Wittgenstein's insight mentioned above came to be integrated into the view of the immense multiplicity of language-games that forms the keystone of *Philosophical Investigations*. To speak language is an act pertaining to a certain form of life. We can assume that what led him to this fundamental cognition was the discussion of intentionality in *Philosophical Grammar*. The task of analyzing in detail the various forms of speech act, including social and institutional conditions, was left to Austin and to others. However, it is undoubtedly Wittgenstein's insight that opened for the first time the horizon of such an investigation. Considering only this respect, the influence the *Logical Investigations* gave him was profound and far-reaching.

IV

We shall return to Husserl. In the latter part of the foregoing paragraph, we referred to the relation between meaning-act of expression and noncognitive intentional experience such as wish or expectation. The interpretation of this relation determines, in fact, all the intentional analysis in the *Logical Investigations*, and had an influence on the further development of phenomenology, so we shall examine it all over again. The kernel of the problem is as follows. As previously stated, Husserl recognized that the intentional experience with a goal has the same structure with the meaning-act of expression. At the same time, he grasped the act of cognition as the unity of meaning-intention and meaning-fulfillment, and analyzed the act of meaning as incorporated into the cognitive process. Then, must we regard all the intentional act as meaning-act, or must we consider that the meaning-act of intention and fulfillment belong only to the class of objectifying acts (i.e., acts of presentation or judgment)? Husserl answered this question in accordance with his epistemological program. That is, what bears meaning is only the act of objectification, and emotional or volitional experience such as wish or expectation does not lead to expression without the aid of the reflective judgment. However, this cannot be a real solution. First of all, if the expression of intentional experience always needs the intermediation of an act of judgment, such an intermediator is equally required for the expression of judgement, and we cannot evade *progressus in infinitum*.[28] On the other hand, if we admit that immediacy of expression is endowed to the act of judgment, why cannot the same immediacy be recognized in the case of noncognitive experience? Next, there remains a problem as to the relation between the reflective act of judgment and noncognitive experience as its object. In order to talk about the adequation between what is meant by reflective judgment and what is actually present as an emotional or volitional experience, we must suppose some kind of isomorphism between them. I mean that the structure corresponding to the intentional construction of judgment must be from the beginning immanent in emotional or volitional experience. We must return right away to the noncognitive experiences the meaning we once took away from them.

If we examine the third section of *Ideas*, Book 1, which analyzes the noesis-noema structure of intentional experience, it is quite evident that Husserl himself was acutely conscious of such a difficulty and sought after the solution with persistence.[29] What deserves special attention is that his phenomenological semantics led, through this investigation, to a definitive turning point. He ways, "The stratum of expression – and this constitutes its peculiarity – apart from the fact that it lends expression to all other intentionalities, is not productive. Or if one prefers: *its productivity, its noematic service, exhausts itself in expressing*, and in the *form of the conceptual* which first comes with the expressing."[30] In short, expression only reflects and copies the meaning and its various moments immanent in preexpressional experience. The relation between expression and what is to be expressed is considered to be applicable to the expression of all the experiences, including such experiences as will or desire. Of course, in the phenomenological description one must discriminate noetic and noematic state of affairs in the experience and bring the latter, i.e., the noematic sense of experience, to the level of logical meaning. However, the shift from the act of will or desire to that of judgment corresponds to the task of transforming the immediate expression of experience into a mediate one. In this stage, Husserl distinguished clearly the relation between experience and its immediate expression from that of the immediate and the mediate expression.[31] The supposition of a judging act that intervenes between expression and experience has been removed once and for all.

Expression which was the main subject of semantical consideration in the *Logical Investigations* has thus been regarded equal to a transparent medium conveying the structure of meaning inherent to experience. Now logical meaning (*Bedeutung*) only reproduces, on the level of the conceptual, the noematic sense (*Sinn*) immanent in the experience. Then the frontier of investigation into meaning moves on together with it. Phenomenological semantics aimed at the *telos* of logicality turns the direction of inquiry to the basis of experience in our life. It must clarify the *genesis* of meaning, that is, the various layers of intentional constitution. Though I am not well informed concerning the later development of Husserl's philosophy, I can imagine that his contemplation on the meaning both of expression and of experience must have motivated his decision to return to the world of life.

Now according to Husserl, the noematic sense of experience belongs to the preexpressive layer of intentional act. Does it come to the same when the noematic sense is definitely of prelinguistic origin? Is preexpressive meaning constituted silently in the absolute absence of language? If the spirit of that semantical reduction dominates Husserl's phenomenology to the last, the answer must be yes. Now that expression and indication have been distinguished from each other, and the nonproductivity of the act of expression thus purified has been confirmed, there remains no room for language to take part in the constitution of noematic sense. So far as I know, there is no sign that Husserl tried a thorough reexamination of the presuppositions of his semantical reduction. For instance Husserl's discussion in *The Origin of Geometry*, a short work belonging to his last period, bears a testimony to my remark. There he asked how the *ideality* of geometry, which is originated from a personal consciousness, can reach the level of the *ideal objectivity*.[32] This questioning can be safely said to lie on an extension of that semantical reduction. The conception that ideal objectivity of geometry came into being by means of language, especially the language of letters and documents, does not move on the boundary between expression and indication. The view that ideal objects are constituted in the enclosure of solitary mental life has not been changed. A word or sentence, not as a token but as a type, which keeps its identity through sensible appearances, is also an ideal-object pertaining to this domain.[33] Roughly speaking, Husserl had just two conceptions of language; one is a language which is constituted as a system of ideal objects in some personal consciousness; the other is a language of physical signs which embodies those ideal objects and leads them into the intersubjective world of human history. The idea of *constituting language* remained quite alien to him. I think that here lies a problem we must take into account.

We must for a moment return to the point where Husserl began the process of his reduction. Expressing our own experiences by words and responding to expressions of others is, needless to say, what we are doing in our everyday life. In this stage, as Husserl himself recognized, expression and intimation are closely united in our speech acts and cannot be separated from each other. Expressing one's own experience is not an autistic activity of copying reflective judgment, but an action directed to the world and others. There language certainly fulfills a productive function. As for this point, Wittgenstein's discussion considered in the pervious paragraph will give us important suggestions. If, for instance, expression of my expectation only describes the experience which fulfills present consciousness, the work of expression will be completed then and there. Even if I experience the feeling of satisfaction (or dissatisfaction) some time later, the two experiences, i.e., expectation and satisfaction which arise at different times should be described as entirely different empirical facts, and only the indicative and external relation of signification can be established between the two. Actually, verbal expression of an expectation combines the two in an internal and intentional connection, as a certain expectation and *its* satisfaction (or dissatisfaction). The first-person expression of one's expectation defines the way of one's description of future experience at the same time. This is to be considered a more fundamental speech act than that of description, because the act of expression underlies the latter, conditioning the possibility of it. Those who understand the expression of my expectation will observe, describe, and explain my behavior from the viewpoint I provided for them. Even if they doubt the authenticity of my expectation, they cannot doubt it without adopting that viewpoint (my expecting so and so). Here consciousness does not constitute language, but conversely language constitutes consciousness. Our consciousness transcends the evidence of present consciousness with the aid of language and reaches a nonpresent, unreal object. It seems to me that Husserl's investigation into time-consciousness or intersubjectivity ended in an unsatisfactory result just because he neglected the creative or constitutive function of language mentioned above. However, I cannot develop this remark further in this article.

I have no objection to dividing the relation between the sign and the thing signified into the internal and expressive one, and external and indicative one. These are in fact the two fundamental forms of signification, and inquiry into the problem of meaning must be carried on always taking this difference into account. Nevertheless, the distinction between expression and indication must be considered to be no other than functional, and one must not combine it, as Husserl did, with the territorial distinction between immanence and transcendence. As shown in this article, most of the difficulties which threaten phenomenological semantics from the inside arise from dividing the domain of expressions and of indications. I would rather argue as follows, following Martin Heidegger's explanation of the distinction between *Phänomen* and *Erscheinung*.[34] Though nonthematically, the expressive or internal sign-relation has shown itself in our everyday speech acts, accompanying or preceding the indicative or external sign-relation. Now the central theme of philosophical semantics is to illuminate this expressive function underlying all the cognitions and actions, and to disclose the transcendental aspect of our speech acts thematically and more conspicuously.

NOTES

The quotations from the works of Husserl and Wittgenstein depend, as a rule, on published English translations. The following abbreviations are used to refer to them

LI — E. Husserl, *Logical Investigations*, trans. J. N. Findlay, Routledge & Kegan Paul, 1970.

ID — E. Husserl, *Ideas*, trans. W. R. Boyce Gibson, George Allen & Unwin, 1931.

TLP — L. Wittgenstein, *Tractatus Logico-Philosophicus*, trans. D. F. Pears and B. F. McGuinness, Routledge and Kegan Paul, 1961.

PR — L. Wittgenstein, *Philosophical Remarks*, trans. R. Hargreaves and R. White, Basil Blackwell, 1975.

PG — L. Wittgenstein, *Philosophical Grammar*, trans. A. Kenny, University of California Press, 1974.

[1] *LI*. II/2, 1, §§1–10.

[2] *Ibid.*, §2. p. 270.

[3] *Ibid.*, §9. p. 280ff. cf. II/2, VI, §14f.

[4] *Ibid.*, §5. p. 275.

[5] *Ibid.*, §§7–8, pp. 276–280.

[6] *Ibid.*, §7. p. 278.

[7] *Ibid.*, §8, pp. 279–280.

[8] They are titled 'Investigations for the Phenomenology and the Theory of Knowledge' and Elements of a Phenomenological Elucidation of Knowledge' respectively.

[9] *LI*. II/2, VI, §39, pp. 765–766.

[10] *LI*. II/1, I, §8, p. 280.

[11] J. Derrida, *La voix et le phénomène*, 1967, p. 109.

[12] A more detailed account of the part played by the *Logical Investigations* in the formation of Wittgenstein's later philosophy can be found in Chapter 8 of my book *Experience and Language* (in Japanese), Tokyo University Press, 1975.

[13] *PR*. 39ff., 76ff.

14 *LI*. II/i, IV, §12, pp. 516-518. §14, pp. 522-524.

15 *LI*, II/1, IV, §14, p. 525.

16 *TLP*. 2.15-2.1515.

17 G. E. M. Anscombe, *An introduction to Wittgenstein's Tractatus*, 1959, p. 68.

18 *PR*. 20, p. 63.

19 Cf. F. Waismann, *Ludwig Wittgenstein und der Wienerkreis*, 1967, p. 167.

20 *PG*. I-45ff.

21 *PG*. I-55, p. 57.

22 *PG*. I-133, p. 184.

23 *PG*. I-90-108.

24 *LI*. II/2, VI, §10, pp. 699-701. cf. II/1, V, §13, p. 563.

25 *PG*. I-92, p. 140.

26 *PG*. I-92, 101.

27 *LI*. II/2, VI, pp. 677-678. cf. *ibid*., pp. 837-851.

28 Wittgenstein also paid attention to this point, and commented that we call intentional an experience, for instance, the wish that that should happen, without any further *meaning*. According to him, saying so is not a description by an *Äusserung* ("utterance" or "manifestation"). See PG. I-103. p. 152.

29 *ID*. §§124-127.

30 *ID*. §124, pp. 348-349.

31 *ID*. pp. 353-356.

32 *Husserliana* Bd. VI, Die Krisis der europäischen Wissenschaften und die transzendentale Phänomenologie, Hrsg. von W. Biemel, 2. Auflage, 1962, Beilage III, S. 369.

33 *Ibid*., S. 368.

34 For me the contrast, between "expression" and "indication" in Husserl, between "phenomenon" and "appearance" in Heidegger, and between "criterion" and "symptom" in Wittgenstein can be considered to represent almost the same situation from the semantical viewpoint. So it was at least with Wittgenstein. We can confirm this in the change of his terminology. He used the word "symptom" in contrast with "phenomenon" at first, when he wrote *Philosophical Remarks*. Next he contrasted it with "expression" in the period of *Philosophical Grammar*, and at last with "criterion" about in 1933 when *The Blue Book* was written. His terminology has been settled without any change since then. The whole process of this change shows how important the contact with phenomenology was in the formation of Wittgenstein's later philosophy. I have tried to trace the process in detail in the eighth chapter of my work mentioned above.

PHÄNOMENOLOGISCHE BETRACHTUNG VOM
BEGRIFF DER WELT

1. DAS THEMA DIESES AUFSATZES

Als Hauptaufgabe der Husserlschen Phänomenologie kann man drei Probleme nennen: erstens die Wesensstruktur aller wissenschaftlichen und vorwissenschaftlichen Erkenntnis und des ihr zugrunde liegenden, transzendentalen Bewußtseins zu erforschen; zweitens die ursprüngliche Seinsweise der gegenständlichen Seienden und die der Welt als des sie umfassenden Gesamthorizontes begreiflich zu machen, und drittens die Seinsweise und transzendentale Leistung des Ich als des erkennenden und handelnden Subjekts gründlich aufzuklären. Im vorliegenden Aufsatz möchte der Verfasser auf Grund der späteren Schriften von Husserl vor allem das oben genannte zweite Problem untersuchen: Wie erfaßt die transzendentale Phänomenologie Husserls die Seinsweise der Welt (hier handelt es sich aber nur um die reale Welt) und deren Verhältnis zum erkennenden Subjekt? Gleichzeitig sollen einige Grundzüge des Husserlschen Denkens erhellt werden.

Wie auch K. Löwith mit Recht herausstellt[1], steht das "Weltproblem", d.h. die Frage nach einer durch Vernunft begründeten Welt, im Mittelpunkt von Husserls Denken. Nach dem Ausdruck von Husserl selbst ist es "die einzige Aufgabe und Leistung des phänomenologischen Idealismus, den Sinn dieser (realen) Welt, genau den Sinn, in welchem sie jedermann als wirklich seiend gilt und mit wirklichem Recht gilt, aufzuklären" (H.V, S.152). Deshalb können wir das Grundanliegen seiner Philosophie darin erblicken, die Welt aus dem letzten Grund ihres Seins her begreiflich zu machen[2]. Hierbei sei noch ganz kurz darauf hingewiesen, warum die neue, radikale Prüfung des Weltbegriffs gerade in der Zeit Husserls für die Philosophie zu einer zentralen Problematik geworden ist. Überblickt man die Hauptprobleme der Philosophie von der Mitte des 19. bis zum Anfang dieses Jahrhunderts, dann zeigt sich eine merkwürdige Tatsache: Das einst so aktive kosmologische Interesse am Weltbegriff tritt gerade in diesem Zeitraum völlig zurück und hat kaum eine größere Bedeutung. Woran liegt das? Wie Landgrebe meint[3], kann man den Grund darin erblicken,

Nitta / Tatematsu (eds.), Analecta Husserliana, Vol. VIII, 109–129. All Rights Reserved.
Copyright © 1978 by D. Reidel Publishing Company, Dordrecht, Holland.

"daß in der damaligen Philosophie ein bestimmter Begriff von Welt so sehr selbstverständlich und allein maßgeblich geworden war, daß seine ausdrückliche thematische Erörterung gar nicht mehr erforderlich schien." Dagegen ist für die heutige Philosophie seit Husserl der Weltbegriff wieder zu einem aktuellen und sogar zu einem zentralen Problem geworden. Diese neue Problematik setzt natürlich voraus, daß durch die sogenannte Krisis der Physik und Mathematik, die sich dem Vordringen sowohl der Relativitätstheorie als auch der nicht-Euklidischen Geometrie verdankt, die alte Selbstverständlichkeit des Weltbegriffs stark erschüttert und eine Neubestimmung dringend nötig geworden ist. In diesem Kontext geht es der Husserlschen Phänomenologie, die ja zunächst als Wissenschaftslehre intendiert war, darum, durch eine radikale Kritik an der Naivität der positiven Wissenschaften das Wesen der Welt als des universalen Horizonts natürlichen Lebens neu zu bestimmen. In der Tat, "von vornherein lebt der Phänomenologe in der Paradoxie, das Selbstverständliche als fraglich, als rätselhaft ansehen zu müssen und hinfort kein anderes wissenschaftliches Thema haben zu können als dieses: die universale Selbstverständlichkeit des Seins der Welt – für ihn das größte aller Rätsel – in eine Verständlichkeit zu verwandeln" (H.VI, S.184). Was nun auf dem Weg der Durchführung dieses Programms gleichzeitig in Frage gestellt und neu gewonnen werden soll, ist die "Lebensbedeutsamkeit" der europäischen Wissenschaften, die es auch mit dem Sein des Menschen zu tun haben.

2.   DIE KRISIS ALLER NEUZEITLICHEN WISSENSCHAFTEN UND
       DES EUROPÄISCHEN MENSCHENTUMS

Die in unserem Jahrhundert immer mehr zutage getretene Krisis der europäischen Wissenschaften geht nach Husserls Meinung auf das Auseinandertreten von Philosophie und Naturwissenschaft zurück, genauer: auf den Riß zwischen dem transzendentalen Subjektivismus der ersteren und dem physikalistischen Objektivismus der letzteren – ein Riß, der schon im 17. Jahrhundert aufgebrochen war. Die klassische Philosophie dagegen hatte noch Wissenschaft sein wollen, d.h. universale und rationale Erkenntnis des Seienden im Ganzen, der es darum ging, die ganze Welt unter einer einheitlichen Ordung zu begreifen. Aber da die neuzeitliche Wissenschaft mit Descartes und Galilei auf ein solches Ideal verzichtet hatte, zerfiel auch die Welt sozusagen dualistisch in zwei Welten: in die Natur als real und theoretisch in sich abgeschlossene

Körperwelt sowie in die seelinsche Welt. Infolgedessen hörte die neuzeitliche Wissenschaft selber damit auf, eine universale Theorie von der Totalität des Seins, bzw. von der einheitlichen Welt zu sein und verzichtete somit auf den Anspruch, die Stellung und Bedeutung des Menschen im Ganzen des Seins zu klären. Auf diese Weise löste sich einerseits die Idee der unabtrennbaren Einheit des Seins im Ganzen oder der Welt auf; andererseits spielte das dem physikalistischen Objektivismus entstammende Natur- und Weltbild eine immer größere Rolle und vertiefte so die gegenseitige Entfremdung von Nature und Geist, Welt und Mensch.

Dieses Auseinanderklaffen hat seine letzte Ursache in der fundamentalen Denkweise der Galileischen Physik, die unter der neuen Idee der mathematischen Naturwissenschaft die ganze, konkrete Welt als mathematisch-objektivierbar ansah. In der Galileischen Mathematisierung der Natur wird diese selbst unter Leitung der neuen Mathematik idealisiert und so interpretiert, daß sie in ihrem "wahren Sein an sich" mathematisch ist (cf. H.VI, S.20 u. 54). So wird diese idealisierte Natur der vorwissenschaftlich anschaulichen Natur untergeschoben, wie denn auch die mathematisch substruierte Welt der Idealitäten für die einzig wirkliche, die wirklich wahrnehmungsmäßig gegebene Welt, nämlich für unsere alltägliche Lebenswelt ausgegeben wird. "Das Ideenkleid, Mathematik und mathematische Naturwissenschaft, oder dafür das Kleid der Symbole, der symbolisch-mathematischen Theorien, befaßt alles, was wie den Wissenschaftlern so den Gebildeten als die 'objektiv wirkliche und wahre' Natur die Lebenswelt vertritt, sie verkleidet. Das Ideenkleid macht es, daß wir für wahres Sein nehmen, was eine Methode ist" (H.VI, S.52). Eben darum ist Galilei, dieser vollendende Entdecker der Physik bzw. der physikalischen Natur, für Husserl ein "zugleich entdeckender und verdeckender Genius". Und seit Galilei nimmt der neuzeitliche Mensch bis heute alles, was er verdeckend entdeckt, für die schlichte Wahrheit.

Wie die oben zitierten Aussagen von Husserl schon deutlich machen, liegt für ihn der Grundzug des physikalistischen Objektivismus darin, daß er von den ursprünglichen Beziehungen des Menschen zur konkreten Dingwelt um ihn herum abstrahiert und die mathematisierte Dingwelt so behandelt, als sei sie an und für sich, d.h. als existiere sie völlig unabhängig von unserem Dasein – und zwar so, als habe sie allein für sich das wahre Sein. Aber ein solches Weltbild beraubt den uns vertrauten Weltbegriff seiner Vertraulichkeit und verändert die Welt zu einer transzendenten Außenwelt, die unserer alltäglichen Lebenswelt völlig ungleichartig ist. Bei einem solchermaßen umgewandelten Weltbegriff erbauen alle Dinge, von jedem Bezug mit unserem Leben losgelöst, an und für sich ein abstraktes und von uns getrenntes Reich. Die physikalische Naturwissenchaft der Neuzeit hat somit ihre Wurzel in der konsequenten Abstraktion, mit der sie an der farbigbunten und bewandtnisreichen Lebenswelt nur blasse Körperwelt sehen will. Und in der Tat sieht sie in jedem Ding bloße Körperlichkeit, obschon es nicht bloß körperlich ist, sondern nur verkörpert ist. Gleicherweise reduziert sich auch die Welt in solcher Abstraktion auf eine abstrakte NATUR. Es sei hier nur ein Beispiel beachtenswerter Kritik an diesem Naturbild genannt, nämlich die bekannte Polemik zwischen der Newton-Schule und Goethes "Farbenlehre". Die Verwandlung des Weltbildes geht so weit, daß zum Schluß auch das menschliche Sein selbst objektiviert und verdinglicht wird. Hier scheint sogar mein eigenes Sein wie ein fremdes Unding zu sein, wie z.B. Kafka in seiner "Verwandlung" dichterisch schildert. Der Gegensatz zwischen der Philosophie und Naturwissenschaft, zwischen dem transzendentalen Subjektivismus und physikalistischen Objektivismus führt auf solche Weise die Entfremdung von Welt und Mensch herbei. Er führt zur Krise nicht nur des Menschentums, sondern auch der Wissenschaft überhaupt. Denn der Verlust der Lebensbedeutsamkeit, die die Wissenschaft eigentlich behalten soll, bedeutet nichts anders als ihre eigene Krise.

3. CHARAKTERISTIK DES OBJEKTIVISMUS UND DES TRANSZENDENTALISMUS

Wie will nun Husserl den oben genannten Gegensatz zwischen Objektivismus und Subjektivismus überwinden? Mit welchem Verfahren will er den objektivistischen Weltbegriff nachprüfen und korrigieren? Er versucht das natürlich in seiner transzendental-phänomenologischen Einstellung, deren Kern die prägnanten Begriffe der "Intention" und "Konstitution" bilden. Wie stellt sich darin der Objektivismus dar, von dem hier die Rede ist? Zunächst einmal muß dieser Begriff genauer bestimmt werden. In der Neuzeit, die unter dem überwältigenden Einfluß der naturwissenschaftlichen Denkweise steht, scheint das Wort "objektiv" die allein herrschende Leitidee der theoretischen und vortheoretischen Erkenntnis zu sein. Objektive Wissenschaft strebt nach der Überwindung jedes Subjektiven und will dieses ganz aus ihrem Forschungsbereich ausschalten. Aus diesem Bestreben ist die Idee der Welt entstanden "als eines Universums durch exakte Methoden, die der mathematisch-physikalischen Naturwissenschaft, beherrschbaren Seins, als eines an sich bestimmten Universums, dessen faktische Bestimmungen dann die Wissenschaft zu ermitteln habe" (EU, S.40). Dementsprechend besteht auch in unserem alltäglichen Weltleben die allgemeine Überzeugung, "die Gegenstände unserer Erfahrung seien an sich bestimmt, und die Leistung der Erkenntnis sei es eben, diese an sich bestehenden Bestimmungen in einer Approximation aufzufinden, sie 'objektiv', wie sie an sich sind, festzustellen" (a.a.O.). Die Objektivität der Welt in diesem Sinne scheint zuerst so selbstverständlich zu sein, daß es völlig sinnlos und vergeblich wäre, nach dem Grund ihrer Gültigkeit zu fragen und ihre Begründung zu versuchen.

Husserl charakterisiert schließlich den Objektivismus folgenderweise: "Das Charakteristische des Objektivismus ist, daß er sich auf dem Boden der durch Erfahrung selbstverständlich vorgegebenen Welt bewegt und nach ihrer 'objektiven Wahrheit' fragt, nach dem für sie unbedingt, für jeden Vernünftigen Gültigen, nach dem, was sie an sich ist. Das universal zu leisten, ist die Sache der Episteme, der Ratio, bzw. der Philosophie. Damit werde das letztlich Seiende erreicht, hinter das zurückzufragen keinen vernünftigen Sinn mehr hätte" (H.VI, S.70). Zweifellos stellt jede objektive Wissenschaft ihre Fragen immer unter der Voraussetzung der ständig vorgegebenen Welt; sie stellt aber deren evidentes Sein (das natürlich nur in einer naiven Einstellung "evident" ist) nie in Frage und hat keine Ahnung davon, daß sie dahinter noch zurückfragen könne oder sogar müsse. M.a.W.: die transzendentale Problematik ist ihr völlig unbekannt. Eben deshalb spricht Husserl von der "transzendentalen Naivität" nicht nur des physikalistischen Naturalismus, sondern auch jeder objektivistischen Philosophie (cf. H.VI, S.99, u. 196). Nach E. Finkes vortrefflicher Darstellung von Husserls Auffassung hat der wissenschaftliche Objektivismus zwar die naive Subjektivität der Lebenswelt überwunden, ist sich jedoch seiner inneren Herkunft aus bestimmten Lebensmotiven dieser Lebenswelt nicht bewußt und begreift außerdem noch nicht, "daß er selber einer subjektiven Aufklärung bedarf, allerdings im Rückgang auf die transzendentale Subjektivität"[4]. In der Tat ist alles Objektive eben als es selbst nie erfahrbar. Mit der Annahme der Objektivität schlechthin oder der Wahrheit an sich macht man, so ist zu sagen, eine Art Hypothese, damit man sowohl in der Theorie als auch in der Praxis alles Subjektiv-Relative ausschließen und einen scheinbar festen Halt gewinnen könne. Aber der Geist des phänomenologischen Radikalismus will nicht auf der Stufe eines solchen Objektivismus stehen bleiben, sondern bis zum letzten und wahren Ursprung des Seinssinnes von allem Objektiven durchdringen und ihn ans Licht bringen.

Nun ist aber der Begriff des Transzendentalen zu klären. In seinem
Berliner Vortrag von 1931 sagt Husserl: "Der cartesianische Rückgang
von dieser vorgegebenen Welt auf die Welt-erfahrende Subjektivität und
so auf die Bewußtseinssubjektivität überhaupt erweckt eine total neue
Dimension wissenschaftlichen Fragens: wir nennen sie im voraus die
transzendentale"[5]. Das Wort "transzendental" ist in seiner Verwendung
durch Kant allgemein bekannt (cf. Kritik der reinen Vernunft, B. S.25 u.
81). Husserl selbst bestätigt, daß er das Wort von Kant übernommen hat
(cf. H.VII, S.230). Aber er selbst gebraucht das Wort "transzendental" in
einem sehr weiten Sinne für das ursprüngliche Motiv, das durch Descartes
in allen neuzeitlichen Philosophien sinngebend geworden ist, nämlich "das
Motiv des Rückfragens nach der letzten Quelle aller Erkenntnisbildungen,
des Sichbesinnens des Erkennenden auf sich selbst und sein erkennendes
Leben, in welchem alle ihm geltenden wissenschaftlichen Gebilde
zwecktätig geschehen, als Erwerbe aufbewahrt und frei verfügbar
geworden sind und werden" (H.VI, S.100f.). Deswegen kreist die ganze
transzendentale Problematik um "das Verhältnis dieses meines Ich – des
'ego' – zu dem, was zunächst selbstverständlich dafür gesetzt wird: meiner
Seele, und dann wieder um das Verhältnis dieses Ich und meines
Bewußtseinslebens zur Welt, deren ich bewußt bin, und deren wahres Sein
ich in meinen eigenen Erkenntnisgebilden erkenne" (H.VI, S.101). Dieser
Begriff des Transzendentalen hat natürlich in dem des Transzendenten
sein Korrelat. In der natürlichen Einstellung, wo man noch keine Ahnung
von der transzendentalen Problematik hat, glaubt man an die
Objektivität   schlechthin,   das   transzendente   An-sich-Sein   alles
vorgegebenen Weltlichen und seiner Wahrheit.

Aber in der transzendentalen Einstellung können wir das objektive An-
sich-Sein   der   transzendenten   Welt   nicht   mehr   ohne   Vorbehalt
anerkennen, denn das "transzendentale" Problem betrifft eben den
bewußtseinsrelativen Seinssinn des "Transzendenten". Die naiv
seinsmäßig vorgegebene "transzendente" Welt wird hier "transzendental"
problematisch, sie kann nicht mehr wie in den positiven Wissenschaften als
Erkenntnisboden dienen (cf. H.IX, S.264). Für den Transzendentalphilosophen wird nun alles Weltliche, alle reale Objektivität, zum Problem
und er muß deswegen auf die Bewußtseinssubjektivität und ihre
"konstitutive Leistung" radikal reflektieren, "in welcher die Welt, die
wissenschaftliche wie die alltäglich-anschauliche, zur Erkenntnis, zu ihrer
Seinsgeltung   für   uns   kommt"   (H.VI,   S.205).   Nur   ein   radikales

Zurückfragen auf die letztlich alle Weltgeltung zustandebringende
Subjektivität kann die objektive Wahrheit verständlich machen und den
letzten Seinssinn der Welt erreichen. Husserl bestimmt also die Transzen-
dentalphilosophie als "eine Philosophie, die gegenüber dem vorwissenschaftlichen und auch wissenschaftlichen Objektivismus auf die
erkennende Subjektivität als Urstätte aller objektiven Sinnbildungen und
Seinsgeltungen zurückgeht und es unternimmt, die seiende Welt als Sinn-
und Geltungsgebilde zu verstehen und auf diese Weise eine wesentliche
neue Art der Wissenschaftlichkeit und der Philosophie auf die Bahn zu
bringen" (H.VI, S.102).

4. DAS WELTPROBLEM DER PHÄNOMENOLOGIE UND IHRE KORRELATIVE BETRACHTUNGSWEISE

In der oben skizzierten transzendentalen Einstellung dürfen wir nicht
mehr schlicht in die Welt hineinleben und auf dem vorausgesetzten Boden
der Welt immer nur objektive Theorien konstruieren. Wenn wir transzendental-phänomenologisch philosophieren wollen, dann müssen wir zuerst
unseren Blick darauf lenken, daß uns allen die Welt und jedes Objekt nicht
nur einfach vorgegeben sind, sondern daß alles in subjektiven Erscheinungsweisen eben als Phänomen uns bewußt wird. Daneben müssen wir
unser universales Interesse ausschließlich und konsequent auf das 'Wie' der
Gegebenheitsweisen richten, also darauf, wie in verschiedenen subjektiven
Erscheinungen und Meinungen die Welt als einheitliches, universales
Geltungsphänomen für uns zustandekommt. Auf diese Weise tut die transzendentale Phänomenologie "gar nichts anderes als die Welt zu befragen,
genau die, die allzeit für uns die wirkliche ist (die uns geltende, uns sich
ausweisende, für uns einzig Sinn habende ist), sie intentional nach ihren
Sinnes- und Geltungsquellen zu befragen, in denen selbstverständlich ihr
wahres Sein beschlossen ist"[6]. Husserl versteht sogar unter dem Titel der
Philosophie "eine universale Wissenschaft von der Welt"[7]. Die Welt zu
befragen bedeutet natürlich keine Verleugnung ihrer Existenz. "Der
phänomenologische Idealismus", so betont Husserl selbst, "leugnet nicht
die wirkliche Existenz der realen Welt (und zunächst der Natur), als ob er
meinte, daß sie ein Schein wäre" (H.V, S.152). Ihre Existenz ist vollkommen zweifellos, es ist aber etwas ganz anderes, diese Zweifellosigkeit zu verstehen und ihren Rechtsgrund aufzuklären. Das phänomenologische Befragen der Welt sieht es auch nicht darauf ab, das Sein und den
Wert jener Objektivität, die von den positiven Wissenschaften erforscht

wird, zu vernichten oder aber zu sichern: "Es gilt nicht, Objektivität zu sichern, sondern sie zu verstehen" (H.VI, S.193). Auf welche Art und Weise soll dann dieses phänomenologische Weltproblem untersucht werden? Um mit einem Wort zu antworten: auf die korrelative Betrachtungsweise.

Das Sein der Welt ist in einem Sinne (nämlich in dem Sinne, daß die Welt kein reelles Stück eines Bewußtseinssubjekts ist) dem Bewußtsein transzendent und bleibt notwendig transzendent. Wir müssen deswegen einerseits anerkennen, daß zum eigenen Sinn der Welt und alles Weltlichen eine so geartete Transzendenz gehört. Aber eine in diesem Sinn objektivistische Auffassung ist eine bloß mögliche, keinesfalls aber eine im letzten radikale Aufklärung der Seinsweise der Welt. Denn sie ist blind dafür, daß alles Weltliche, jede Objektivität, nur f ü r u n s Sinn und ausweisbares Sein haben kann. Ohne die Einsicht, "daß wahre Objektivität etwas ist, das nur im Bewußtsein Sinn und ursprüngliche realisierende Bewährung erfahren kann" (H.VII, S.77), ist es unmöglich, das wahre Sein alles Weltlichen phänomenologisch zu befragen und im letzten radikal aufzuklären. Eben deshalb behauptet Husserl: "unsere Wissenschaft behandelt jederlei Objektives als Objektives des Bewußtseins und als in subjektiven Modis sich Gebendes; Bewußtseinssubjekt und Bewußtsein selbst wird nicht vom bewußten Gegenständlichen getrennt betrachtet, sondern im Gegenteil, Bewußtsein trägt Bewußtes in sich, und so, wie es das in sich trägt, ist es Forschungsthema" (H.VII, S.50). "Cartesianisch gesprochen: Die Untersuchung wird ebensowohl das cogito an sich selbst wie das cogitatum qua cogitatum betreffen. Selbstverständlich ist das eine und das andere, wie im Sein, so in der Forschung untrennbar verflochten"[8]. Der Phänomenologe nimmt also alle Objektivitäten "nur als Bewußtseinskorrelate, rein nach dem Was und Wie ihrer aus den betreffenden Bewußtseinserlebnissen und -zusammenhängen zu entnehmenden Phänomene"[9].

Diese korrelative Betrachtungsart, die zuerst schon in der "Philosophie der Arithmetik" (1891) auftrat, ist der Phänomenologie so wesentlich, daß Husserl sie selbst "das philosophische Grundmotiv" nennt. So kann gesagt werden: "Die Spannungseinheit des in eins subjektiv und objektiv gerichteten Fragens ist der eigenste und fruchtbare Ansatz Husserls, der in seiner wachsenden Vertiefung und Wandlung die Etappen bestimmt, in welchen schließlich eine neue Idee der Philosophie zum Durchbruch kommt"[10]. Alles Seiende ist in dieser korrelativen Betrachtung (entgegen dem falschen Ideal eines absoluten Seienden und seiner absoluten

Wahrheit) letztlich relativ auf die transzendentale Subjektivität (cf. FTL, S.241). Die gesamte Problematik der Korrelation von erkennendem und sonstigem Bewußtsein einerseits und seiner Gegenständlichkeit andererseits ist einer naturalistischen Psychologie und Erkenntnistheorie völlig fremd (cf. H.VII, S.122). Husserl behauptet mit Stolz, daß vor dem ersten Durchbruch seiner Phänomenologie in den "Logischen Untersuchungen" (1900/1901) die Korrelation von Welt und subjektiven Gegebenheitsweisen von ihr das philosophische Staunen nie erregte (cf. H.VI, S.168). Wir können also ohne Übertreibung sagen, daß die Korrelationsforschung zwischen dem Bewußtsein als Erlebnis und dem Bewußten als solchem den ersten Ausgang aller phänomenologischen Forschungen bildet und daß die Möglichkeit aller transzendentalen Untersuchungen ausschließlich vom richtigen Verständnis dieser Forschungsmethode abhängt.

Unser Forschungsthema in dieser phänomenologischen Einstellung kann also nicht mehr die Objektivität oder die Welt schlechthin, sondern nur diejenige sein, die den sie bestimmenden Sinn und ihre Seinsgeltung aus meiner bzw. unserer Erfahrung und Erkenntnis gewinnt. Eben in diesem Sinne gehören alle Objektivitäten, die sich normalerweise in objektive Wissenschaften einordnen, zugleich auch in die universale Wissenschaft von der Bewußtseinssubjektivität (cf. H.VII, S.49). M.a.W.: Die Phänomenologie als Wissenschaft von jederlei Gegenstandsphänomenen muß zugleich notwendig auch die Wissenschaft vom reinen Bewußtsein und von reinen Bewußtseinsphänomenen sein: "Die grundsätzliche Klärung der Notwendigkeit des Rückgangs auf das Bewußtsein, die radikale und ausdrückliche Bestimmung des Weges und der Schrittgesetze dieses Rückgangs, die prinzipielle Umgrenzung und systematische Durchforschung des auf diesem Rückgang sich erschließenden Feldes der reinen Subjektivität heißt Phänomenologie" (H.IX, S.256). Andererseits umfaßt doch dieselbe Phänomenologie auch "eine universale Ontologie erweiterten Sinnes", weil sie in ihrer Korrelationsforschung die mögliche Welt und ihre ontischen Strukturen nach den Seiten möglicher Sinngebung und Seinsbegründung erforscht, ohne die die Welt undenkbar ist (cf. H.IX, S.251). In diesem Sinne kann man mit Recht behaupten, daß für die Phänomenologie die Logik als rationale Wissenschaft von der Objektivität überhaupt und die des Erkennens immer in notwendiger Korrelation stehen (cf. H.VII, S.45). Dieses Verhältnis ist schon von vornherein bestimmt, weil Husserl seinen "Ausgang von der notwendigen Korrelation von Gegenstand, Wahrheit und Erkenntnis"[11] genommen hat. Zum Schluß dieses Paragraphen sei noch erwähnt, daß die korrelative Denkweise schon von Anfang an im zentralen Begriff der Intentionalität eingeschlossen ist. Unter Intentionalität, durch deren Entdeckung die Phänomenologie allein möglich geworden ist, versteht Husserl die Eigenschaft von Erlebnissen, "Bewußtsein von etwas zu sein", und bezeichnet sie als das Generalthema der "objektiv" orientierten Phänomenologie (cf. H.III, S.203f. u. H.V, S.155).

5.   DIE TRANSZENDENTALE EPOCHÉ HINSICHTLICH DER
     OBJEKTIVEN WELT UND IHR RESIDUUM

Womit wäre nun anzufangen, um die ursprüngliche Seinsweise und Seinsgeltung der Welt transzendental-phänomenologisch aufzuklären? Dazu ist vor allem eine universale Veränderung unserer Einstellung, unserer Interessenrichtung unentbehrlich. In der natürlichen Einstellung ist uns die Welt, wie schon erwähnt, immer notwendig als Gesamthorizont jedes weltlich Seienden und als Universalfeld wirklicher und möglicher Praxis vorgegeben und alle Dinge und Vorgänge existieren und geschehen in diesem Horizont. Dementsprechend gibt es natürlich einen grundsätzlichen Unterschied zwischen der Seinsweise der Welt und jener der Dinge, obwohl andererseits beide eine untrennbare Einheit bilden. Doch herrscht bei dieser natürlichen Einstellung unbewußt eine schwer auszurottende Naivität des Objektivismus, derzufolge die transzendentale Leistung der erfahrenden und erkennenden Subjektivität unbefragt bleibt. Ja, Husserl selbst betont: "Alle unsere Wissenschaften – die positiven – sind naiv. Sie entbehren letzter Begründung aus einer Enthüllung all der Geltungsfundierungen, durch die ihr universales Thema 'Welt' naiv vorgegeben ist" (H.VIII, S.464).

Die Grundfrage der transzendentalen Phänomenologie ist aber – im radikalen Gegensatz zum naiven Objektivismus – nichts anderes als die Frage nach dem Seins- und Geltungsursprung der Welt und jeder Objektivität. Darum ist es dem Transzendentalphilosophen nicht mehr erlaubt, dieses Problem in derselben Naivität zu behandeln, mit der die Welt ständig als die Allheit an sich seiender Dinge vorausgesetzt und auf deren Boden positiv-wissenschaftliche Fragen aufgerollt werden. Vielmehr muß er, solange er eben als Transzendentalphilosoph philosophiert, die Grundtatsache immer genau im Auge behalten, "daß uns allen die Welt bzw. die Objekte nicht nur überhaupt vorgegeben sind, in einer bloßen Habe als Substrate ihrer Eigenschaften, sondern daß sie (und alles ontisch Vermeinte) in subjektiven Erscheinungsweisen, Gegebenheitsweisen uns bewußt werden, ohne daß wir eigens darauf achten und während wir zum größten Teil überhaupt nichts davon ahnen" (H.VI, S.147). Und dann muß er sein ständiges Interesse darauf richten, "wie im Wandel relativer Geltungen, subjektiver Erscheinungen, Meinungen die einheitliche, universale Geltung Welt, die Welt für uns zustande kommt" (H.VI, S.147). Wenn wir die radikale Enthüllung des letzten Seinsursprungs der Welt erzielen wollen, dürfen wir nicht mehr von der scheinbaren Selbstverständlichkeit der vorgegebenen Welt geblendet sein, weil gerade hinter einer solchen Selbstverständlichkeit noch eine tiefer liegende Unverständlichkeit verborgen ist.

Für uns, die in der "Zeit der Naturwissenschaft" leben, ist aber die objektivistische Einstellung zusammen mit ihrer Naivität ein so tief eingewurzelter Habitus geworden, daß wir ihn nur mit Hilfe einer radikalen Methode aufheben und so in die transzendental-phänomenologische Einstellung eintreten können. Die dazu erforderliche Methode ist eben die der phänomenologischen (oder transzendentalen) Epoché bzw. Reduktion, die Husserl selbst "die prinzipiellste aller Methoden" nennt (H.VII, S.234). Er hat diese Methode durch die Radikalisierung des cartesianischen Zweifels etabliert und fordert damit universale "Ausschaltung" bzw. "Einklammerung" des naiven Glaubens an das Sein von all bloß objektiv Seiendem und sogar der Welt überhaupt, um diese der natürlichen Einstellung fundamentale Voraussetzung einmal im letzten radikal zu hinterfragen. Solange wir uns mit der transzendentalen Problematik beschäftigen, wird von uns gefordert, die zum Wesen der natürlichen Einstellung gehörige Generalthesis außer Aktion zu setzen, alles und jedes, was sie in ontischer Hinsicht umspannt, in einem Schlag in Klammer zu setzen (cf. H.III, S.67). Um uns schließlich zur unnatürlichen Einstellung der phänomenologischen Reflexion durchdringen zu können, müssen wir zur radikalen Veränderung unserer Einstellung jederzeit bereit sein, wodurch wir unseren thematischen Blick von den jeweils objektiv vorgegebenen Dingen und Gedanken, Werten und Zwecken zu unseren subjektiven Bewußtseinserlebnissen selbst unwenden können, in denen die ersteren uns erscheinen und bewußt werden. Husserl selbst spricht darum von der "Epoché der Berufszeit": "In der Zeit, in der ich transzendentaler oder reiner Phänomenologe bin, bin ich ausschließlich im transzendentalen Selbstbewußtsein und bin ausschließlich als transzendentales ego nach allem darin intentional

Implizierten mein Thema. Hier gibt es nichts von Objektivität schlechthin, hier gibt es Objektivität, Dinge, Welt und Weltwissenschaften (also alle positiven Wissenschaften und Philosophien inbegriffen) nur als meine, des transzendentalen ego Phänomene" (H.VI, S.261f.).

Manchmal wendet man gegen diese Methode der Epoché und Reduktion ein, daß sie das ganze Sein der Welt und jedes Weltlichen vernichte. In der Tat gebraucht Husserl selbst die Ausdrücke wie Weltvernichtung oder Weltentsagung, und sie könnten eventuell unbedachte und unangenehme Mißverständnisse hervorrufen. Aber wie er selbst ausdrücklich betont, stellt uns die Epoché – dieses universale Außergeltungsetzen aller Stellungnahmen zur objektiven Welt und so zunächst der Seinsstellungnahmen (betreffend Sein und Nichtsein, Wirklichsein und Vermutlichsein etc.) – nicht einem Nichts gegenüber (cf. H.I, S.60). Durch die Epoché haben wir die Welt nicht einfach verloren; "wir behalten sie ja qua cogitata" (H.I, S.75). Auch wenn Husserl selber von der Weltverlorenheit spricht, meint er: "Man muß erst die Welt durch *Epoché* verlieren, um sie in universaler Selbstbesinnung wieder- zugewinnen" (H.I, S.183). "Alles preisgeben heißt, alles gewinnen; radikale Weltentsagung ist der notwendige Weg, die letztwahre Wirklichkeit zu erschauen und damit, ein letztwahres Leben zu leben" (H.VIII, S.166). Um ein mögliches Mißverständnis hinsichtlich des Wesens der Epoché zu vermeiden, muß man unbedingt folgendes bedenken: "Alles phänomenologisch Ausgeschaltete gehört doch in einer gewissen Vorzeichenveränderung in den Rahmen der Phänomenologie. Nämlich die realen und idealen Wirklichkeiten, die der Ausschaltung verfallen, sind in der phänomenologischen Sphäre vertreten durch die ihnen entsprechenden Gesamtmannigfaltigkeiten von Sinnen und Sätzen" (H.III, S.329). In der nunmehr eröffnezten transzendentalen Erfahrung ist "alles transzendente Sein", verstanden im normalen Sinne als wahrhaftes Sein, ausgeschaltet, und das, was als Reisduum nach der Epoché einzig und allein übrig bleiben soll, ist das Bewußtsein selbst in seinem eigenen Wesen, und an Stelle des sogenannten transzendenten Seins das "Vermeintsein von Transzendentem" und somit alle Arten von Bewußtseinskorrelaten, die Noemata (cf. H.V, S.76). Nun kommt hier also ständig beides in Betracht: Bewußtsein und sein Korrelat, Noesis und Noema.

So haben wir nun als Residuum nach der Durchführung der universalen Epoché "mein reines Leben mit all seinen reinen Erlebnissen und all seinen reinen Gemeintheiten, das Universum der Phänomene im Sinne der

Phänomenologie" (H.I, S.60) gewonnen. Es kann nichts anderes sein als dieses reine Bewußtseinsleben, "in dem und durch das die gesamte objektive Welt für mich ist, und so, wie sie eben für mich ist" (a.a.O). Hier erscheint mir also nicht mehr die schlechthin an sich seiende Welt, sondern "die so und so bewußte (wahrgenommene, erinnerte, beurteilte, gedachte, gewertete etc.) Welt als solche" (H.IX, S.282). Die Welt ist für mich ursprünglicherweise nur als die in meinen cogitationes (in meiner Wahrnehmung, Erinnerung, Beurteilung usw.) bewußt seiende und mir geltende da. M.a.W.: Die Welt gilt in phänomenologischer Einstellung nicht mehr als Wirklichkeit schlechthin, sondern nur als Wirklichkeits- phänomen, und zwar rein als Korrelat der ihr Seinssinn gebenden Subjektivität. Das Wort "Phänomen" bedeutet hier einen gewissen, dem betreffenden Bewußtsein selbst einwohnenden Gehalt, der das Substrat der jeweiligen Wirklichkeitsbewertung ist[12]. Hierbei müssen wir noch den Grund genau angeben, warum Husserl den sogenannten Bewußtseins- immanentismus zum Ausgang seiner philosophischen Forschungen gewählt hat. Der Grund dazu liegt letztlich im "Prinzip apodiktischer Evidenz als philosophischem Leitprinzip" (H.VIII, S.126). Eben aus dem Grund, daß die objektive Welt nicht apodiktisch gegeben ist, hat Husserl als Phänomenologe, der in seinem Philosophieren unbedingt die absolute Evidenz erreichen will, absichtlich gewagt, selbst das in unserem natürlichen Leben so selbstverständliche Sein der Welt provisorisch auszu- schalten, und zum reinen Bewußtsein des je eigenen Ego zurückzugehen, das einzig und allein absolut evident ist. "Prinzipiell kann die Philosophie", so behauptet Husserl, "nur von absolut einsichtigen Urgründen ausgehen und in einem absolut vorurteilslosen, in jedem Schritte aus evidenten Prinzipien sich rechtfertigenden Begründungswege emporsteigen" (H.VII, S.190f.).

6. DER RÜCKGANG ZUR LEBENSWELT

In diesem Paragraphen wollen wir über einen beachtenswerten Unter- schied nachdenken, der zwischen der "Krisis", Husserls letztem, hauptsächlich zwischen 1935 und 1936 entstandenem Werk und seinen früheren Schriften aufscheint. Der Unterschied besteht darin, daß in der "Krisis" zum ersten Mal das Problem der Lebenswelt als ein Haupt-thema diskutiert und die Notwendigkeit der Reduktion der objektiv-wissenschaft- lichen Welt und Erkenntnis auf die sie fundierende und vorwissenschaftlich uns geltende Lebenswelt deutlich eingesehen und hervorgehoben wird[13], während in den früheren Schriften die objektiv-wissenschaftliche Welt und Erkenntnis, gar nicht oder wenigstens nicht ausdrücklich durch die Zwischenstufe der Lebenswelt hindurch, direkt auf die reinen Bewußtseinserlebnisse des je eigenen Subjekts reduziert wurde. In der Tat erwähnt Husserl selbst den "viel kürzeren Weg zur transzendentalen Epoché" d.h. den unmittelbaren Rückgang von der objektiv-theoretischen Welt zum transzendentalen ego, den er z.B. in den "Ideen" (1913) gemacht hatte, und erkennt dessen großen Nachteil selbstkritisch an, wenn er feststellt, daß jener Weg "wie in einem Sprung schon zum transzendentalen ego führt, dieses aber, da jede vorgängige Explikation fehlen muß, in einer scheinbaren Inhaltsleere zur Sicht bringt, in der man zunächst ratlos ist, was damit gewonnen sein soll, und gar, wie von da aus eine neue und für eine Philosophie entscheidende, völlig neuartige Grundwissenschaft gewonnen sein soll" (H.VI, S.158). Es ist zweifellos von größter Bedeutsamkeit, "daß Husserl hier (in der "Krisis") den Begriff des Wissens nicht mehr ausschließlich an den Wissenschaften aufgreift, sondern auf die vorwissenschaftliche Lebenswelt des Menschen, auf seine alltägliche und vortheoretische Erfahrung zurückgeht und dort die Motive aufsucht, welche schlißlich zur wissenschaftlichen Theorie hindrängen"[14].

Außerdem können wir in der Thematisierung der Lebenswelt mindestens noch zwei sehr wichtige Vorhaben Husserls ausfindig machen: Zum einen will Husserl den Begriff der Subjektivität, die den Ichpol der Intentionalität bildet, sowie jenen der Welt, die als Korrelat der ersteren einen einheitlichen Gegenstandspol ausmacht, in einer inhaltsreicheren Konkretion erfassen. Zum anderen soll das Subjektiv-Relative neu bewertet werden. Wie bekannt, lehnt Husserl in der "Philosophie als strenger Wissenschaft" (1911) den Relativismus und Historismus entschieden ab[15] und stellt dagegen die Erkenntnisleistung der "absoluten" Subjektivität in den Mittelpunkt seines Philosophierens. Auch in der "Krisis" behauptet er ausdrücklich, daß das transzendentale Urmotiv aus der Forderung der Apodiktizität entsprungen ist (H.VI, S.274) und daß "die Reduktion auf das absolute ego als das letztlich einzige Funktionszentrum aller Konstitution" hinfort ganz die Methode der transzendentalen Phänomenologie bestimmt (H.VI, S.190). In diesem Sinne kann man sagen, daß ein konsequent durchgeführter Relativismus mit dem echten Ausgangspunkt von Husserls Philosophie von Grund aus unverträglich ist. Trotzdem versucht Husserl in der "Krisis" – auf dem Wege zur radikalsten Aufklärung der Gegebenheits- und Erscheinungs-

weisen von allem Objektiven – die bedeutsame Rolle des Subjektiv-
Relativen, worin sich jedes Objektive eben als solches gibt, neu zu
bewerten. Ja, Husserl bringt diese neue Problematik mit den Worten zum
Ausdruck, "daß der rechte Rückgang zur Naivität des Lebens, aber in
einer über sie sich erhebenden Reflexion, der einzig mögliche Weg ist, um
die in der 'Wissenschaftlichkeit' der traditionellen objektivistischen
Philosophie liegende philosophische Naivität zu überwinden" (H.VI,
S.60). Was steht hinter dieser neuen Problematik?

Für den Wissenschaftler, dem es ausschließlich um die objektive
Wahrheit geht, ist sicher alles bloß Subjektiv-Relative zu überwinden und
zu beseitigen. Bedeutet das aber, daß es für ihn bloß ein irrelevanter
Durchgang sei? Nein, im Gegenteil: Das mir ursprünglich und unmittelbar
Gegebene sind meine eigenen subjektiven Erlebnisse und mein erfahrendes
Leben, und dann meine darin sich gebende relative Lebenswelt. Der
Seinssinn dieser primordialen Lebenswelt ist "subjektives Gebilde",
Leistung des erfahrenden, des vorwissenschaftlichen Lebens. Diese
Lebenswelt ist freilich in stetem Wandel der Relativitäten auf
Subjektivität bezogen (cf. H.VI, S.176). Ist aber diese subjektiv-relative
Lebenswelt, also diese aller Abstraktion vorgängige, konkrete Welt – für
die "objektive" bzw. "wissenschaftlich wahre" Welt wie auch für das
"objektiv-wissenschaftliche" Wissen – nicht eben der sie begründende und
fundierende Boden! Umfaßt sie nicht die letztere in ihrer eigenen
Konkretion – auch wenn die objektive Wissenschaft dem Subjektiv-
Relativen "ein hypothetisches An-sich-Sein, ein Substrat für logisch-
mathematische Wahrheit an sich" zuordnet und das letztere bevorzugt!
Wenn man sich diesen Einsichten stellt, dann wird man auch damit
zusammenhängende, weitere Aussagen Husserls akzeptieren, daß nämlich
die Lebenswelt ein Reich ursprünglicher Evidenzen sei und alles objektive
Apriori im subjektiv-relativen Apriori der Lebenswelt gründe (cf. H.VI,
S.130 u. 143); daß alle theoretische Leistung objektiver Wissenschaft auf
dem Boden der vorgegebenen Welt – der Lebenswelt – statthabe und die
schlichte Erfahrung, in der die Lebenswelt gegeben ist, die letzte
Grundlage aller objektiven Erkenntnis sei (H.VI, S.229). Aufgrund dieser
Einsichten zeigt Husserl, daß sich das naturwissenschaftliche Weltbild
erst auf Grund der subjektiv-relativen Lebenswelt konstituieren kann –
und damit bringt er die verdeckte Naivität der Naturwissenschaft ans
Licht. Nun wäre die Gegebenheitsweise der Lebenswelt und eines jeden in
ihr gegebenen Objekts, also die transzendentale Konstitution jedes
gegenständlichen Seienden zu betrachten.

7. DIE KONSTITUTIVE LEISTUNG DER TRANSZENDENTALEN
SUBJEKTIVITÄT

Im vorigen Paragraphen ergab sich, daß die objektiv-wissenschaftliche
Welt zunächst einmal zur vorwissenschaftlichen Lebenswelt reduziert
werden muß. Das bedeutet aber gar nicht, daß die Erreichung dieser
Lebenswelt selbst schon das Endziel der phänomenologischen Forschung
sei. Das echte Ziel der transzendentalen Reduktion besteht nicht im
bloßen Rückgang zur Lebenswelt, sondern eben in der "Entdeckung der
universalen, in sich absolut geschlossenen und absolut eigenständigen
Korrelation von Welt selbst und Weltbewußtsein" (H.VI, S.154). Nach der
Epoché gilt uns auch die Lebenswelt nur als transzendentales Phänomen,
also bloß als Korrelat der dieser Lebenswelt Sinn und Seinsgeltung
verleihenden transzendentalen Subjektivität. "In der reinen Korrelativ-
einstellung, die die Epoché schafft, wird die Welt, das Objektive, selbst zu
einem besonderen Subjektiven" (H.VI, S.182). Wie schon oft erwähnt, ist
die Welt im gewöhnlichen Verstand uns allen im voraus gegeben als
universaler Horizont, der alles und jedes Seiende umfaßt und in dem wir
selber geboren sind und leben. Wenn unser Bewußtseinsleben Innenwelt
genannt wird, dann wird die Dingwelt für eine an sich seiende Außenwelt
gehalten, die unserem Bewußtseinsleben transzendent ist. Wie verhalten
sich nun diese zwei Welten? Haben sie beide in einer einheitlichen
umfassenden Raumwelt je ihre eigene Hälfte? In transzendental-phäno-
menologischer Einstellung darf man das Verhältnis zwischen "Innen und
Außen" bzw. "Immanenz und Transzendenz" nicht im Bild z.B. der
sogenannten Zweiweltentheorie verstehen, als ob sie beide räumlich
nebeneinander abgegrenzt und voneinander unüberbrückbar getrennt
wären. Wenn man es so begriffe, dann würde der Sinn von "Subjektiv-
Werden eines Objektiven" nicht richtig verstanden. Diese Redeweise muß
in dem Sinn aufgefaßt werden, daß sich das in objektivistischer Einstellung als Transzendentes oder als Äußerlich-Objektives Genommene
jetzt in transzendentaler Einstellung als Immanentes oder Subjektives ausweist.

Um diesen Sachverhalt besser zu verstehen, empfiehlt es sich, sich an die
Husserlsche Erläuterung des Begriffes "Intentionalität" zu erinnern. Auf
die Frage, was das unmittelbar Gegebene im Bewußtseinsleben ist,
antwortet Husserl, daß dies keineswegs bloße Farbendaten, Tondaten und
sonstige Empfindungsdaten sind, sondern eben das "Ich sehe einen Baum,
der grün ist"; "Ich höre das Rauschen einer Blätter, ich rieche seine Blüten"

usw. (cf. H.VI, S.236). Das unmittelbar Gegebene ist also nichts anders als der gesamte einheitliche Bereich der dreigliedrigen Struktur "ego-cogito-cogitatum" oder genauer der viergliedrigen Struktur "ego-cogito-cogitatum-qua-cogitatum". Deshalb wird jedes "transzendente, an sich seiende Objekt", sei es die Welt oder sei es einzelnes Ding, hier in transzendentaler Einstellung notwendigerweise als ein intentionales aufgefaßt, das nur in subjektiver Korrelation seinen eigenen Seinssinn und seine je bestimmte Seinsgeltung haben kann. Eben in diesem Sinn will Husserl das Objektive "nur als Intentionales des Aktes, der ihm Geltung zumißt", gelten lassen (cf. H.VIII, S.111 u. 445). Für ihn ist das Objektive nichts anderes als die "der transzendentalen Subjektivität eigenwesentlich zugehörige synthetische Einheit aktueller und potenzieller Intentionalität" (FTL, S.242). Dementsprechend faßt Husserl den allgemeinen Begriff des Subjektiven in der Epoché in der Weise, daß in ihm alles, Ichpol, Universum der Ichpole und Universum der Gegenstandspole umfaßt werden (cf. H.VI, S.183). In diesem Kontext bedeutet auch die Transzendenz der Welt "Transzendenz in Relation zum transzendentalen Ich und mittels seiner zu der offenen Ichgemeinschaft als der seinen" (FTL, S.222). Auch die objektive Welt transzendiert die Intersubjektivität nicht mehr im eigentlichen Sinne, sondern wohnt ihr als "immanente Transzendenz" ein (cf. H.I, S.137f.). Eine objektive Welt rein als Welt möglichen Bewußtseins wird so transzendental zu einer rein im transzendentalen ego sich bewußtseinsmäßig konstituierenden Welt.

In Bezug auf den Begriff der "Konstitution" müssen wir nun den sehr trefflichen Hinweis von E. Fink beachten, daß das wahre Thema der Phänomenologie weder die Welt selbst, noch die ihr gegenüberzustellende transzendentale Subjektivität für sich, sondern eben das "Werden der Welt in der Konstitution der transzendentalen Subjektivität" ist[16]. Husserl selbst bestätigt, daß die Probleme der Konstitution der Bewußtseinsgegenständlichkeit im transzendentalen Bewußtsein die allergrößten Probleme der Phänomenologie sind (cf. H.III, S.212 u. 364). Was bedeutet "Konstitution"? Diese Frage ist schon lange genug unter Fachleuten diskutiert worden und hat verschiedene, manchmal kritische Interpretationen hervorgerufen. Eine große Schwierigkeit, diesen Begriff eindeutig zu bestimmen, liegt natürlich daran, daß Husserl selbst das Wort "Konstitution" in verschiedenen Bedeutungsnuancen verwendet. Einmal gebraucht er das Wort "sich Konstituieren" im Sinn von "sich Beurkunden" von Erfahrungsgegenständlichkeit (z.B. H.XVI, S.8) und andermal im Sinn von "Erzeugen" (besonders in der "Formalen und transzendentalen Logik"). Beim letzteren Fall entseht danach die neue Frage, was dann dabei erzeugt wird. Wenn man darunter "ein Erzeugen oder ein handelndes Hervorbringen von realen Dingen und Vorgängen" verstehen würde, dann wird der Grundsinn dieses Begriffs völlig mißverstanden. In der phänomenologischen Einstellung ist es unsinnig, vom Erzeugen eines real Daseienden zu sprechen, weil hier schon bekannterweise alle Existenzialsetzung, alles reale Dasein, bei der unentbehrlichen Durchführung der transzendentalen Epoché völlig außer Geltung, außer Aktion gesetzt ist. Darum spricht Husserl meist sehr vorsichtig von der Konstitution des "Seinssinnes", z.B. in folgenden Formulierungen: "das in sich selbst objektiven Sinn konstituierende Bewußtsein" (FTL, S.13); "Denken im weitesten Sinn als sinn-konstituierendes Erlebnis" (FTL, S.19); "Intersubjektivität, die für die objektive Welt als sinnkonstituierende fungiert" (FTL, S.209). Auch andere Redeweisen wie "transzendentale Subjektivität als die Urstätte aller Sinngebung und Seinsbewährung" (H.V, S.139), "Subjektivität als Quelle aller Sinngebung und Geltung" (H.VII, S.122) etc. machen dieses Sinn-Konstituieren begreiflich. Das transzendentale Problem im Husserlschen Sinne betrifft ja eigentlich "den Seinssinn der Welt überhaupt als nur aus Bewußtseinsleistungen Sinn und Geltung gewinnender" (H.IX, S.273).

Die phänomenologische Auslegung ist im oben erwähnten Sinne die "reine Sinnesauslegung". Sie tut also auch "hinsichtlich der objektiven Welt der Realitäten nichts anders als den Sinn auszulegen, den diese für uns alle vor jedem Philosophieren hat und offenbar nur aus unserer Erfahrung hat" (H.I, S.177). Die phänomenologische Konstitution bedeutet demnach die Erzeugung von Sinn und Geltung jedes Seienden oder die Sinngebung zum intentionalen Gegenstand. Nämlich kann sich diese Sinngebung nur im Wechselspiel mit dem Sich-selbst-Geben des Gegenständlichen vollziehen (cf. H.VII, S.30).

8. WELT UND INTERSUBJEKTIVITÄT

Für den Phänomenologen, der durch die Epoché eine Reduktion auf reines cogito und cogitatum als solches vollzieht, beziehen sich notwendigerweise die Prädikate Sein und Nichtsein und ihre modalen Abwandlungen "nicht auf Gegenstände schlechthin, sondern auf gegenständlichen Sinn" (H.I, S.91). Schließlich werden bei Husserl auch "Realität und Welt", obwohl mit einiger Vorsicht im Wortgebrauch, eben doch Titel für "gewisse Sinneseinheiten", bezogen auf gewisse sinngebende und Sinnesgültigkeit ausweisende Zusammenhänge des absoluten, reinen Bewußtseins (cf. H.III, S.134). Kurz gesagt erforscht die Phänomenologie das konstitutive Problem gerade als das der Sinngebung. Deswegen kann man darin mit Recht und ohne weiteres den Gedanken des "Primats des Sinnes gegenüber dem Sein" erkennen, bzw. den des Mitimpliziertseins des Daseins im Sosein oder auch den der Identifizierung des Sinnes mit dem Sein[17]. Wir begegnen jedenfalls gerade in diesem Gedanken einem der wesentlichen Grundzüge der phänomenologischen Denkweise. Für sie sind die transzendentale Idealität und die empirische Realität eines Seienden gar nicht unverträglich, sondern auch die Realität kann nur dann als eine sinnhafte, als eine je bestimmte Seinsgeltung habende, bestehen und existieren, wenn sie von der Idealität sinngemäß begründet ist. Husserl behauptet in diesem Sinne, daß der transzendentale Idealismus der Phänomenologie den natürlichen Realismus ganz in sich birgt (H.IX, S.254). Wir leben sicher in der fundamentalen Paradoxie, daß der Mensch im Verhältnis zur Welt Subjektivität ist und zugleich in ihr objektiv weltlich sein soll. Aber insofern die sich uns erschließende Welt immer die in unserem menschlichen Leben Sinn habende und neuen Sinn gewinnende ist, ergibt es sich, daß, was Erkennen betrifft, für uns Menschen unser eigenes Sein dem Sein der Welt vorangeht, obgleich es der Seinswirklichkeit nach anders ist (cf. H.VI, S.265f.).

Eben in diesem Sinne spielen in der Philosophie des späteren Husserl das Prinzip des "Ich bin" bzw. die apodiktische Evidenz des "Ich bin" (H.VII, S.166) eine höchst bedeutsame Rolle. Husserls Meinung nach muß der Satz "Ich bin" das wahre Prinzip aller Prinzipien und der erste Satz aller wahren Philosophie sein (H.VIII, S.42), und dieses "Ich bin" ist für mich der intentionale Urboden für meine Welt (FTL, S.209). Dieses "Ich", von dem hier die Rede ist, darf nicht ein bloßer Mensch im natürlichen Sinn sein, der selber als ein real und weltlich Seiendes zur vorgegebenen, objektiven Welt gehört, sondern es soll das transzendentale ego sein, und zwar, das in seiner vollen Konkretion, also mit all seinen intentionalen Korrelaten (nämlich nicht nur mit einzelnen Gegenständen, sondern auch mit seiner Welt) erfaßt wird. Mit welchem Recht kann nun dieses "Ich bin" der erste Satz alles Philosophierens sein? Erstens, weil nur dieses "Ich bin" wegen der absoluten Zweifellosigkeit des Selbstbewußtseins die erste und absolut apodiktische Evidenz ist (cf. H.VI, S.79) und zweitens – natürlich im Zusammenhang mit dem ersten Grund – weil für mich das Sein des transzendentalen Ich erkenntnismäßig allem objektiven Sein vorangeht und dazu noch in gewissem Sinne der Grund und Boden ist, auf dem sich alle objektive Erkenntnis abspielt (cf. H.I, S.66). So ist das Ich nur als ein apodiktisches und transzendentales gegenüber dem Sein der Welt das an sich Frühere und kann dieses Weltsein nach seinem Was und Wie letztlich begründen. Nur so kann der Philosoph, der die eigene radikale Selbstverantwortung auf sich nehmen soll, das Sein seiner Welt und alles Weltlichen letztlich verantworten.

Zum Schluß müssen wir auch den Zusammenhang der Intersubjektivität mit dem Weltproblem noch etwas erwägen. Die phänomenologische Philosophie soll unter der Leitidee des "Prinzips der Apodiktizität" und der "radikalen Selbstverantwortung des autonomen Philosophen" als reine, transzendentale Egologie anfangen, die naturgemäß den Schein einer solipsistischen Philosophie mit sich führt. Aber die Phänomenologie darf keineswegs konsequent solipsistisch bleiben, sondern muß sich notwendigerweise zu einer Phänomenologie der transzendentalen Intersubjektivität entfalten. Die Welt als Lebenswelt des je eigenen Subjekts hat als Untergrund für den Weltbegriff von höherer Ordnung den Charakter der "Je-meinigkeit". Die Welt bleibt aber nicht für immer die Welt für mich allein, obwohl sie objektiv die eine und selbe Welt für Jedermann, mich selbst eingeschlossen, ist. So ergibt sich das Problem einer diese objektive Welt konstituierenden Monadengemeinschaft, bzw. der die "Welt für alle" konstituierenden, transzendentalen Intersubjektivität, von der Husserl ausdrücklich sagt, daß die transzendentale Intersubjektivität, diese durch mögliche Wechselverständigung verbundene Allgemeinschaft der transzendentalen Einzelsubjekte das an sich erste Sein sei, das jeder weltlichen Objektivität vorangehe und sie in sich selbst trage (cf. H.I, S.158, 182 usw.). Selbstverständlich dürfen wir auch hierbei nicht übersehen, daß sowohl die objektive Welt als auch andere Iche im letzten radikal sich nur in mir, in diesem einsamen transzendentalen ego konstituieren und absolut gerechtfertigt werden können. Die Intersubjektivität konstituiert sich zwar von mir aus und sogar in mir selbst, aber zugleich ist die je meinige Subjektivität nur in der Intersubjektivität, was sie ist, also ein konstitutiv fungierendes Ich. So erhebt sich wieder vor uns die neue schwierige Frage: Wie kann ich über mein individuelles Selbstbewußtsein hinaus ein transzendental-intersubjektives Bewußtsein haben? Der Husserlsche Versuch, dieses intersubjektive Problem und das Erkenntnisproblem des anderen Ich mit Hilfe des Begriffs "Einfühlung" zu lösen, scheint noch nicht ein theoretisch vollkommenes und recht befriedigendes Resultat gebracht zu haben. Aber immerhin haben das Problem der Intersubjektivität und die daraus

hervorgehenden Probleme der Sozialität und Geschichtlichkeit neue
Aspekte und große Entwicklungsmöglichkeiten im Forschungsbereich der
Phänomenologie aufgezeigt.

Nagoya

ANMERKUNGEN

Die Quellenbelege für die gebrachten Husserl-Zitate sind an den betreffenden Stellen in
folgenden Abkürzungen angegeben: H = Husserliana; FTL = Formale und transzendentale
Logik, Halle a.d.S., 1929; EU = Erfahrung und Urteil, Prag, 1939.

[1] K. Löwith: 'Eine Erinnerung an E. Husserl', in Van Breda, "Husserl 1859–1959", La Haye,
1959, S.51; cf. H.VI, S.11f.

[2] cf. E. Fink: *Studien zur Phänomenologie* 1930–1939, Den Hagg, 1966, S.101, 103f. u. 139.

[3] L. Landgrebe: *Philosophie der Gegenwart*, Bonn, 1952, S.56f.

[4] E. Fink: 'Die Spätphilosophie Husserls in der Freiburger Zeit', in *Husserl 1859–1959*, S.107
oder E. Fink: Nähe und Distanz, Freiburg/München, 1976, S.216.

[5] E. Husserl: 'Phänomenologie und Anthropologie', Philosophy and Phenomenological
Research, Vol.II (1941), No.1. p.4.

[6] E. Husserl: ibid. p.13.

[7] E. Husserl: ibid. p.5.

[8] E. Husserl: 'Die reine Phänomenologie, ihr Forschungsgebiet und ihre Methode'
(Freiburger Antrittsvorlesung), *Tijdschrift voor Filosofie*, Jaargang 38 (1976), S.368.

[9] E. Husserl: a.a.O. S.374f.

[10] *Philosophenlexikon*, hrsg. von Ziegenfuss, 1949/50, Bd.I, S.570.

[11] E. Husserl: a.a.O. (*Tijdschrift*) S.365.

[12] E. Husserl: a.a.O. S.366.

[13] Das Wort "Life World" wird schon im Vorlesungsmanuskript von 1924 verwendet (H.VII,
S.232: "die wirkliche Lebenswelt, die Welt im Wie der Erlebnisgegebenheit"). Außerdem
könnte man in dem Begriff "Umwelt", der schon in den früheren Schriften Husserls (z.B. in
den *Ideen* Bd.I) verwendet wird, einen Vorgänger des Begriffs "Lebenswelt" sehen.

[14] E. Fink: in 'Husserl 1859 1959'. S.106.

[15] Auf der ersten Seite der *Krisis* beklagt Husserl, daß die Philosophie "in unserer Gegenwart
der Skepsis, dem Irrationalismus, dem Mystizismus zu erliegen droht". Aber im übrigen
äußert er in diesem Werk, absichtlich oder zufällig, an keiner anderen Stelle eine ausdrückliche Kritik am Relativismus und Historismus.

[16] E. Fink: *Studien zur Phänomenologie* 1930–1939, S.108f., 139 u.178.

[17] E. Fink: a.a.O. S.96; A. Diemer: *Edmund Husserl*, Meisenheim, 1956, S.86; The. De Boer:
De ontwikkelingsgang in het denken van Husserl, Assen, 1966, S.597.

JIRO WATANABE

WAHRHEIT UND UNWAHRHEIT
ODER
EIGENTLICHKEIT UND UNEIGENTLICHKEIT

Eine Bemerkung zu Heideggers Sein und Zeit

INHALT

Nitta / Tatematsu (eds.), Analecta Husserliana, Vol. VIII, 131–203. *All Rights Reserved.*
Copyright © 1978 by D. Reidel Publishing Company, Dordrecht, Holland.

I. EINLEITUNG

Das Problem, das wir im folgenden zu behandeln suchen wollen, bezieht sich auf *einen* Aspekt der existenzialen Analytik des Daseins im *Sein und Zeit* Heideggers.

Dieses in seiner früheren Periode verfaßte Werk hat nach einem Ausdruck G. Mischs "wie der Blitz eingeschlagen" und "eine ungewöhnliche philosophische Erregung hervorgerufen".[1] Es besteht daran kein Zweifel, daß es den Gang der ganzen Philosophie des 20. Jahrhunderts tief beeinflußt hat. Es bleibt allerdings heutzutage, da wir eine Menge anderer wichtiger Veröffentlichungen Heideggers kennen, nur eine frühere Etappe seiner Hauptschriftenreihen, in denen sich seine ungeheuren Gedanken niedergelegt haben. Sein gigantischer Denkweg hat sicher im Laufe seines Lebens Wandlungen durchgemacht. Jedoch kann dieses Werk gewissermaßen, losgelöst von dem historischen Gedankengang Heideggers, fast als ein selbständigeres vollendetes Werk angesehen werden, und wie bei jedem klassischen Werk, so auch bie diesem, könnte ein philosophischer Anfänger einen sicheren Ausgangspunkt seines eigenen Philosophierens finden, indem er sich damit philosophisch-systematisch auseinandersetzt. Freilich ist dieses Werk im historischen Sinne nicht vollendet. Seine siebente Auflage bestätigte sogar, daß sich die lange erwartete zweite Hälfte davon nicht mehr der ersten direkt anschließen sollte. Und über die damit zusammen-hängende sogenannte Kehre-Problematik gehen ja die Meinungen vielfältig auseinander. Aber wir haben hier nicht die Absicht, uns in dieses zwar wichtige, doch in einem gewissen Sinne historische Problem einzulassen. Unsere Absicht ist hier durchaus systematisch orientiert. Wir werden im folgenden versuchen, *ein* in der existenzialen Daseinsanalytik Heideggers *verstecktes* wichtigstes Problem möglichst deutlich an den Tag zu bringen, und dadurch hoffentlich eine Notwendigkeit und Richtung einer *neuen weiteren* Daseinsanalytik anzudeuten.

Nun, bei den philosophischen Gedanken kommt es alles darauf an, wie sie tief in die Wahrheit ihrer jeweiligen Sachen des Denkens eingedrungen sind. Alle nachkommende philosophische Untersuchungen müssen mehr oder weniger an die durch vorangehende philosophische Systeme zustandegebrachten Denkergebnisse anknüpfen und davon ausgehen, kritisch zu beurteilen, wie weit und tief diese Systeme den wahren Sachverhalt der jeweiligen Sachen ans Licht gebracht haben. Erst durch diese kritischen Auseinandersetzungen können philosophische

Untersuchungen ihren eigenen Ausgangspunkt ergreifen. Das soll auch für unseren Fall gelten. So müßte es auch hier im Grunde genommen um die Überprüfung der Wahrheit der existenzialen Daseinsanalytik Heideggers gehen. Aber dabei soll beachtet werden, daß man bei einer solchen kritischen Überprüfung nicht von vornherein eine Position nehmen darf, was bloß entweder zu einer einfachen Zustimmung oder zu einer einfachen Ablehnung führte. Bei einer echten Kritik sollte man im Gegenteil immanent dem Gedankengang eines Systems getreu nachfolgen und es gemäß der Forderung des darin entfalteten Sachverhaltes sich von sich selber her kritisieren, eventuell sich widersprechen lassen. Auf diese Weise allein könnte es unseres Erachtens einer Kritik eine gerechte positive Auseinandersetzung gelingen. Bevor wir aber jetzt gleich diese Arbeit in Angriff nehmen, sollte es wahrscheinlich angebracht sein, den Begriff der Wahrheit und die darauf begründete phänomenologische Methode der Daseinsanalytik Heideggers selbst zunächst in Betracht zu ziehen, damit gezeigt wird, daß Heideggers Ansicht darüber uns erlaubt, ja sogar von uns verlangt, ihm gegenüber eine immanente Kritik durchzuführen.

II. METHODEN- UND WAHRHEITSBEGRIFF

§1. Methodenbegriff

1. Das Thema von Sein und Zeit

Die Absicht Heideggers in *Sein und Zeit* ist bekanntlicherweise "die konkrete Ausarbeitung der Frage nach dem Sinn von 'Sein'" (SZ 1).[2] Um diesen "erfragten" Sinn von Sein als dem "Gefragten" klar zu machen, muß aber ein "exemplarisches" Seiendes als das "Befragte" gewählt werden, an dem der Sinn von Sein abgelesen werden sollte. Dieses Seiende ist nun nach Heidegger das Dasein, das wir selbst je sind (SZ 7). In dem veröffentlichten Teil von *Sein und Zeit* ist dieses exemplarische Seiende, d. h. das Dasein, hinsichtlich seines Seins eingehenderweise expliziert; sein Sein wird als Sorge bestimmt, und dessen Sinn wird als Zeitlichkeit herausgestellt.

Natürlich ist diese Explikation eigentlich zum Zweck der Fragestellung nach dem Sinn von *Sein überhaupt und als solchem* durchgeführt. Diese ist aber, obwohl als eine Problematik der "Temporalität des Seins" (SZ 19) schon antizipiert, doch am Ende nicht in vollem Maße ausgeführt worden. Hier liegt schon ein Problem der Unvollendung des Werks nahe, das wierderum mit der oben genannten Kehre-Problematik zu einem Teil zusammenhängt. Diese Kehreproblematik stünde vermutlich mit derjenigen Zirkelstruktur im engeren Zusammenhang, die hierin bereits zum Vorschein kommt; nämlich die Zirkelstruktur, die zwischen dem zuvor zu explizierenden *Sein* des Daseins und dem danach erneut zu erklärenden Sinn von *Sein überhaupt* zu bestehen scheint. Gewiß verleugnet dabei Heidegger das Bestehen eines eigentlichen Zirkels im Beweis, gesteht aber in der Seinsfrage die Unvermeidlichkeit der Zirkelhaftigkeit des Problems zu (SZ8, 152f., 314ff.), und fordert, in den Zirkel "nach der rechten Weise hineinzukommen" (SZ 153) und "zu springen" (SZ 315). Heidegger selbst ist auf diese Weise in dem veröffentlichten Teil von *Sein und Zeit* in die Analytik *des Seins des Daseins* hineingesprungen, die aber von vornherein eine spätere, erneute Erörterung *des Seins überhaupt* und eine darauf gegründete, nochmalige Überprüfung der gewonnenen Analysen *des Seins des Daseins* verlangte; was wohl einer der Gründe gewesen sein müßte, die "Kehre" später zu veranlassen. Doch, wir müssen, wie gesagt, wegen unseres beschränkten jetzigen Themas all diese wichtige Problematik zunächst einmal beiseite lassen.

Wir müssen uns augenblicklich damit begnügen, festzustellen, daß was Heidegger in dem veröffentlichten Teil von *Sein und Zeit* wirklich analysierte, das Sein des Daseins (Sorge) und dessen Sinn (Zeitlichkeit) war, obwohl er immer wieden darauf hinweist, daß die durchgeführten Analysen einst "eine erneute Wiederholung im Rahmen der grundsätzlichen Diskussion des Seinsbegriffes" (SZ 333), d.h. "das Licht aus der zuvor geklärten Idee des Seins überhaupt" (SZ 436) verlangen.

2. *Die phänomenologische Methode*

Nun, dieses Dasein wird bei seiner existenzialen Analytik in *Sein und Zeit* mit der Behandlungsart der *hermeneutischen Phänomenologie* aufgewiesen und erschlossen, so daß Sorgestruktur und Zeitlichkeit, wie oben gesagt, freigelegt worden sind.

Der methodologische Sinn der *Phänomenologie* ist von den beiden Begriffen von "Phänomen" und "Logos" zu erschöpfen. Das *Phänomen* bedeutet bei Heidegger "das Sich-an-ihm-selbst-zeigende" (SZ 28), kurzum "das Sichzeigende". Übrigens, davon streng unterschieden werden müssen sowohl "der Schein", d.h. "das so Aussehende wie" als auch "die Erscheinung", d. h. "das Sichmelden von etwas, das sich nicht zeigt, durch etwas, was sich zeigt", sowie "die bloße Erscheinung", d. h. "das Meldende", was "das wesenhaft selbst nie Offenbare" anzeigt und "von diesem ausstrahlt" (SZ 28 ff.); alle diese sind verschiedenerweise im Phänomen, dem Sichzeigenden, fundiert. Andererseits bedeutet *Logos* bei Heidegger "ἀποφαίνεσθαι": das, "wovon die Rede ist", "von ihm selbst her" (ἀπό), "sehen lassen" (φαίνεσθαι) (SZ 32). So sagt nun die *Phänomenologie*: "Das was sich zeigt, so wie es sich von ihm selbst her zeigt, von ihm selbst her sehen lassen" (SZ 34).

Aber dieser Titel der Phänomenologie bezeichnet nur das Wie der direkten, sachgemäßen Aufweisung und Ausweisung der Sachen selbst. Was soll denn aber eigentlich in dieser phänomenologischen Behandlungsart aufgewiesen werden? Heideggers Antwort darauf ist die folgende: das, was in einem "ausgezeichneten" Sinne, "seinem Wesen nach notwendig", einer "ausdrüklichen" Aufweisung benötigt, ist "offenbar solches, was sich zunächst und zumeist gerade nicht zeigt, was gegenüber dem, was sich zunächst und zumeist zeigt, verborgen ist, aber zugleich etwas ist, was wesenhaft zu dem, was sich zunächst und zumeist zeigt, gehört, so zwar, daß es seinen Sinn und Grund ausmacht" (SZ 35). Dieses, was in einem ausnehmenden Sinne verborgen, verdeckt oder verstellt sein kann, was also in einem ausgezeichneten Sinne *phänomenologisch* aufgezeigt und sehen gelassen werden soll, ist nach Heidegger "nicht dieses oder jenes Seiende, sondern das Sein des Seienden" (SZ 35), das ja heutzutage, völlig vergessen, nicht mehr der Gegenstand der Frage geworden sei.

Also, bei Heidegger wird der *formale* Phänomenbegriff als das *Sichzeigende* bestimmt, und dieser *formale* Phänomenbegriff kann gewiß auf irgend ein *Seiendes* rechtgemäß angewandt werden. Dann kommt der *vulgäre* Phänomenbegriff zustande, wobei unter dem *Sichzeigenden* das *Seiende* verstanden wird. Aber der *phänomenologische* Phänomenbegriff meint als das Sichzeigende "*das Sein* des Seienden, seinen *Sinn*, seine Modifikationen und Derivate" (SZ 35, gesperrt vom Verfasser). Da es sich nun in dem veröffentlichten Teil von *Sein und Zeit* vor allem um das Dasein handelt, so ist ganz klar, daß es in erster Linie nichts anderes sein kann, als das *Sein* und der *Sinn* des Daseins, welche dort gegen die "Verdeckungstendenz" (SZ 311) mit der *phänomenologischen* Methode von sich selbst her aufgewiesen, ja sogar "gewaltsam" abgewonnen werden sollen. Aber dabei muß zu beachten sein, daß "es für das Absehen auf eine Freilegung des Seins zuvor einer rechten Beibringung des

Seienden selbst bedarf", weil "das Sein je das Sein von Seiendem ist" (SZ 37). Das Seiende selbst, das hier das Dasein ist, soll nämlich nicht nur in der Seinsfrage als ein "exemplarisches" Seiendes "fixiert" werden (SZ 15, 37), sondern auch zuerst vor der Herausstellung seines Seins gerade als das Seiende in der Sicherung einer "rechten Zugangsart" (SZ 6, 7, 15, 16, 37) gezeigt werden. "So wird der vulgäre Phänomenbegriff phänomenologisch relevant" (SZ 37). Mit anderen Worten: Bevor nämlich das *Sein* des Daseins phänomenologisch intensiv untersucht werden soll, muß zuvor das Dasein selbst als das *Seiende*, nachdem es als ein exemplarisches Seiendes gewählt ist, nach der rechten Weise, d. h. in der Sicherung der gewissen Zugangsart, beigebracht werden, so denkt Heidegger. Und gerade hierin besteht unserer Meinung nach der Grund dafür, daß die phänomenologische Untersuchungsart Heideggers ihren *hermeneutischen* Charakter, den sie allerdings ursprünglich an sich hat, um so mehr verschärfen muß.

3. *Hermeneutik*

Das Dasein, das wir selbst sind, ist "dadurch ontisch ausgezeichnet, daß es diesem Seienden in seinem Sein *um* dieses Sein selbst geht" (SZ 12). Es versteht sich irgendwie in seinem Sein. Das Dasein bewegt sich immer schon in einem Seinsverständnis. "*Seinsverständnis ist selbst eine Seinsbestimmtheit des Daseins*" (SZ 12). Daß die phänomenologische Aufweisung als solche von Hause aus einen *hermeneutischen* Charakter haben muß, kommt daher, daß sie gerade diesem "zum Dasein gehörigen *Seinsverständnis* den eigentlichen *Sinn von Sein* und die *Grundstrukturen seines Seins* kundgeben" soll (SZ 37, gesperrt vom Verfasser). Die *phänomenologische* Untersuchung des Daseins muß eine *Auslegung* und *Hermeneutik* des Daseins werden, gerade deshalb, weil das in Frage stehende Dasein im wesentlichen das *Seinsverständnis* hat, so daß die Frage nach dem Dasein hinsichtlich seines Seins unumgänglich das zum Dasein gehörige *Seinsverständnis* selbst zum Problem machen muß. Daß sich die *phänomenologische* Aufweisung stets auf diese Weise eine *Auslegung des Seinsverständnisses* des Daseins zur Aufgabe machen muß, kennzeichnet   gerade   das   Eigentümliche   der   Heideggerschen Phänomenologie.

Aber darüber hinaus ist sehr wichtig und keineswegs zu übersehen, daß diese auslegende Hermeneutik Heideggers ständig "den Charakter einer Gewaltsamkeit" (SZ 311) haben muß. Der Grund dafür besteht darin, daß sich das Dasein *immer schon* in einem Seinsverständnis hält, das aber zunächst und zumeist nicht nur vage und durchschnittlich ist, sondern vielmehr eine hartnäckige Verdeckungstendenz seiner selbst in sich enthält. Das heißt, daß das Dasein, das wir selbst je sind, zwar uns bereits vertraut, ontisch sogar das nächste, doch ontologisch das fernste ist, weil diese vorontologische Seinsauslegung, in der sich das Dasein zunächst und zumeist immer schon bewegt, keineswegs "einer thematisch ontologischen Besinnung auf die eigenste Seinsverfassung entsprungen" ist und so nicht ohne weiteres als angemessener ontologischer Leitfaden übernommen werden darf (SZ 15). Nach Heidegger hat nämlich das Dasein gemäß seiner Seinsart des Verfallens die unüberwindbare Verdeckungstendenz, sich aus dem nichtdaseinsmäßigen Seienden, der "Welt" her, zu verstehen (SZ 15). Gerade durch diese alltägliche Daseinsauslegung bleibt meistens das eigenste Sein des Daseins verdeckt. Überdies kann diese öffentliche Ausgelegtheit des Seins des Daseins mit den verkehrten überlieferten Daseinsauslegungen durchsetzt sein, was die Verdeckungstendenz immer mehr verstärkt und die eigentliche Auslegung verwirrt und erschwert. Beispielsweise darf nach Heidegger das Dasein, das durch die Existenz und Jemeinigkeit gekennzeichnet werden soll, keineswegs durch die bisher übliche Betrachtungsweise behandelt werden, die "auf die Seinsart des innerhalb der Welt nur Vorhandenen" (SZ 43) orientiert ist, die aber gemäß der Verfallenstendenz des Daseins an die "Welt" eine vorherrschende Macht hat: nämlich, die Vorhandenheit.

Aus diesem Grund ist schon eine ursprüngliche, rechte, phänomenale *Vorgabe des Daseins selbst* als eines durch Existenz und Jemeinigkeit bestimmten *Seienden*, geschweige denn sein Sein und dessen Sinn, nichts weniger als selbstverständlich. Gerade im Gegenzug zur verfallenden Auslegungstendenz an den Horizont der Vorhandenheit muß also das Dasein als *ein Seiendes*, erst recht *sein Sein* und dessen *Sinn*, allererst phänomenal recht vorgegeben und aufgrund der genuinen Vorgabe dann näher ontologisch erforscht werden. Eben deshalb ist bereits die Sicherung der genuinen rechten Zugangsart zu dem Dasein als einem Seienden sehr wichtig und so der vulgäre Phänomenbegriff des Daseins auch relevant. Mit anderen Worten: bei dem Versuch der existenzialen Analytik des Daseins soll allererst das Dasein als eine phänomenale Sache im Gegenzug zur verfallenden ontisch-ontologischen Auslegungstendenz auf die rechte

Begegnungsart hin vorgegeben werden, und somit sollen auch sein Sein und dessen Sinn in der eigentlichen auslegend-hermeneutischen sowie phänomenologischen Aufweisung und Explikation den phänomenalen Tatbeständen des Daseins abgerungen bzw. abgewonnen werden, und auf diese Weise ständig mit Gewaltsamkeit erschlossen, schließlich in einer angemessenen Begrifflichkeit festgestellt werden.

Bei dieser Gewaltsamkeit aber darf nun selbstverständlicherweise keine "beliebige Idee" des Seins an das Dasein "konstruktiv-dogmatisch" (SZ 16) herangebracht und ihm keine "'aprioristische' Konstruktion" (SZ 50 Anm.) aufgezwungen werden. Im Gegenteil soll sich das Dasein "an ihm selbst von ihm selbst her" (SZ 16) *phänomenologisch* zeigen, eben dadurch, daß es *hermeneutisch* den vulgären Ausgelegtheiten gewaltsam abgewonnen und so interpretiert wird. Die *Hermeneutik* des Daseins soll ebenso *phänomenologisch* vorgehen, wie die *Phänomenologie* gerade *hermeneutisch* durchgeführt werden soll. Das bedeutet bei Heidegger, daß sich das Dasein in dieser hermeneutischen Phänomenologie gerade "in dem zeigen soll, wie es *zunächst und zumeist* ist, in seiner durchschnittlichen *Alltäglichkeit*" (SZ 16), daß "das Dasein im Ausgang der Analyse gerade nicht in der Differenz eines bestimmten Existierens interpretiert, sondern in seinem indifferenten Zunächst und Zumeist aufgedeckt werden soll" (SZ 43). An dieser Indifferenz der durchschnittlichen Alltäglichkeit, die eine "rechte Bereitung des phänomenalen Bodens" (SZ 50 Anm.) ausmacht, sollen die "wesenhaften Strukturen" (SZ 17) des *Seins* des Daseins in Hinblick auf die sogenannten "Existenzialien", also im Gegenzug zur verfallenden Auslegung aus den Vorhandenheitskategorien, herausgestellt werden, und schließlich der Seins*sinn* des Daseins durch eine ursprüngliche Interpretation des eigentlichen Ganzseinkönnens des Daseins ans Licht gebracht werden. Daß aber dieser methodische Ansatz bei der indifferenten durchschnittlichen Alltäglichkeit einige Probleme in sich enthält, wird später erwähnt werden.

§2. *Wahrheitsbegriff*

1. *Wahrheitsbegriff als die Grundlage des Methodenbegriffs*

Nun, wir müssen jetzt mit großem Nachdruck darauf hinweisen, daß dem oben kurz dargelegten hermeneutisch-phänomenologischen Methodenbegriff der existenzialen Daseinsanalytik Heideggers in Wirklichkeit sein eigener *Wahrheitsbegriff* zugrund liegt, und zwar als das, was das Ganze jener hermeneutisch-phänomenologischen Methode von Grund aus trägt. Die Rechtmäßigkeit jener Methode gründet also ursprünglich in einer bestimmten Auffassung Heideggers von der Wahrheitsstruktur. Jetzt soll diese Seite in unseren Blick kommen.

Die Phänomene der Wahrheit sind, sofern wir uns hier auf ihre Erörterung in *Sein und Zeit* (SZ 212-230) beschränken müssen, von zwei Seiten her betrachtet: nämlich, Entdecktheit und Erschlossenheit. Jene bezieht sich auf die Wahrheit des innerweltlichen Seienden, diese auf die Wahrheit des Daseins. Das Wichtigste ist der von Heidegger dargelegte Inhalt der Erschlossenheit (Wahrheit) des Daseins. Aber zunächst einmal, um die Struktur des Wahrheitsphänomens überhaupt begreiflich zu machen, soll sich unsere Aufmerksamkeit dem Wahrheitsphänomen der Entdecktheit zuwenden, an dem ja doch die Wahrheitsstruktur im allgemeinen am deutlichsten an den Tag gebracht zu werden vermag.

Heidegger hält den traditionellen Wahrheitsbegriff als Überein-stimmung (*adaequatio*) für ungenügend und abkünftig; er fordert vielmehr, in den "Seinszusammenhang" (SZ 216) und die "Seinsart des Erkennens selbst" (SZ 217) einzudringen. Nach Heidegger hat eine erkennende Aussage bei einer Ausweisung ihrer Wahrheit gar nichts zu tun mit der Übereinstimmung zwischen Erkennen und Gegenstand. Die Aussage ist einzig und direkt auf ein Seiendes "bezogen" (SZ 217), das sie meint. "Das Aussagen ist ein Sein zum seienden Ding selbst" (SZ 218). "Das Erkennen bleibt im Ausweisungsvollzug einzig auf das Seiende selbst bezogen" (SZ 218). Dies ist das fundamentale Kennzeichen der Heideggerschen Auffassung der Wahrheitsstruktur, daß er nämlich das aussagende Erkennen gerade als ein *Seinsverhältnis des Daseins zu dem Seienden selbst* ansieht. Jede Interpretation, die hier irgend etwas anderes als den Seinsbezug des Daseins zum Seienden einschiebt, soll nach ihm als eine totale Verkennung der phänomenalen Tatbestände schier abgelehnt werden.

Nun, was tut aber diese erkennende Aussage an dem Seienden? "Sie *entdeckt* das Seiende an ihm selbst. Sie sagt aus, zeigt auf, sie 'läßt sehen' (ἀπόφανσις) das Seiende in seiner *Entdecktheit*" (SZ 218, gesperrt vom Verfasser). Die Aussage als eine wahre ist *entdeckend*. "Wahrsein (Wahrheit) der Aussage muß verstanden werden als *ent-deckend-sein*" (SZ 218). Dieses Wahrsein als entdeckend-sein ist, wie gesagt, "eine Seinsweise des Daseins", d. h. "des In-der-Welt-seins"

(SZ 220), so zwar, daß das Dasein gerade in dieser Seinsweise entweder "umsichtig" oder "verweilend hinsehend" (SZ 218) das innerweltliche Seiende entdeckt. Doch, diese objektive Seite des Entdeckens sei vorläufig beiseite gelassen. Jedenfalls ist das Wahrsein der erkennenden Aussage "das ἀληθεύειν in der Weise des ἀποφαίνεσθαι: Seiendes – aus der Verborgenheit herausnehmend – in seiner Unverborgenheit (Entdeckt-heit) sehen lassen" (SZ 219). "Das 'Wahrsein' des λόγος als ἀληθεύειν besagt: das Seiende, *wovon* die Rede ist, im λέγειν als ἀποφαίνεσθαι aus seiner Verborgenheit herausnehmen und es als Unverborgenes (ἀληθές) sehen lassen, *entdecken*" (SZ 33). Die ἀλήθεια bedeutet "die 'Sachen selbst', das, was sich zeigt, *das Seiende im Wie seiner Entdecktheit*" (SZ 219).

Die Wahrheit ist auf diese Weise die Unverborgenheit, ἀ-λήθεια. Dabei muß darauf aufmerksam gemacht werden, daß der Begriff ἀ-λήθεια "in einem *privativen* Ausdruck" (SZ 222) gemeint ist. "Die Wahrheit (Ent-decktheit) muß dem Seienden immer erst abgerungen werden. Das Seiende wird der Verborgenheit entrissen. Die jeweilige faktische Entdecktheit ist gleichsam immer ein *Raub*" (SZ 222). Das bedeutet, daß sich das Dasein, mit dem Schein, der Verstellung, Verdeckung und Verborgenheit kämpfend, stets bemühen muß, das Seiende, so wie es an ihm selbst ist, "sich ausdrücklich zuzueignen und sich der Entdecktheit immer wieder zu versichern" (SZ 222). Meistens "vollzieht sich alle Neuentdeckung nicht auf der Basis völliger Verborgenheit, sondern im Ausgang von der Entdecktheit im Modus des Scheins" (SZ 222). Im Gegensatz dazu muß die Wahrheit als Entdecktheit fast wie ein Raub mit Gewaltsamkeit abgewonnen werden.

Aus dieser Darstellung könnte man nun sehr leicht ersehen, daß die bereits erwähnte hermeneutisch-phänomenologische Behandlungsart der existenzialen Daseinsanalytik, die ja gerade im Gegenzug zur verfallenden ontisch-ontologischen Auslegungstendenz mit Gewaltsamkeit die Abgewinnung einer rechten Vorgabe des Daseins, die deutliche Herausstellung und die begriffliche Fixierung von seinem Sein und dessen Sinn bezweckt, durch nichts anderes als die vorstehende Auffassung Heideggers der Wahrheitsstruktur ermöglicht ist und aufgrund deren erst zustande gekommen ist. Allerdings bezieht sich alles, was über die Wahrheitsstruktur als Entdecktheit und Entdeckend-sein gesagt ist, genauer besehen, auf das innerweltlich Seiende und keineswegs auf das Dasein. Indessen läßt sich unserer Ansicht nach die Wahrheitsstruktur als solche, die aus der Entdecktheit und dem Entdeckend-sein zu entnehmen ist, mutatis mutandis auf die hermeneutisch-phänomenologische Untersuchung des Daseins durch sich selbst übertragen. Diese hermeneutisch-phänomenologische Untersuchung des Daseins kann nämlich gar nicht anders durchgeführt werden, als daß das Dasein, das in einem ausnehmenden Sinne das innerweltliches Seiende *entdeckt*, in einem *übertragenen* Sinne gleichsam sich selbst hinsichtlich seines Seins zu "*entdecken*" sucht, d. h. sich selbst, im Gegenzug zum verfallenden Daseins- und Seinsverständnis, aus der Verborgenheit herausnehmend, ans Licht zu bringen versucht. Eben dadurch kann das Dasein in einem *strengen* Sinne sich selbst "*erschlossen*" werden. Mit der "Erschlossenheit" des Daseins muß ja doch wenigstens, so müssen wir annehmen, als eine ihrer fundamentalen Bedeutungen, dieses übertragene Verhältnis der "Entdecktheit" des Seins des Daseins durch sich selbst mit gemeint sein. Die Erschlossenheit muß also für die im übertragenen Sinne "entdeckte", d.h. streng genommen, hermeneutisch-phänomenologisch "erschlossene" Erschlossenheit (wahres Verhältnis) des Seins des Daseins gehalten werden.

2. Das Problem der Bewährung

Bevor wir aber die erschlossene Erschlossenheit des Daseins zum Problem machen, soll unsere Diskussion einigermaßen nochmals auf den Ausgangspunkt der Wahrheit als Entdecktheit und Entdeckend-sein zurückgehen. Denn die Feststellung, daß bei Heidegger die Wahrheit überhaupt in ihrer Struktur als die Unverborgenheit des Seienden erfaßt ist, genügt ja zur Überprüfung seines Wahrheitsbegriffes gar nicht. Das Allerwichtigste ist, zu wissen, wie sich diese Entdecktheit des Seienden, bzw. das Entdeckend-sein des Daseins gerade als wahr bewähren läßt. Es gilt vor allem, deutlich zu erfassen, wie sich beispielsweise ein aussagendes Erkennen als wahres ausweist. Diese Frage würde sich natürlich nicht erheben, wenn das Dasein ohne weiteres gewissermaßen mit einem Schlag das Seiende in seiner wahren Entdecktheit entdecken könnte, was doch leider nicht der Fall ist; denn Heidegger sagt selbst, daß die Entdecktheit gerade gegen die Verstellung und Verdeckung dem Seienden erst "abgerungen" und "abgewonnen" werden muß. Wenn dem so ist, müßte es ja irgendwie ein Kriterium geben, das von der "verstellten" Entdecktheit die "wahre", "echte", entdeckte" Entdecktheit zu unterscheiden vermag. Oder bedarf es denn für das Entdeckend-sein des Daseins gegenüber dem innerweltlichen Seienden und insbesondere für die Erschließung des

Daseins durch sich selbst gar keiner "Regelung" oder "Leitung" (SZ 312) oder keines "Beweises", daß sich der mit der methodologisch zwar erforderlichen Gewaltsamkeit durchgeführte Versuch von ἀληθεύειν doch irgendwie "den freien Belieben" (SZ 313) entziehen lassen kann? Auch bei der Erschließung des Daseins durch sich selbst muß, so liegt es nahe, wenn auch nicht ein exakter Beweis, doch irgend eine "Bezeugung" vorgebracht werden. Jedenfalls ist es klar, daß es doch auf eine gewisse Weise Möglichkeiten der Ausweisung, Bewährung und Bezeugung der entdeckenden bzw. erschließenden Aussage geben muß. Sonst wäre es ja möglich, daß eine vermeintlich wahre Entdeckung des innerweltlichen Seienden oder eine angeblich tiefgreifende Erschließung des Daseins in der Tat in ein sich undurchsichtiges Mißverständnis und Verkennen eines wirklichen Sachverhaltes geraten wäre. Sicher ist dieses Problem der Bewährung sehr schwer zu lösen; ferner muß ja wohl die Art der Bewährungen je nach dem Problembereich, wie etwa bei der Entdeckung des umsichtigen Besorgens des innerweltlich zuhandenen Seienden oder des verweilenden Hinsehens des innerweltlich vorhandenen Seienden oder bei der hermeneutisch-phänomenologischen Erschließung des Seins des Daseins durch sich selbst, unterschiedlich sein. Hier müssen wir aber auf eine näher eingehende Erörterung dieses Problems verzichten. Wir beabsichtigen nur, es in einer allgemeineren Form der Überprüfung dessen, was Heidegger dazu sagt, soweit zu bedenken, als es unser vorstehendes Thema angeht und seiner Exposition dienlich ist.

Nach Heidegger ist das Wahrsein der erkennenden Aussage, wie oben gesagt, das Entdeckend-sein. Durch dieses Entdeckend-sein der Aussage wird das Seiende entdeckt, und so wird das Entdeckt-sein des Seienden, es im Wie seiner Entdecktheit zugänglich gemacht. Aber dabei muß es irgendwie bewährt und ausgewiesen werden. Das heißt, daß von der Seite der Aussage her gesehen das Entdeckend-sein der Aussage bewährt und ausgewiesen wird, daß von der Seite des Seienden her gesehen das Entdeckt-sein des Seienden, es im Wie seiner Entdecktheit zur Ausweisung kommt. Nun, wie geschieht das denn? Heidegger sagt: An dem Seienden selbst spielt sich die Bewährung ab (SZ 218). Aber wie? Die Bewährung erfolgt zweiseitig. *Zum einen:* "Die Bewährung vollzieht sich auf dem Grunde eines *Sichzeigens des Seienden*" (SZ 218, gesperrt vom Verfasser). "Das gemeinte Seiende selbst zeigt sich *so, wie* es an ihm selbst ist, d. h., daß *es* in Selbigkeit so ist, als wie seiend *es* in der Aussage aufgezeigt, entdeckt wird" (SZ 218). "*Bewährung* bedeutet: *sich zeigen des Seienden* in *Selbigkeit*" (SZ 218). *Zum anderen* aber: "Das ist nur so möglich, daß das aussagende und sich bewährende Erkennen seinem ontologischen Sinne nach ein *entdeckendes Sein zum* realen Seienden selbst ist" (SZ 218). "Das aussagende Sein zum Ausgesagten" ist "ein Aufzeigen des Seienden"; "es *entdeckt* das Seiende, zu dem es ist" (SZ 218).

Also, die Wahrheitsstruktur ist folgende: (1) das Dasein verhält sich als entdeckend-seiend zum Seienden selbst, und läßt das Seiende an ihm selbst offenbaren in seiner Entdecktheit. (2) Damit aber diese Entdeckt-heit des Seienden selbst eigentlich zustande kommen kann, muß sich das Seiende selbst gerade so, wie es an ihm selbst ist, zeigen, und aufgrund dessen entdeckt ja das Dasein das Seiende gerade in seiner Entdecktheit, so zwar, daß eben dieses Entdeckend-sein sich bewährt. Erst gerade aufgrund eines Sichzeigens des Seienden kann das Dasein es entdecken. (3) Aber gleichzeitig damit muß beachtet werden, daß sich das Seiende wiederum erst aufgrund des Entdeckend-seins des Daseins zeigen kann.

Wenn dem so ist, so liegt hier offensichtlich ein Zirkel. Das Sichzeigen des Seienden ist der Grund der Bewährung einer entdeckenden Aussage; das Entdeckend-sein des Daseins ist aber umgekehrt die Voraussetzung für das Sichzeigen des Seienden; beide Momente nehmen sich in Anspruch, bedürfen sich füreinander. Was bedeutet aber dieses zirkelnde Verhältnis?

Es deutet sicher auf einen enormen Komplex schwieriger Wahrheits-probleme; doch scheint mindestens soviel zwiefellos zu sein, daß die Wahreit aus dem tiefen Verwobensein beider Momente, d.h. des Sichzeigens des Seienden und des Entdeckend-seins des Daseins besteht, daß sich die faktische Wahrheitsgewinnung nur in einem *dynamischen* Prozeß, der das Verflochtensein beider Momente in sich beschließt, entwickeln kann. Das Dasein verhält sich nämlich zu einem Seienden entdeckend, und dieses zeigt sich dementsprechend in einer gewissen Weise seiner Entdecktheit. Aber dieses Seiende kann sich als Schein oder in einer verdeckten, verstellten Weise zeigen. So könnte es prinzipiell nicht für ausgeschlossen gehalten werden, daß das Dasein, wenn es sich nur entdeckend benimmt, Fehler, Irrtümer und Täuschungen nicht beginge. "Daher muß das Dasein wesenhaft das auch schon Entdeckte *gegen* den Schein und die Verstellung sich ausdrücklich zueignen und sich der Entdecktheit immer wieder versichern" (SZ 222). Indessen, trotz der wiederholten selbstbewußten Versicherungen kann das Dasein möglicher-weise immer wieder der Verstellung verfallen. Um so vorsichtiger muß die Bewährung durchgeführt werden. Wahrscheinlich wird die Entdecktheit

eines Seienden faktisch nicht einfach dadurch, daß sich das Dasein
aufgrund eines Sichzeigens des Seienden entdeckend verhält, ohne
weiteres mit einem Zug erzielt werden, sondern in einem mühevollen
dynamischen, wenn man so sagen darf, in einem gewissen Sinne
dialektischen Prozeß erreicht werden, in dem ein anstrengender Schritt
von der niedrigen Wahrheit zur höheren getan wird, oder wo die höhere
Wahrheit als die manche überwundene niedrige Unwahrheiten in sich
enthaltende besteht. "Die Wahrheit (Entdecktheit) muß dem Seienden
immer erst abgerungen werden. Das Seiende wird der Verborgenheit
entrissen. Die jeweilige faktische Entdecktheit ist gleichsam immer ein
Raub" (SZ 222). Dieser Raub aber wird wohl nicht ein für allemal
vollzogen werden, sondern vielmehr "immer wieder" seine eigenen
Versicherungen in Anspruch nehmen. Heidegger erwähnt als Beispiel
einer Bewährung nur einen äußerst einfachen Fall, daß sich nämlich die
Aussage, das Bild an der Wand hänge schief, bewährt (SZ 217); wobei sich
der obengenannte dynamische Bewährungsprozeß anscheinend leicht
abspielen kann. Er spricht gewiß auch von den Gesetzen Newtons, aber in
einem ganz anderen Zusammenhang (SZ 226 f.), so daß es nicht leicht
einleuchtet, wie er bei diesem Fall die Bewährung präzisieren will. Diese
beiden Beispiele gelten gewissermaßen der Entdecktheit des innerweltlich
zuhandenen oder vorhandenen Seienden. Aber wenn anders die aus der
Entdecktheit erschöpfte Wahrheitsstruktur, wie gesagt, auch auf die Erschlossenheit des Seins des Daseins in einem *übertragenen* Sinne angewandt werden kann, so liegt es nahe, daß man auch bei dieser hermeneutisch-phänomenologischen Erschließung der Erschlossenheit des
Daseins die gewonnenen Resultate "gegen den Schein und die Verstellung
sich ausdrücklich zueignen" und sich der erschlossenen Erschlossenheit
"immer wieder versichern" muß. Das bedeutet, daß die existenziale
Analytik des Daseins, wie sie von Heidegger wirklich ausgeführt worden
ist, nicht ein für allemal als ein schon vollendetes System zu gelten vermag,
sondern gerade wegen des darin entwickelten Wahrheitsbegriffes eine
mögliche Überprüfung ihrer selbst, ja sogar eine Möglichkeit ihres Weiter-
denkens grundsätzlich einräumen muß. Das ist ja gerade, was uns am
Anfang dieser Abhandlung auf eine mögliche immanente Kritik zu
sprechen hat kommen lassen. Abgesehen davon, müssen wir aus der
zirkelnden Wahrheitsstruktur offensichtlich den Schluß ziehen können,
daß die Wahrheit genau besehen eben das sein muß, was in dem Kampf mit
der Unwahrheit besteht, was ständig der Gefahr, in die Unwahrheit zu
geraten, ausgesetzt ist, was also erst durch die Bemühung, sie zu
überwinden, erzielt werden kann, und was diesen mit Unwahrheit

verflochtenen dynamischen Prozeß wohl nicht einfach hinter sich verwerfen kann, sondern ihn im wesentlichen in sich selbst als ein überwundenes Moment enthalten muß, ja sogar ohne ihn nicht bestehen kann.

3. Husserls Wahrheitsbegriff

Bei Heidegger wird auf diese Weise das Sichzeigen des Seienden als der Grund der Bewährung von Entdeckend-sein des Daseins angesehen, und umgekehrt gilt das Entdeckend-sein des Daseins als die Voraussetzung für das Sichzeigen des Seienden. Aus dieser zirkelnden Sachlage hat sich der dynamische Charakter der Wahrheitsstruktur, wie oben erwähnt, ergeben.

Nun, damit man sich aber über diesen Charakter noch klarer wird, wollen wir hier, obwohl es von unserem eigentlichen Thema ein bißchen abzukommen scheint, doch Husserls Ansicht darüber kurz heranziehen, weil in Heideggers Wahrheitsbegriff ein großer Einfluß von Husserl zu finden ist. Man könnte wahrscheinlich berechtigt sein, anzunehmen, daß Heideggers Wahrheitsbegriff erst dadurch entstanden sei, daß er sich Husserls Wahrheitsgedanken mit Hilfe der Interpretation der griechischen Gedanken von ἀλήθεια angeeignet hätte. Hier aber beschränken wir uns nur auf einen skizzenhaften Vergleich mit Husserls Wahrheitsgedanken, der aber auch bloß mit Bezugnahme auf deren verständlicheren Darstellungen in "Ideen I"³ vorgenommen werden muß.

Bei Husserl wird ein Seiendes bzw. Gegenstand in seinen noetisch-noematischen Analysen als ein intentionales Korrelat des Bewußtseins angesehen. Dieses Korrelat heißt das Noema, und dieses setzt sich schließlich aus zwei Momenten zusammen: nämlich, Sinn und Sein. Der noematische Sinn besteht aus einem Kern (x) und den bestimmenden Inhalten, und wird aus der Hyle durch die korrelativ entsprechenden synthetisierenden Akten der Noesis intentional konstituiert, indem die Noesis die Hyle beseelt, auffaßt und ihr Sinn gibt. Die Hyle und die noetischen Akte der Synthesis sind in dem Bewußtsein reell vorfindbar, während der noematische Sinn nicht darin als reelles Moment zu finden ist, sondern erst intentional konstituiert wird. Der noematische Sinn besagt den Gegenstand im Wie seiner Bestimmtheiten. Andererseits hat das Noema außer diesem noematischen Sinn noch das Seinsmoment, und besonders wichtig sind verschiedene Seinscharaktere, die den Gegenstand im Wie seiner Gegebenheiten kennzeichnen. Diesen Seins-charakteren, die auch thetische Charaktere heißen, entsprechen gemäß der Korrelation zwischen Bewußtsein und Gegenstand die Glaubens-charaktere auf der noetischen Seite. Indem nun der noematische Sinn mit den Seinscharakteren verbunden wird, wird das volle Noema erst als der Satz konstituiert. Das volle Noema ist ein Komplex von Sinn und Sein.

Was hat nun dieses mit Sinn und Sein ausgestattete Noema mit der Wahrheitsstruktur zu tun? Für Husserl ist die Wahrheit "die Einheit einer Vernunftsetzung mit dem sie wesensmäßig Motivierenden".[4] Der noematische Sinn wird nämlich jeweils, den Fall der Neutralisierung aus-genommen, mit verschiedenen Seinscharakteren gesetzt. Die Seins-charaktere modifizieren sich in Korrelation der Glaubenscharaktere als z. B. gewiß, möglich, wahrscheinlich und zweifelhaft usw. Wenn beispiels-weise ein noematischer Sinn mit dem Seinscharakter von "zweifelhaft" gesetzt wird, so bedeutet es, daß die bestimmenden Inhalte eines noematischen Sinnes, die ein bestimmter Kern trägt, mit dem "zweifel-haften" Charakter gesetzt sind, daß sie also noch nicht zu dem evidenten Sachverhalt der betreffenden Sache gelangt sind. Falls hingegen ein noematischer Sinn, als Korrelat "der Urdoxa, der Glaubensgewißheit", mit der höchsten Seinsgewißheit thetisch gesetzt wird, so wird das auf einen Gegenstand intentional bezogene Bewußtsein zum Vernunftbewußt-sein, dessen Seinssetzung Vernunftsetzung ist. Diese Vernunftsetzung enthält die vernünftige Ausweisung, setzt also einen noematischen Sinn mit dem Seinscharakter der Gewißheit und trifft die Evidenz bzw. Wahr-heit der betreffenden Sache. Aber wie geschieht diese Vernunftsetzung? Denn ohne Ausweisung kann ein noematischer Sinn nicht mit dem Seins-charakter der Gewißheit gesetzt werden. Für die Vernunftsetzung muß ja irgend eine Ausweisung ev. ein Rechtsgrund vorhanden sein. Nun, Husserl sagt: "Die Setzung hat in der originären Gegebenheit ihren *ursprünglichen Rechtsgrund*".[5] Das heißt: die Vernunftsetzung muß durch "*die originäre* Gegebenheit" der betreffenden Sachen selbst "vernünftig motiviert" sein. Erst dadurch kann der noematische Sinn zur Erfüllung kommen und so mit der Seinssetzung der Gewißheit versehen werden. Die Wahrheit, d. h. die Evidenz heißt also "die Einheit einer Ver-nunftsetzung mit dem sie wesensmäßig Motivierenden", d. i. mit der originären Gegebenheit, in der die betreffende Sache, ohne Ansehen der individuellen oder Wesensanschauung, zu ihrer anschaulichen Erfüllung kommt.

Nun, Heideggers Auffassung der Wahrheitsstruktur hat offensichtlich eine enge Beziehung zu dieser Husserlschen Auffassung der Evidenz. Aber natürlich ist alles, was hier im Zusammenhang mit der transzendental-phänomenologischen Bewußtseinsanalyse stand, dort aus der Richtung der existenzialen Analytik des Daseins und der Seinsfrage her subtilerweise umgeändert und weiter entwickelt. Der Bewußtseinsstandpunkt ist überwunden; das Wahrheitsproblem ist, jetzt in der existenzialen Daseinsanalytik umgelagert und darin verankert, aus dem Horizont des Seinsproblems her tiefer umgedeutet worden. Dabei muß im übrigen Heideggers Auseinandersetzung mit dem griechischen Wahrheitsbegriff ἀλήθεια eine große Rolle gespielt haben. Abgesehen davon, liegt es nahe, daß nämlich *einmal* die auf das Noema gerichteten noetischen Akte der Synthesis des *intentionalen Bewußtseins* Husserls in das entdeckende Sein *des Daseins* zum realen Seienden selbst als eine Seinsweise des Daseins Heideggers umgewandelt worden sind. *Zum anderen* verwandelt sich daher auch *die originäre Gegebenheit der Sache selbst,* die bei Husserl die Vernunftsetzung motivieren soll, bei Heidegger in ein *Sichzeigen des Seienden* in Selbigkeit. *Zum dritten* ist also das volle Noema, das durch die motivierte Vernunftsetzung als ein seiendes Sinnhaftes *im Wie seiner Bestimmtheiten* und zugleich *im Wie seiner Gegebenheiten* konstituiert wird, bei Heidegger umgeändert worden in das Entdeckt-sein des Seienden selbst, d. h. *das Seiende im Wie seiner Entdecktheit.* Daran, daß auf diese Weise die Terminologie der Bewußtseinsphilosophie in die der Fundamentalontologie umgewandelt worden ist, besteht unserer Meinung nach kein Zweifel.

Doch, wir müssen hier auf das weitere Eingehen in eine vertiefende Vergleichung dieser Ansätze Verzicht leisten. Wir müssen vielmehr im vorstehenden darauf achtgeben, daß der Husserlschen Wahrheitsstruktur, die vermutlich zur Gestaltung des Heideggerschen Wahrheitsbegriffes einen entscheidenden Antrieb gegeben hat, eine Idee der "Motivierung" als ein wichtiges Moment zugrunde liegt. Damit nämlich ein noematischer Sinn vernünftig, d. h. mit der Glaubensgewißheit gesetzt werden kann, muß diese Setzung durch die *originäre Gegebenheit,* eventuell durch die *originär gebende Anschauung* "motiviert" sein. Ohne diese Motivierung läßt sich keine gewisse Vernunftsetzung vollziehen. Aber die ursprünglichen Vernunftsetzungen können sicher dabei im Fortgang einstimmiger Erfüllung eine positive phänomenologische Steigerung in Hinsicht auf ihre motivierende Kraft erfahren, aber auch unter Umständen im Gegenteil durch das Auftauchen der Gegenmotive ihre Kraft verlieren und endlich durchstrichen werden. In diesem letzten Fall müssen durch ein genaueres Achtgeben auf die originäre Gegebenheit neue Vernunftsetzungen nochmals herbeigeführt werden, aber selbstverständlicherweise unter derselben Wahrheitsstruktur. Die Wahrheitsstruktur als solche hält sich überall durch, die Evidenz kann immer nur als "die Einheit der Vernunftsetzung mit dem sie wesensmäßig Motivierenden" gelten. Jedoch kann diese "*Einheit*" der Seinssetzung der Gewißheit *mit* dem sie motivierenden, originär Gegebenen *nicht immer* so ohne Schwierigkeit leicht verwirklicht werden, wie man meinen möchte, sondern in einem Fortgang der Erfahrungen durch das ständige Prüfen gewonnener Ergebnisse erst gestiftet werden.

Durch die Rücksichtnahme auf die Husserlsche Auffassung der Wahrheitsstruktur hat sich hoffentlich immer deutlicher herausgestellt, daß die Evidenz oder die Entdecktheit keineswegs *mit einem Schlag* erzielt werden kann, sondern daß sie erst dadurch zu gewinnen ist, indem sich das intentionale Bewußtsein anstrengt, auf die originäre Gegebenheit genau achtgebend, und dadurch motiviert, einen noematischen Sinn mit dem ausgezeichneten Seinscharakter der Gewißheit zu bekleiden; oder aber erst dadurch, indem sich das Dasein bestrebt, dem Seienden selbst, gegen den Schein und die Verstellung, die Entdecktheit zu entreißen, und es selbst so, wie es an ihm selbst ist, sehen zu lassen. Indessen bleibt trotz aller Bemühung stets die Möglichkeit der Fehler, Irrtümer und Täuschungen. Um so mehr bedarf es für die Entdecktheit sorgfältiger Überprüfung der schon gewonnenen Resultate. In diesem Sinne muß die Wahrheit einen Werdegang in sich beschließen, der ein verflochtener Entwicklungsprozeß von Wahrheit und Schein ist. Wahrheit und Unwahrheit sind unzertrennlich miteinander verwoben.

4. *Erschlossenheit als Wahrheitsproblem*

Alles, was oben, insbesondere bei Heidegger, für die Wahrheitsstruktur gesagt wurde, ist eigens im Hinblick auf die Wahrheit als Entdecktheit herausgearbeitet, und die Entdecktheit ist ja, wie gesagt, einzig und allein auf das innerweltliche Seiende bezogen. Jedoch, die allgemeine Wahrheitsstruktur, vor allem ihr Charakter des Werdegangs, soll, obwohl er aus der ausschließlichen Orientierung an der Wahrheit als Entdecktheit erschöpft ist, trotzdem auch für die Erschlossenheit des Daseins gelten. Denn diese ist in Wirklichkeit nicht anders freigelegt worden, als daß das Dasein hinsichtlich seines Seins gerade durch die aufgrund der oben erörterten Auffassung der Wahrheitsstruktur ermöglichte hermeneutisch-phänomenologische Methode gleichsam "entdeckt", im strengen Sinne erschlossen wurde. Dieselbe Sachlage trifft auch für Husserl zu. Seine Auffassung der Evidenz und deren sozusagen offener Charakter haben ebenso wie bei Heidegger seine transzendental-phänomenologische Philosophie selbst im Ganzen, in der dieser Wahrheitsbegriff entwickelt wurde, im Hinblick auf ihren Wahrheitscharakter bedingt. Gerade deswegen ist Husserls Phänomenologie nicht in einer und derselben Stelle stehengeblieben, sondern vielmehr befand sie sich durch sein ganzes Leben hindurch in einer dauernden Metamorphose. Abgesehen davon, ist es ganz klar, daß die philosophischen Untersuchungsergebnisse überhaupt grundsätzlich wegen ihres solchen Wahrheitscharakters einer kritischen Überprüfung ihrer selbst keineswegs entgehen können; denn die gewonnene Wahrheit kann sich prinzipiell nicht der Gefahr entheben, möglicherweise den Mißdeutungen der betreffenden Sachverhalte verfallen zu sein. Das läßt sehen, daß die von Heidegger durchgeführten hermeneutisch-phänomenologischen Analysen wegen der dort von ihm selbst dargelegten Wahrheitsbegriffe von sich selbst her ihre eigene Überprüfung verlangen und eine immanente Kritik unumgänglich machen. Für unmöglich muß in diesem Sinne gehalten werden, daß die von Heidegger selbst vollzogenen Analysen allein von diesem Wahrheitscharakter des Werdegangs enthoben wäre. Das bedeutet aber nicht, daß seine Daseinsanalytik irregeführt wäre, sondern daß sie als ausgezeichnete Anhalte gilt, bei denen man sein eigenes Nachdenken über Dasein ansetzen könnte, um erneut das Problem von Mensch und Welt gründlich zu bedenken, so zwar, daß dadurch die von dem vorangegeangenen System erreichten Ergebnisse so wenig vernichtet würde, wie vielmehr eben damit vertieft, weitergedacht und entwickelt worden wäre.

Wenn man nun aber an diejenigen Stellen in *Sein und Zeit* herankommt, in denen, auf die Darlegungen der Wahrheit als Entdecktheit folgend, endlich die Wahrheit als Erschlossenheit des Daseins zur Sprache kommt, hat es den Anschein, als ob damit die endgültige Wahrheit des Daseins erreicht worden wäre. Wir müssen jetzt also bedächtig auf diejenige Weise, in der die Wahrheit als Erschlossenheit von Heidegger eingeleitet wird, achtgeben.

Nach Heidegger "ist Wahrsein als entdeckend-sein eine Seinsweise des Daseins", und "was dieses Entdecken selbst möglich macht, muß notwendig in einem *noch ursprünglicheren Sinne* 'wahr' genannt werden" (SZ 220, gesperrt vom Verfasser). Das Entdecken bezieht sich, wie gesagt, auf das innerweltliches Seiende. Dieses ist aber "'wahr' in einem zweiten Sinne". "Primär 'wahr', das heißt entdeckend, ist das Dasein"

(SZ 220). "Die Entdecktheit des innerweltlichen Seienden gründet in der Erschlossenheit der Welt", die aber die "Grundart des Daseins" ist (SZ 220). "Die Struktur der Sorge" "birgt in sich Erschlossenheit des Daseins" (SZ 220). "*Mit* und *durch* sie", d. h. "erst mit der *Erschlossenheit des Daseins*" wird "das *ursprünglichste* Phänomen der Wahrheit erreicht" (SZ 220 f.). Alles, was in Bezug auf die Seinsstruktur des Daseins "aufgezeigt" worden ist, "betraf nichts anderes als das ursprünglichste Phänomen der Wahrheit" (SZ 221). "Sofern das Dasein wesenhaft seine Erschlossenheit *ist*, als erschlossenes erschließt und entdeckt, ist es wesenhaft 'wahr'. *Dasein ist 'in der Wahrheit'*" (SZ 221). So ungefähr geht Heidegger vor.

Wir müssen über diesen Gedankengang Heideggers sehr bedachtsam nachdenken. Heideggers Argument, daß die Erschlossenheit als Bedingung der Möglichkeit der Entdecktheit vom innerweltlichen Seienden ursprünglicher ist, als diese, und daß sie, vorausgesetzt, daß die Entdecktheit eine Art der Wahrheit ist, verglichen mit dieser, das "*ursprünglichere*", ja sogar "*ursprünglichste*" Phänomen der Wahrheit "*betrifft*", hat gewiß vollkommen recht. Denn: solange die Entdecktheit als eine Art Wahrheit nur auf das innerweltliches Seiende bezogen ist, das aber erst aufgrund der Erschlossenheit der Welt wirklich entdeckt wird, liegt es doch ganz nahe, daß die Erschlossenheit die "*ursprünglichere*", bzw. "*ursprünglichste*" Dimension ausmacht, die die Wahrheit der Entdecktheit ermöglicht. Das soll aber jedoch nicht bedeuten, daß Heideggers Analyse dieser Dimension, d. h. der Erschlossenheit als das, was die Entdecktheit als eine Wahrheit möglich macht, ohne weiteres auch *inhaltlich* bereits "wahr" wäre. Es ist ein Ding, auf den Ort, wo das "ursprünglichere" bzw. "ursprünglichste" Phänomen der Wahrheit zu suchen ist, hinzuweisen, und es ist ein anderes Ding, über den bloßen Hinweis dieses Ortes hinaus dieses Phänomen selbst inhaltlich *konsequenterweise* an den Tag zu bringen, bzw. *zu erschließen*. Wenn man sich so ausdrücken darf, muß ja die *Erschlossenheit des Daseins* als der ursprünglichste Ort, in dem das Wahrheitsphänomen verwurzelt, in ihrer Wahrheit nun erst gerade hermeneutisch-phänomenologisch von jetzt an *erschlossen* werden. Ob diese Untersuchungen wirklich es zu einer im wahrhaften Sinne *erschlossenen Erschlossenheit* des Daseins gebracht hätte, kann "erst *nach dem Gang*" (SZ 437) entschieden werden. Nun aber hat die Darlegung Heideggers an dieser Stelle den Anschein, als ob der von ihm durchgeführte Versuch, diese Erschlossenheit des Daseins herauszustellen, das "ursprünglichste" Phänomen der Wahrheit *nicht nur* "*beträfe*", sondern auch "*erreichte*", und zwar dermaßen, daß damit auch inhaltlich bereits dieses Phänomen auf eine *wahre* Weise *erschlossen* worden wäre. Sicher betreffen die von ihm vollzogenen Analysen der Erschlossenheit insofern das "ursprünglichste" Phänomen der Wahrheit, als die Erschlossenheit des Daseins den "ursprünglicheren", oder "ursprünglichsten" Boden ausmacht, auf dem die Entdecktheit erst als eine sekundäre Art der Wahrheit zustande kommt. Jedoch kann das nicht ohne weiteres besagen, daß sie auch in *inhaltlicher* Hinsicht schon wahr wären, d. h. die wahren Verhältnisse des Seins des Daseins *erschlossen* hätten. Allerdings kann es auch *nicht* bedeuten, daß sie die wahren Verhältnisse *nicht* erschlossen hätte. Ob es diesen Analysen wirklich gelungen hat, muß erst nach einer eingehenden Überprüfung entschieden werden. Jedenfalls, soviel ist klar, daß mit dem Hinweis auf den "ursprünglichen" Ort des Wahrheitsphänomens nicht ohne weiteres bereits die Rechtmäßigkeit der betreffenden Analysen gewährleistet ist. Zwischen beiden Dingen muß genau unterschieden werden. Daraus folgt aber auch, daß der Begriff der Erschlossenheit zwei Seiten in sich enthält und eigentlich die *erschlossene Erschlossenheit* bedeuten soll; "erschlossen" — das bezieht sich auf die hermeneutisch-phänomenologische Erschließungsweise, und "Erschlossenheit" —, das bedeutet die durch diese Methode erschlossene Erschlossenheit als die Seinsweise des Daseins, auf deren Grund erst auch die Entdecktheit des innerweltlichen Seienden möglich wird. Und solange die erste hermeneutisch-phänomenologische Erschließungsweise, wie gesagt, die Möglichkeit der Überprüfung ihrer eigenen Ergebnisse grundsätzlich einräumen muß, muß die letztgenannte Erschlossenheit im Sinne der Seinsweise des Daseins als die durch diese Methode erreichten Ergebnisse inhaltlich genau durchgegangen und auf ihre Wahrheit hin überprüft werden können.

Bei Heidegger kommt die erste Bedeutung der Erschlossenheit, die doch die methodische Bedingung für die ganzen Untersuchungen bildet, indessen nicht deutlich zum Vorschein, und mit der Erschlossenheit ist *in erster Linie* die letztere Seinsweise des Daseins gemeint, gemäß der es sein Da ist. Das Dasein "*ist* selbst je sein 'Da'" (SZ 132), "es *ist* in der Weise, sein Da zu sein" (SZ 133). "Der Ausdruck 'Da' meint diese wesenhafte Erschlossenheit" (SZ 132). Das Dasein ist "'erleuchtet'", "an ihm selbst *als* In-der-Welt-sein gelichtet", so zwar, daß "es selbst die Lichtung *ist*" (SZ 133). "Das Dasein ist seine Erschlossenheit" (SZ 133). Da das Dasein auf diese Weise sein Da ist und in seinem Da sich die Lichtung und Unverborgenheit des Seienden ereignet, so sagt Heidegger, wie angeführt, gerade in diesem

Sinne, daß "das Dasein 'in der Wahrheit' ist" (SZ 221). Diese Wahrheit besagt nicht die erschlossene bzw. entdeckte Wahrheit als Ergebnisse, sondern meint, daß das Dasein als erschlossenes erschließt und entdeckt und so mitten in der Lichtung und Unverborgenheit des Seienden existiert. So sagt Heidegger in der Tat, daß diese Aussage nicht meine, das Dasein sei ontisch immer oder auch nur je 'in alle Wahrheit' eingeführt (SZ 221). Sie sagt nur, daß das Dasein sein Da ist, aufgrund der Erschlossenheit existiert, und seiner existenzialen Verfassung nach vor allem im Hinblick auf sein eigenstes Sein erschlossen existiert.

Es kommt alles auf die Überprüfung des hermeneutisch-phänomenologisch erschlossenen Inhalts dieser Erschlossenheit an. An dieser Stelle aber, wo die Wahrheit erörtert wird, gibt uns Heidegger darüber hinaus einen entscheidenden Anhaltspunkt für diesen kritischen Versuch, weil er hier einen seiner zentralsten Gedanken zum Ausdruck bringt. Es ist nämlich folgendes. Zur Seinsverfassung des Daseins gehöre das erschließende Sein zu seinem Seinkönnen. Aber das Dasein könne sich dabei als verstehendes *entweder* aus der "Welt" und den Anderen her verstehen *oder* aus seinem eigensten Seinkönnen. Die letztgenannte Möglichkeit besage: das Dasein erschließe sich ihm selbst im eigensten und als eigenstes Seinkönnen. Diese *eigentliche* Erschlossenheit zeige das Phänomen der ursprünglichsten Wahrheit im Modus der Eigentlichkeit. *Die ursprünglichste* und zwar eigentlichste Erschlossenheit, in der das Dasein als Seinkönnen sein könne, sei die *Wahrheit der Existenz*. Dagegen sei bei der erstgenannten Möglichkeit das Dasein an seine "Welt" verloren und in das Man aufgegangen. Bei dieser Seinsart des Verfallens stehe das Entdeckte und Erschlossene im Modus der Verstelltheit und Verschlossenheit. Das Dasein sei, weil wesenhaft verfallend, seiner Seinsverfassung nach in der 'Unwahrheit'. Zur Faktizität des Daseins gehören Verschlossenheit und Verdecktheit. Der volle existenzial-ontologische Sinn des Satzes: "Dasein ist in der Wahrheit" sage gleichursprünglich mit: "Dasein ist in der Unwahrheit". *Das Dasein sei gleichursprünglich in der Wahrheit und Unwahrheit* (Vgl. SZ 221 ff.).

Es muß bei dieser Stelle beachtet werden, daß es sich hier einzig und allein um die Erschlossenheit als die Seinsweise des Daseins handelt. Die hermeneutisch-phänomenologische Erschließungsweise, ja sogar die echte Entdecktheit des innerweltlichen Seienden stehen jetzt völlig außer Frage. Es geht hier Heidegger allein um den Tatbestand, wie das Dasein sein Da ist, wie es faktisch existieren kann. Es macht allerdings ein Problem aus, ob man bei der Betrachtung der Existenz völlig z. B. die Entdecktheit des innerweltlichen Seienden außer Acht lassen kann. Aber wir müssen hier diese Frage offen dahingestellt bleiben lassen. Jedenfalls geht es hier eigens um die Existenzweise des Daseins in seiner Erschlossenheit. Dabei gibt es, wie aus der oberen Zusammenfassung hervorgeht, zwei Grundmöglichkeiten: das Dasein existiert, *entweder* sich aus seinem eigensten Seinkönnen her verstehend, *oder* aus der "Welt" und den Anderen her, mit anderen Worten, *entweder* eigentlich *oder* uneigentlich, d. h. verfallend. Bei dem ersteren Fall kommt die Erschlossenheit als die ursprünglichste Seinsweise des Daseins *eigentlichsterweise* zustande, was die *Wahrheit der Existenz* ausmacht. Hierin ist das Dasein sozusagen *ursprünglichst-eigentlichsterweise* in der Erschlossenheit (Wahrheit), d. h. in der Wahrheit der Existenz. Hingegen ist das Dasein bei dem letzteren Fall *verfallend-uneigentlicherweise* in der Erschlossenheit (Wahrheit) seines Seins. Aber wichtig ist, daß nach Heidegger beide Existenzweisen miteinander verwoben sind, so daß das Dasein gleichursprünglich in der *Wahrheit* und *Unwahrheit* ist. Das, was wir früher aus der Betrachtung der allgemeinen Wahrheitsstruktur als Schluß gezogen haben, taucht hier, in die Dimension der Existenz umgelagert, in einem ähnlichen Ton wieder auf. Wenn dem aber so ist, ist es denn auch hier bei der Existenz so, daß sie sich nur in einem dynamischen, gewissermaßen dialektischen Werdegang verwirklicht, in dem die Wahrheit und Unwahrheit der Existenz ineinander verstrickt sind? Oder aber kommt die Wahrheit der Existenz nur so zustande, daß sie entschieden die verfallende Uneigentlichkeit von sich abwehrt? Wie steht es denn bei Heidegger mit dem Verhältnis der Eigentlichkeit zur Uneigentlichkeit, oder der Wahrheit zur Unwahrheit in der Existenz? Der Gedanke, daß das Dasein *gleichursprünglich* in der Wahrheit und Unwahrheit ist, nimmt, genau besehen, in *Sein und Zeit* die zentralste Stellung, und tritt auch in der späteren Periode Heideggers, obwohl in den vertieften Formulierungen, doch als Grundgedanke auf. Das Problem von Eksistenz und In-sistenz, die Zwiefalt von Sein und Seiendem, und nicht zuletzt der Grundgedanke, daß sich das Sein, indem es sich als Seiendes zuschickt, selber entzieht usw., stehen im engen Zusammenhang mit der obigen These, allerdings durch das tiefbohrende Seinsdenken in unermeßlichem Ausmaße und unvergleichbarer Schärfe weitaus weitergedacht. Auf alle Fälle müssen wir umso mehr, uns auf diesen springenden Punkt konzentrierend, jedoch unter der Einschränkung, uns allein mit

dem Problem der Beziehung der Eigentlichkeit zur Uneigentlichkeit in
Sein und Zeit zu befassen, nun die hermeneutisch-phänomenologisch
erschlossene Erschlossenheit des Daseins gründlich durchgehen.

III. EIGENTLICHKEIT UND UNEIGENTLICHKEIT

§1. *Problemstellung*

Das Werk *Sein und Zeit* setzt sich, abgesehen von seiner "Einleitung", die
das Thema und die Methode der Untersuchung in ihren allgemeinen
Zügen erläutert, in seiner jetzigen unvollendeten Form aus zwei großen
Abschnitten zusammen. In seinem ersten Abschnitt setzt die existenziale
Analytik des Daseins bei der indifferenten durchschnittlichen Alltäglichkeit des Daseins an, auf daß bei der Analytik, wie erwähnt, nicht eine
beliebige Idee konstruktiv-dogmatisch an das Dasein unbesehen herangebracht wird, sondern daß das Apriori der Daseinsstruktur gerecht an ihr
selbst herausgestellt wird. So werden dort zunächst drei Aspekte der Seinsweise des Daseins nacheinander aufgegriffen und analysiert: nämlich, die
Weltlichkeit der Welt, das Wer des Daseins in seiner Alltäglichkeit und das
In-sein als solches. Erst nach diesen Analysen kommt der abschließende
Teil des ersten Abschnitts, in dem das Sein des Daseins deutlich als Sorge
bestimmt wird. Da es aber bis hierher bei der Analytik in erster Linie um
das Dasein in seiner indifferenten bzw. uneigentlichen und zwar unganzen
Seinsweise geht, entsteht eine Notwendigkeit, das Sein des Daseins in
seiner Eigentlichkeit und Ganzheit existenzial ans Licht zu bringen, und
so beginnt der zweite Abschnitt, der sucht, sich dadurch des ursprünglichen Seins des Daseins zu versichern, daß er das eigentliche Ganzseinkönnen des Daseins aufweist. So wird zuerst das Ganzseinkönnen des
Daseins aufgrund des Seins zum Tode erörtert, so zwar, daß zugleich eine
ontologische Möglichkeit eines existenziellen eigentlichen Seins zum Tode
als Vorlaufen sichtbar wird. Als Bezeugung dieser Möglichkeit eines
existenziellen eigentlichen Seins zum Tode wird dann das Phänomen des
Gewissens exponiert, und so ergibt sich als die existenziale Struktur des im
Gewissen bezeugten existenziellen eigentlichen Seinkönnens die Entschlossenheit. Die zu Ende gedachte Entschlossenheit verbindet sich notwendig
mit dem Vorlaufen. An dem eigentlichen Ganzsein des Daseins als der so
freigelegten vorlaufenden Entschlossenheit wird nun als der Seinssinn des
Daseins die Zeitlichkeit erfahren und in ihrer Struktur herausgestellt. Der
Seinssinn bedeutet das, was das Sein des Daseins ermöglicht. Alle Seins-

weisen des Daseins gründen in der Zeitlichkeit; die Zeitigung der Zeitlichkeit ermöglicht verschiedene Existenzweisen des Daseins in ihrer Gesamtheit. Die Zeitlichkeit, die an der vorlaufenden Entschlossenheit in ihrem ausgezeichneten Modus erfahren wird und zugleich diese möglich macht, heißt die ursprüngliche und eigentliche Zeitlichkeit. Aber die Sorge überhaupt wird ferner als in der Zeitlichkeit gründend erwiesen. Wenn aber die Zeitlichkeit alle Seinsweisen des Daseins konstituiert, so muß aufgewiesen werden, daß nicht nur das eigentliche Ganzseinkönnen des Daseins, sondern auch seine Alltäglichkeit in ihrer Konkretion aufgrund der Zeitlichkeit ermöglicht wird. Auf diese Weise wird eine zeitliche Interpretation der indifferenten bzw. uneigentlichen Alltäglichkeit durchgeführt, so daß diese als durch die Zeitigung der uneigentlichen Zeitlichkeit ermöglicht ausgewiesen wird. Dieser Interpretation schließt sich dann die Analyse der Geschichtlichkeit an, und schließlich wird die echte Genesis des vulgären Zeitbegriffs aus der uneigentlichen Zeitlichkeit verfolgt und erklärt.

In Bezug auf diesen ungeheueren Gedankengang der existenzialen Daseinsanalytik können und sollen sicher von verschiedensten Gesichtspunkten aus eine Menge der Probleme, denen Heidegger auf eine hervorragende Weise nachgegangen ist, die aber weithin weitergedacht zu werden verlangen, als Thema einer tiefgreifenden Untersuchung jeweils für sich getrennt ausführlich erörtert werden. Aber hier beschränken wir uns im vornherein nur auf *einen* besonderen Aspekt, der aber, obwohl er scheinbar *ein* Aspekt aussieht, doch im Grunde genommen allen diesen von Heidegger durchgeführten existenzialen Analysen zugrunde liegt und sich als Grundproblem durch das Ganze seiner Darlegungen durchhält. Es ist nämlich das Problem von Eigentlichkeit und Uneigentlichkeit des Daseins, genauer gesagt, das von deren Verhältnis zueinander. Wie oben gesagt, beginnt in der Tat der erste Abschnitt von *Sein und Zeit* mehr oder weniger mit der Daseinsanalyse in seiner *indifferenten bzw. uneigentlichen* durchschnittlichen Alltäglichkeit, während der zweite Abschnitt sie in Richtung des ursprünglichen Aspekts, d. h. der Eigentlichkeit und Ganzheit des Daseins hin zu vertiefen sucht. Hieraus allein könnte man schon klar einsehen, wie das Problem der Eigentlichkeit und Uneigentlichkeit das Ganze der existenzialen Daseinsanalytik in *Sein und Zeit* beherrscht. In der Tat sagt Heidegger in dem letzten Paragraphen dieses Werks: "Die Aufgabe der bisherigen Betrachtungen war, das *ursprüngliche Ganze* des faktischen Daseins hinsichtlich der Möglichkeiten des eigentlichen und uneigentlichen Existierens existenzial-ontologisch *aus seinem Grunde* (d. h. aus dem Seinssinn des Daseins als Zeitlichkeit – angemerkt vom Verfasser) zu interpretieren" (SZ 436). Eigentlichkeit und Uneigentlichkeit sind also die Grundmöglichkeiten des Existierens. Sorge und nicht zuletzt Zeitlichkeit können sich am Ende nicht anders als mit Rücksicht auf diese zwei Möglichkeiten des Existierens erst ausführlich präzisieren. Es ginge nicht zu weit, wenn man sagte, das Ganze der in *Sein und Zeit* vollzogenen Analysen sei einzig darauf aus, Seinsstruktur und Seinssinn des Daseins im Hinblick auf diese zwei Möglichkeiten zu verfolgen und klar herauszuarbeiten. Überdies bezieht sich gerade dieses Problem auf das, was wir uns am Anfang zur Aufgabe gemacht haben: nämlich, die Aufgabe der Überprüfung der gewonnenen Ergebnisse. Ferner, wie oben schon gezeigt, fordert denn auch der Wahrheitsbegriff Heideggers selbst dem Problem von Eigentlichkeit und Uneigentlichkeit nachzugehen; denn das Dasein existiert eigentlich und zugleich uneigentlich, d. h. gleichermaßen in der Wahrheit und Unwahrheit. So soll dieses Problem von Eigentlichkeit und Uneigentlichkeit jetzt näher in Betracht gezogen werden, damit dadurch das Ganze der existenzialen Daseinsanalytik von diesem Gesichtspunkt her im Hinblick auf die hermeneutisch-phänomenologische Wahrheit ihrer eigenen Ergebnisse an den Tag gebracht wird.

§2. *Eigentlichkeit als Modifikation der Uneigentlichkeit*

Nun, Eigentlichkeit und Uneigentlichkeit bilden die zwei Grundmöglichkeiten des Existierens des Daseins. Inhaltlich gesehen, ist die Eigentlichkeit im Grunde genommen nichts anderes als die vorlaufende Entschlossenheit und die darauf gegründete eigentliche Geschichtlichkeit, während die Uneigentlichkeit in ihrem besonderen Sinne die Seinsart des Verfallens (vgl. SZ 175 f.) bedeutet. Natürlich beschließt das Sein des Daseins mannigfaltige Strukturmomente in sich, die alle ihrerseits jeweils auf diese zwei Grundmodi von Eigentlichkeit und Uneigentlichkeit bezogen sind. So betreffen alle Analysen der Seinstrukturen des Daseins mehr oder weniger das Problem der Eigentlichkeit und Uneigentlichkeit. Ferner werden diese beiden Seinsmodi schließlich aufgrund der Zeitigung der Zeitlichkeit erst deutlich interpretiert; alle Seinsweisen des Daseins sind, wie gesagt, am Ende von der Zeitlichkeit her und zwar hinsichtlich dieser Grundexistenzmöglichkeiten ausführlich interpretiert. So soll der zeitliche Sinn dieser beiden Seinsmodi auch nicht übersehen werden. Wir haben aber hier nicht die Absicht, die *gesamten* Ergebnisse, die

Heidegger in Bezug auf diese Grundmöglichkeiten dargelegt hat, zu betrachten, damit die ganzen Phasen der Eigentlichkeit und Uneigentlichkeit erklärt werden. Sondern wir wollen uns hier in erster Linie gerade zumal *dem Verhältnis dieser beiden Möglichkeiten des Existierens zueinander* zuwenden.

Zuvor aber müssen einige Bemerkungen zu dem Begriff von Eigentlichkeit und Uneigentlichkeit vermerkt werden. *Zum ersten* hat dieser Begriff mit einem Werturteil der Seinsweise des Daseins gar nicht zu tun. Zum Beispiel darf man die Uneigentlichkeit nicht dahingehend verstehen, daß sie etwa "ein weniger Sein" oder "eine niedriger Seinsgrad", ja sogar "seines-Seins-verlustig-gehen" bedeutet (SZ 43, 176). Die Uneigentlichkeit ist vielmehr "eine positive Möglichkeit des In-der-Welt-seins", in der es sich zunächst und zumeist hält (SZ 176); "sie kann vielmehr das Dasein nach seiner vollsten Konkretion bestimmen in seiner Geschäftigkeit, Angeregtheit, Interessiertheit, Genußfähigkeit" (SZ 43). Sie ist eine Seinsweise, die vom Dasein selbst "als 'Aufstieg' und 'konkretes Leben'" ausgelegt wird (SZ 178). Die Interpretation der Uneigentlichkeit ist "von einer moralisierenden Kritik des alltäglichen Daseins und von 'kulturphilosophischen' Aspirationen weit entfernt" (SZ 167). Sie ist durchaus rein ontologisch orientiert. *Zum zweiten* muß dieser Begriff vielmehr aus dem Tatbestand her verstanden werden, daß das Dasein im Grunde genommen "seinem Wesen nach mögliches, *eigentliches*, d. h. sich zueigen" ist (SZ 42). Die Eigentlichkeit des Daseins besagt also, daß das Dasein auf sein *eigenes* Da-sein, d. h. In-der-Welt-sein hin existiert; dagegen meint die Uneigentlichkeit des Daseins, daß es sein *eigenes* In-der-Welt-sein vergißt, verliert, und vom dem *Un-eigenen*, d. h. *Un-eigentlichen* benommen und so darin aufgegangen ist. Es handelt sich dabei einzig darum, ob das Dasein sein *eigenes* In-der-Welt-sein aus seinem *eigensten* Seinkönnen her sein läßt oder nicht. Die Eigentlichkeit ist die Seinsart, in der sich das Dasein, dem es in seinem Sein um dieses geht, gerade aus seinem *eigensten* Seinkönnen her versteht und entwirft; hingegen die Uneigentlichkeit die Seinsart, in der es sich, indem es sich selbst verliert, von der "Welt" und den Anderen her versteht. Jene findet hinsichtlich ihrer tiefgehenden Inhalte ihren Höhepunkt in der vorlaufenden Entschlossenheit und der eigentlichen Geschichtlichkeit mit ihren reichhaltigen zeitlichen Interpretationen, und diese "erfährt durch die Interpretation des Verfallens eine schärfere Bestimmung" (SZ 175 f.). *Drittens* darf man ferner nicht übergehen, daß sowohl die Eigentlichkeit wie auch die Uneigentlichkeit entweder "echt" oder "unecht" sich zeigen können (vgl. SZ 58, 146, 148, 326, 333). "Echt" meint unserer Ansicht nach, daß das betreffende Phänomen sich unverstellt zeigt, anders ausgedrückt, daß das "echte" Phänomen dem Wesen nach zu dem wirklichen Sichzeigen des betreffenden Phänomens selbst gehört und darin verwurzelt. Heidegger will die eigentliche wie uneigentliche Seinsweise des Daseins in ihren "echten", unverstellten, dem Phänomen selbst notwendig gehörigen, wesenhaften Strukturen aufweisen. So bezweckt die existenziale Analytik des Daseins, so könnte man sagen, mit der rein ontologischen Absicht die Grundmöglichkeiten des Existierens von Eigentlichkeit und Uneigentlichkeit in ihren unverfälschten echten Phänomenen herauszustellen, in der Art, daß es dadurch klar wird, daß das Dasein in der Eigentlichkeit auf sein eigenes Selbstseinkönnen hin existiert, während es in der Uneigentlichkeit sich selbst verliert und in dem Un-eigensten, d. h. der "Welt" und dem Man aufgeht.

Nun, unsere Frage ist nicht, was für Strukturmomente jeweils die Eigentlichkeit und Uneigentlichkeit als solche wesensmäßig weiter in sich beschließen, wie sie je in ihrer Konstitution näher aufgebaut werden, sondern, wie sie sich zueinander verhalten, in welcher Beziehung sie sich zueinander befinden. Noch genauer, präziser gewendet, sieht es unsere Frage einzig darauf ab, wie sich die Eigentlichkeit *zur* Uneigentlichkeit *verhält*, auf welche Weise jene diese *modifiziert*, wie die Uneigentlichkeit durch die Eigentlichkeit *verändert wird*. Alle je einzelne Strukturmomente der beiden Grundmöglichkeiten sollen nur insoweit, als sie diese unsere Frage betreffen und deren Erläuterungen dienlich sind, herangezogen werden. Indessen mag unsere Fragestellung gegebenenfalls einigermaßen befremdlich erscheinen, weil sie von Heidegger selbst, zumindest in dem Werk, nirgendwo deutlich als eine Frage zum Ausdruck gebracht wird. Jedoch beruht unsere Fragestellung auf einer in der Sache selbst gründenden Notwendigkeit.

Eigentlichkeit und Uneigentlichkeit sind die "existenziellen" Grundmöglichkeiten des Daseins (SZ 350). Aber sie stehen nicht nebeneinander, sozusagen bloß parallel, ohne Beziehung zueinander. Denn "die *eigentliche* Existenz ist nichts, was über der verfallenden Alltäglichkeit schwebt, sondern existenzial nur ein modifiziertes Ergreifen dieser" (SZ 179). "Das eigentliche Selbstsein bestimmt sich als eine existenzielle Modifikation des Man", d. h. der Uneigentlichkeit (SZ 267). "Das *eigentliche Selbstsein* beruht nicht auf einem vom Man abgelösten Ausnahmezustand des Subjekts, sondern *ist eine existenzielle Modifikation des Man als eines wesenhaften Existenzials*" (SZ 130). Und umgekehrt ist das Man-selbst, d. h. die Uneigentlichkeit "eine existenzielle Modifikation des eigentlichen Selbst" (SZ 317). Also, beide Existenzmöglichkeiten stehen keineswegs ohne Verhältnis zueinander, bloß parallel, sondern nehmen einander in Anspruch, so zwar, daß die eine Möglichkeit ohne die andere nicht möglich ist, daß die eine Möglichkeit jeweils eine existenzial begründete existenzielle *Modifikation* der anderen ist.

So ist es ganz klar, daß die Eigentlichkeit eine Modifikation der Uneigentlichkeit ist und umgekehrt die Uneigentlichkeit eine Modifikation der Eigentlichkeit ist. Nun, was die letztere anbelangt, so hat Heidegger gerade diese existenzielle Modifikation der Eigentlichkeit zur Uneigentlichkeit in *Sein und Zeit* als die Seinsart des Verfallens erläutert; denn die existenzielle Modifikation des eigentlichen Selbstseins zum uneigentlichen Man-selbst ist ja in ihrem existenzial-ontologischen Sinne nichts anderes als das Phänomen des Verfallens. Hingegen, wie steht es mit der "existenziellen Modifikation des Man-selbst zum *eigentlichen* Selbst-sein" (SZ 268) in ihrer existenzialen Struktur? Diese Struktur hat Heidegger gewiß, so könnte man sagen, als die vorlaufende Entschlossenheit klar gemacht. Aber ob hierbei wirklich eine völlige Modifikationsstruktur ans Licht gebracht ist, ist eine Frage. Denn bei dem Übergang von der Uneigentlichkeit zur Eigentlichkeit ist die Sachlage nicht so einfach, wie bei dem umgekehrten Übergang, bei dem der Zerfall und Zusammenbruch des Selbst nur beobachtet werden darf. Bei dem Übergang von der Uneigentlichkeit zur Eigentlichkeit reicht der bloße Hinweis auf eine Umwendung des Selbst von der uneigentlichen Seinsweise zur eigentlichen durchaus nicht aus. Wenn auch das Dasein nämlich wirklich von der Uneigentlichkeit, d. h. von der Selbstverlorenheit und Selbstvergessenheit, sich eigentlich zurückhält, so muß es doch als eigentliches Selbstsein auch noch in einer Welt existieren, so zwar, daß es irgendwie als In-der-Welt-sein mit der "Welt" und den Anderen zu tun haben muß; denn das In-der-Welt-sein, gemäß dessen Grundverfassung das Dasein existiert, heißt immer zugleich mit der Seinsweise des Daseins als geworfener Entwurf Mitsein mit den Anderen in der Fürsorge und Zu-tun-haben-mit der "Welt" in dem umsichtigen Besorgen. Aber wie verhält sich das Dasein dann eigentlich als In-der-Welt-sein? Die Modifikation soll nicht einfach der Art sein, daß sich das eigentlich gewordene Dasein von der selbstverlorenen Uneigentlichkeit zurückholt, sondern, daß das eigentliche Selbst gleichzeitig als konkretes In-der-Welt-sein eigentlicherweise sich zu der "Welt" und dem anderen Mitdasein in dem besorgend-fürsorgenden

Umgang verhalten kann. Sozusagen muß sich die Freiheit *von* der Uneigentlichkeit zur Freiheit *für* die Eigentlichkeit entwickeln, und zwar dermaßen, daß diese die Uneigentlichkeit, indem sie diese in sich einnimmt als ein überwundenes Moment, völlig verändert und modifiziert. Wichtig ist, aufgrund des so erreichten eigentlichen Selbstseins sich als eigentliches In-der-Welt-sein durchzusetzen und so das In-der-Welt-sein zum vollen Modus einer höheren Eigentlichkeit erheben zu suchen. Das soll indes selbstverständlich rein und genuin existenzial-ontologisch gemeint sein. Das heißt, daß eine existenzial-ontologische Möglichkeit des Seins des Daseins, sich auf eine eigentliche Weise zu der "Welt" und den Anderen verhalten zu können, zumindest als eine Möglichkeit aufgezeigt werden sollte; oder gar wenigstens deutet das darauf, sich zu fragen, wie über diese ontologische Möglichkeit als solche von Heidegger selbst nachgedacht worden ist.

Heidegger selbst meint in der Tat auch immer wieder das gleiche, was wir annehmen. Z. B. sagt er: Sicher, "wenn es um das eigenste Seinkönnen geht", "versagt alles Sein bei dem Besorgten und jedes Mitsein mit Anderen", weil "das Dasein nur dann *eigentlich es selbst* sein kann, wenn es sich von ihm selbst her dazu ermöglicht". Aber "das Versagen des Besorgens und der Fürsorge bedeutet jedoch keineswegs eine Abschnürung dieser Weisen des Daseins vom eigentlichen Selbstsein. Als wesenhafte Strukturen der Daseinsverfassung gehören sie mit zur Bedingung der Möglichkeit von Existenz überhaupt. Das Dasein ist eigentlich es selbst nur, sofern es sich *als* besorgendes Sein bei ... und fürsorgendes Sein mit ... primär auf sein eigenstes Seinkönnen, nicht aber auf die Möglichkeit des Man-selbst entwirft" (SZ 263). Oder: "Die Entschlossenheit löst als *eigentliches Selbst-sein* das Dasein nicht von seiner Welt ab, isoliert es nicht auf ein freischwebendes Ich. Wie sollte sie das auch – wo sie doch als eigentliche Erschlossenheit nichts anderes als das *In-der-Welt-sein eigentlich* ist. Die Entschlossenheit bringt das Selbst gerade in das jeweilige besorgende Sein bei Zuhandenem und stößt es in das fürsorgende Mitsein mit den Anderen" (SZ 298). Oder: "Die Angst", in deren Grundbefindlichkeit das Dasein vor sein eigenstes nacktes Da-sein gebracht wird und so sich in seinem eigensten Seinkönnen versteht, "vereinzelt und erschließt so das Dasein als 'solus ipse'. Dieser existenziale 'Solipsismus' versetzt aber so wenig ein isoliertes Subjektding in die harmlose Leere eines weltlosen Vorkommens, daß er das Dasein gerade in einem extremen Sinne vor seiner Welt als Welt und damit es selbst vor sich selbst als In-der-Welt-sein bringt" (SZ 188) usw. Aber die

Frage bleibt die, auf welche Weise sich dieses besorgend-fürsorgende Sein-bei, das jetzt auf die Eigentlichkeit hin entworfen wird, in der Welt verwirklichen läßt. wie sich das entschlossene, angstbereitete Sein-bei und Mitsein eigentlich durchsetzen kann, wie die existenzial-ontologische Möglichkeit dafür gedacht werden soll, was für Zeitigungen der Zeitlichkeit diese ontologische Möglichkeit ermöglichen sollten.

Alle diese Fragen aber lassen sich, wenn man ihnen genauer nachsieht, nicht so einfach beantworten, wie es beim ersten Blick zu sein scheint, weil sie sich an manche Schwierigkeiten, die in dem Vorgehen Heideggers in *Sein und Zeit* versteckt liegen, unvermeidlich stoßen müssen. Wenn wir beispielsweise schon jetzt eine dieser Schwierigkeiten vorausnehmend andeuten dürfen, so wird das Besorgen, in dem "auch die eigentliche Existenz des Daseins sich hält, selbst dann, wenn es für sie 'gleichgültig' bleibt" (SZ 352), wenigstens in *Sein und Zeit* in seiner zeitlichen Interpretation merkwürdigerweise einzig und allein als durch die uneigentliche Zeitlichkeit konstituiert aufgewiesen (vgl. SZ 352 ff.). Wenn dem so ist, müßte das Dasein, wenn es auch noch so eigentlich in der Welt zu existieren entschlossen sein mag, wieder, sobald es in das für es unentbehrliche Besorgen eingeht, in die Uneigentlichkeit geraten. Oder gibt es eine andere ontologische Möglichkeit, daß das Besorgen durch die eigentliche Zeitlichkeit ermöglicht wird? Müßte die Eigentlichkeit, wenn sie über einen höchsten Augenblick der Entschlossenheit hinaus mit dem Zuhandenen befaßt wird, gleich immer wieder der Uneigentlichkeit verfallen, oder gibt es gar die Möglichkeit, in der die Eigentlichkeit, die Uneigentlichkeit völlig überwindend, in der Welt sich durchzusetzen imstande ist?

Wir müssen aber jetzt mit einer voreiligen Antwort zurückhalten. Zunächst einmal soll ein Zugang zur Lösung dieser Frage von einer anderen Seite her Schritt für Schritt, an mehrere Schwierigkeiten wieder anstoßend und sie verwindend, freigelegt werden.

§3. *Indifferenz*

Wir haben bisher immer nur von den zwei Grundmöglichkeiten der Existenz gesprochen. Aber wir müssen augenblicklich unseren Schritt anhalten und uns überlegen, ob es denn außer diesen Grundmöglichkeiten sonst gar keine Möglichkeiten gibt. Wir sehen aber doch, daß diese noch in der Tat vorhanden sind. Denn Heidegger selbst sagt es an mehreren Stellen; z. B. "Dasein existiert je in einem dieser Modi (d. h. Eigentlichkeit und Uneigentlichkeit), bzw. in der modalen Indifferenz ihrer" (SZ 53, die Worte in Klammern sind vom Verfasser ergänzt). "Als je *meines* aber ist das *Seinkönnen* frei für Eigentlichkeit oder Uneigentlichkeit oder die modale Indifferenz ihrer" (SZ 232). Sofern auf diese Stellen Bezugnahme genommen wird, muß gesagt werden, daß eine Existenzmöglichkeit einer modalen Indifferenz noch außer oder neben den beiden Seinsmodi von Eigentlichkeit und Uneigentlichkeit bestehen soll, weil hier offensichtlich von diesen drei Möglichkeiten in einem gleichwertigen Nebeneinander die Rede ist.

Nun, was für eine Existenzweise ist sie? Wenn man mit dieser Frage Heideggers Darlegungen verfolgt, so bleibt nichts übrig, als anzunehmen, daß unter dieser Indifferenz gerade diejenige Seinsweise des Daseins verstanden werden muß, von der die Daseinsanalytik ausgegangen ist, auf daß sie nicht dogmatisch-konstruktiv vorgehen darf. Dort heißt es: "Das Dasein soll im Ausgang der Analyse gerade nicht in der Differenz eines bestimmten Existierens interpretiert, sondern in seinem indifferenten Zunächst und Zumeist aufgedeckt werden. Diese Indifferenz der Alltäglichkeit des Daseins ist *nicht nichts*, sondern ein positiver phänomenaler Charakter dieses Seienden. Aus dieser Seinsart heraus und in sie zurück ist alles Existieren, wie es ist. Wir nennen diese alltägliche Indifferenz des Daseins *Durchschnittlichkeit*" (SZ 43). Die Indifferenz als eine Seinsweise meint also, aus dieser Stelle vermutet, nichts anderes als die durchschnittliche Alltäglichkeit des Daseins, in der es, ohne sich in ein bestimmtes Existieren zu differenzieren, modal indifferent existiert.

Wenn man aber den Darlegungen in *Sein und Zeit* genau nachsieht, so sind andere Stellen leicht zu finden, in denen diese Seinsweise der alltäglichen Indifferenz, die doch ursprünglich mit jeder der Grundmöglichkeiten von Eigentlichkeit und Uneigentlichkeit nichts zu tun haben sollte, trotzdem merkwürdigerweise sehr oft wieder mit der Uneigentlichkeit gleichgesetzt wird. Z. B.: "Die bisherige Interpretation beschränkte sich, ansetzend bei der durchschnittlichen Alltäglichkeit, auf die Analyse des indifferenten bzw. uneigentlichen Existierens" (SZ 232). Hier ist klar, daß die alltägliche Indifferenz mit der Uneigentlichkeit, mit dem Wort "beziehungsweise" verbunden, in einem engen Zusammenhang steht. Oder: "Auch in ihr (d. h. in der durchschnittlichen Alltäglichkeit – ergänzt vom Verfasser) und selbst im Modus der Uneigentlichkeit liegt a priori die Struktur der Existenzialität" (SZ 44). Auch hier ist die enge Beziehung der Alltäglichkeit zur Uneigentlichkeit zweifellos. Aber viel deutlicher zeigen es die Analysen über das Wer der Alltäglichkeit (vgl. SZ 126–130) an, in denen es z. B. heißt: "Das Selbst des alltäglichen Daseins ist das *Man-selbst*, das wir von dem *eigentlichen*, d. h. eigens ergriffenen *Selbst* unterscheiden" (SZ 129), und dieses "Man ist in der Weise der Unselbständigkeit und Uneigentlichkeit" (SZ 128) und es ist gerade "das 'realste Subjekt' der Alltäglichkeit" (SZ 128). Darüber hinaus wird der Teil der Darlegungen (SZ 166–180), in dem "die Alltäglichkeit des Daseins" (SZ 166) hinsichtlich ihrer Erschlossenheit sichtbar gemacht wird, unverkennbar der Herausarbeitung des Verfallens als einer uneigentlichen Seinsweise gewidmet. "Die Weise, in der das Dasein alltäglich sein 'Da' ist", d. h. "die Grundart des Seins der Alltäglichkeit", stellt sich dort als das durch Gerede, Neugier und Zweideutigkeit gekennzeichnete "Verfallen" des Daseins heraus, durch das gerade "die Uneigentlichkeit des Daseins" "eine schärfere Bestimmung erfahren" kann (SZ 175). Ferner wird in dem vierten Kapitel des zweiten Abschnitts (SZ 334–372), das "Zeitlichkeit und Alltäglichkeit" überschrieben ist, der Versuch gemacht, eine zeitliche Interpretation der Alltäglichkeit vorzunehmen, und diese Aufgabe wird damit gleichgesetzt, "die *Uneigentlichkeit* des Daseins in ihrer spezifischen Zeitlichkeit sichtbar zu machen" (SZ 331), und "eine Zeitigungsmöglichkeit der Zeitlichkeit, in der die Uneigentlichkeit des Daseins ontologisch gründet", zu "enthüllen" (SZ 335). In der Tat werden dort mannigfaltige Phänomene, die im ersten Abschnitt aufgrund der durchschnittlichen Alltäglichkeit zergliedert worden sind, so interpretiert, daß sie eigens durch die uneigentliche Zeitlichkeit ermöglicht werden, und es ergibt sich überdies, daß es gewiß außer dieser die eigentliche Zeitlichkeit gibt, doch keineswegs eine besondere Seinsweise der Indifferenz, die weder eigentlich noch uneigentlich wäre und zwar durch irgend eine Zeitlichkeit konstituiert würde, welche wiederum auch weder eigentlich noch uneigentlich wäre. Wenn man außerdem dem Gedankengang Heideggers in *Sein und Zeit* im Ganzen nachblickt, leuchtet es ein, daß der erste Abschnitt mehr oder weniger, wie gesagt, bei der durchschnittlichen Alltäglichkeit ansetzt, während der zweite gerade deswegen in die ursprüngliche Analyse des Daseins vordringt, die die Eigentlichkeit in den Blick einnimmt, und daß schließlich außer diesen beiden Aspekten eine besondere Seinsweise der Indifferenz gar nicht ausdrücklich als solche zum Thema gemacht worden ist.

Aus diesen Gründen müßte nun die Indifferenz als eine Seinsweise inhaltlich am Ende, wie es scheint, auf die Uneigentlichkeit des Daseins zurückgeführt werden, die durch die Zeitigung der uneigentlichen Zeitlichkeit konstituiert wird, in der sich das Dasein alltäglich durchschnittlich gemäß der Seinsart des Verfallens hält. Dürfte man aber, so erhebt sich hier eine Frage, so einfach die Indifferenz mit der Uneigentlichkeit gleichsetzen? Kann sie ohne weiteres durch diese völlig substituiert werden? Denn wenn dem so wäre, müßte die alltägliche Indifferenz, die ursprünglich zum Leitfaden der vorurteilsfreien sachgemäßen Daseinsanalytik ergriffen wurde, in Wirklichkeit keine echte Indifferenz, sondern ein Modus der Uneigentlichkeit sein, und so Heideggers Ansatz bei der indifferenten Alltäglichkeit kein echt neutrales Vorgehen, sondern nur ein Vorwand sein, scheinbar seine Analysen als neutral hinzustellen. Täte er aber wirklich nur so, als ob er neutral vorginge? Oder aber gibt es tatsächlich eine indifferente Seinsweise? Wie aber sollte diese gedacht werden, wo sie doch von Heidegger selbst anscheinend nicht so deutlich inhaltlich zum Problem gemacht worden ist? Sicher scheint aus vielen Stellen eine Interpretation, daß die indifferente Alltäglichkeit schließlich auf die Uneigentlichkeit zurückgeht und sonst weiter nichts ist, unvermeidlich zu sein. Jedoch, wie soll man manche unterstrichene, häufig wiederholte Ausdrücke Heideggers der Art zurechtlegen, daß "alles Existieren, wie es ist, aus dieser Seinsart (d. h. der Indifferenz der Alltäglichkeit – ergänzt vom Verfasser) heraus und in sie zurück ist" (SZ 43), daß "die Alltäglichkeit das Dasein auch dann, wenn es sich nicht das Man als 'Helden' gewählt hat, bestimmt" (SZ 371), daß die Existenz den Alltag "nie auslöschen" kann (SZ 371) usw. Diese Alltäglichkeit, von der gesagt wird, daß ohne sie auch die Eigentlichkeit nicht wäre, – was ist sie denn? Ist sie ohne weiteres mit der Uneigentlichkeit gleichzusetzen? Dann aber taucht wiederum unser Problem des Verhältnisses der Eigentlichkeit zur Uneigentlichkeit auf. Wie verhält sich eigentlich jene zu dieser? Könnten sich denn diese beiden scharf zueinander im Gegensatz stehenden˙ Seinsmodi von Eigentlichkeit und Uneigentlichkeit ohne irgend einen gemeinsamen Boden, ohne neutrales Medium, zueinander verhalten? Oder aber verhält es sich anders?

So nachdenkend, sind wir jetzt an den Punkt gelangt, wo uns erst eine Richtung einer angemessenen konsequenten Interpretation über dieses Problem aufgeht. Es geht nämlich der anscheinend einzige Ausweg, diesen verworrenen Verhältnissen zu entgehen, aus den folgenden Sätzen Heideggers hervor: "Die durchschnittliche Alltäglichkeit des Daseins darf aber nicht als ein bloßer 'Aspekt' genommen werden. Auch in ihr und selbst im Modus der Uneigentlichkeit liegt a priori die Struktur der Existenzialität. Auch in ihr geht es dem Dasein in bestimmter Weise um sein Sein, zu dem es sich im Modus der durchschnittlichen Alltäglichkeit verhält und sei es auch nur im Modus der Flucht *davor* und des Vergessens *seiner*. Die Explikation des Daseins in seiner durchschnittlichen Alltäglichkeit gibt aber nicht etwa nur durchschnittliche Strukturen im Sinne einer verschwimmenden Unbestimmtheit. Was ontisch in der Weise der Durchschnittlichkeit *ist*, kann ontologisch sehr wohl in prägnanten Strukturen gefaßt werden, die sich strukturell von ontologischen Bestimmungen etwa eines *eigentlichen* Seins des Daseins nicht unterscheiden" (SZ 44).

Nach diesen Stellen ist die alltägliche Indifferenz keineswegs ein flüchtiges Phänomen. Sie tritt sicher zunächst und zumeist in dem Modus der Uneigentlichkeit auf. Aber sie birgt in sich sozusagen eine *"apriorische Struktur der Existenzialität"*, ja sogar die *"prägnanten Strukturen, die sich strukturell von ontologischen Bestimmungen etwa eines eigentlichen Seins des Daseins nicht unterscheiden"*. Das besagt, daß mit der *Indifferenz* als Seinsweise gerade die *formal-indifferenten Strukturen* gemeint werden sollen, die zwar zunächst in der *indifferenten Alltäglichkeit* anzutreffen sein muß, die aber sowohl der Eigentlichkeit wie auch der Uneigentlichkeit als formale Strukturen gemeinsam zugrunde liegen. Unter der Indifferenz sollte also nicht eigentlich die durchschnittliche Alltäglichkeit, die sich nachher als Uneigentlichkeit herausstellt, verstanden werden, sondern vielmehr und vorwiegend, vorausgesetzt, daß man Heideggers Darlegungen *konsequent* interpretieren wolle, *die formale Struktur der Existenz* verstanden werden, an der sowohl die Eigentlichkeit als auch die Uneigentlichkeit beteiligt sind, und die man aber zunächst bei der durchschnittlichen Alltäglichkeit ansetzend zu finden suchen muß. In der Tat ist der Begriff der Indifferenz eigens im Hinblick darauf eingeführt worden, daß sich das Dasein an ihm selbst von ihm selbst her, und zwar ohne Ansehen der Eigentlichkeit und Uneigentlichkeit, in seinen "wesenhaften Strukturen" (SZ 17) überhaupt zeigen kann. Aus eben demselben Grund ist immer wieder betont worden, daß dem Dasein eine bestimmte Idee der Existenz nicht dogmatisch aufgezwungen werden darf (SZ 16 f, 43, 50 Anm., 313). Wenn man dieser Sachlage Rechnung trägt, besteht daran kein Zweifel, daß die Indifferenz ursprünglich und eigentlich die *formale Struktur* des Existierens meinen sollte.

Nun, was ist aber diese formale Struktur der Existenz? Sie hat sich bereits in dem obigen Zitat durchblicken lassen. Es steht darin, daß "es

dem Dasein in bestimmter Weise um sein Sein geht", gleichgültig ob es
eigentlich oder uneigentlich existiert. "Das Dasein bestimmt sich als
Seiendes je aus einer Möglichkeit, die es ist und in seinem Sein irgendwie
versteht. Das ist der *formale* (gesperrt vom Verfasser) Sinn der Existenz-
verfassung des Daseins" (SZ 43). "Dasein ist Seiendes, das sich in seinem
Sein verstehend zu diesem Sein verhält. Damit ist der *formale* (gesperrt
vom Verfasser) Begriff von Existenz angezeigt" (SZ 52 f.). "Der Titel (der
Existenz - ergänzt vom Verfasser) besagt in *formaler* (gesperrt vom
Verfasser) Anzeige: das Dasein *ist* als verstehendes Seinkönnen, dem es in
seinem Sein um dieses selbst geht" (SZ 231). Aber was bedeutet es, daß es
dem Dasein in seinem Sein um dieses geht? "Diese Seinsbestimmungen des
Daseins" müssen "a priori auf dem Grund der Seinsverfassung gesehen
und verstanden werden, die wir das *In-der-Welt-sein* nennen" (SZ 53).
So beginnen die ausführlichen Analysen des In-der-Welt-seins und
gelangen schließlich an das Ergebnis, daß "die *formal* (gesperrt vom
Verfasser) existenziale Ganzheit der ontologischen Strukturganzen des
Daseins" in der Sorgestruktur "gefaßt werden muß" (SZ 192). Ferner
gehen die Analysen über die Exposition des Seins zum Tode und des
Gewissens so vonstatten, daß sich endlich herausstellt: "die ursprüngliche
Einheit der Sorgestruktur liegt in der Zeitlichkeit" (SZ 327). "Der
ursprüngliche ontologische Grund der Existenzialität des Daseins aber ist
die *Zeitlichkeit.* Die gegliederte Strukturganzheit des Seins des Daseins
als Sorge wird erst aus ihr existenzial verständlich" (SZ 234). "Die Grundmöglichkeiten der Existenz, Eigentlichkeit und Uneigentlichkeit des
Daseins, gründen ontologisch in möglichen Zeitigungen der Zeitlichkeit"
(SZ 304).

Also kann und muß das Verhältnis so zusammengefaßt werden: die
formal-indifferente Struktur der Existenz heißt, daß es dem Dasein
in seinem Sein um dieses geht, was das In-der-Welt-sein bedeutet, daß
diese Struktur in der Sorge zentriert, und daß diese Sorge durch die
Zeitlichkeit ermöglicht wird, die erst die Differenz von Eigentlichkeit und
Uneigentlichkeit aus sich selbst her zeitigt. Nun, die Frage ist, ob damit das
Problem der Indifferenz vollkommen aufgelöst worden ist. Im Gegenteil,
scheint es doch leider nur so zu sein. Wir stoßen, genauer gesehen,
wiederum auf schwierige Probleme, die wir weiter im folgenden etwas
bedenken müssen.

§4. *Schwierigkeiten um die Sorgestruktur*

Wir knüpfen jetzt an die Sorgestruktur an. Diese soll, wie gesagt, die
"formal existenziale", d. h. *indifferente* "Ganzheit des ontologischen

Strukturganzen des Daseins" (SZ 192) bezeichnen. Der Titel Sorge besagt: "Sich-vorweg-schon-sein-in-(der-Welt-) als Sein-bei (innerweltlich begegnendem Seienden)" (SZ 192). Sorge beschließt in sich die drei Momente; das erste ist "Sich-vorweg", das zweite "Schon-sein-in-(der-Welt)" und das dritte "Sein-bei (innerweltlich begegnendem Seienden)".

Nun, ein schwieriges Problem taucht auf, wenn man darauf aufmerksam wird, daß Heidegger sehr oft, statt von dieser langwierigen Formulierung der Sorge Gebrauch zu machen, die je den obengenannten drei Momenten entsprechenden, sogar damit gleichzusetzenden drei Begriffe benutzt und jene durch diese vertreten zu lassen pflegt. Diese drei Begriffe heißen: Existenzialität (Existenz), Faktizität und Verfallen (Verfallensein, Verfallenheit) (SZ 181, 191 ff., 231, 249 f., 252, 284, 314, 316, 350). An diesen Begriffen orientiert, soll das erste Moment die Existenzialität, das zweite die Faktizität, und das dritte das Verfallen genannt werden. Wollen wir aber jetzt vor allem auf das dritte Moment achtgeben. Dann springt uns ein merkwürdiger Widerspruch sogleich in die Augen. Denn die Sorgestruktur soll ursprünglich "die *formal* existenziale Ganzheit des ontologischen Strukturganzen des Daseins" bedeuten, die das Sein des Daseins *ohne Ansehen seiner Eigentlichkeit und Uneigentlichkeit* konstituiert. Wenn aber ihr drittes Moment "Sein-bei", das ursprünglich völlig eine *formal-indifferente* Seinsverfassung des Daseins bezeichnen sollte, mit der Seinsart des Verfallens gleichgesetzt wird, die doch nichts anderes als die Uneigentlichkeit ist, so müßte das Sein des Daseins als Sorge im vornherein ein solches verfallendes uneigentliches "Sein-bei" in sich beschließen und somit notwendigerweise von Hause aus mit der Uneigentlichkeit durchsetzt sein; infolgedessen wäre das Dasein, solange es durch eine so strukturierte Sorge konstituiert ist, auf die Dauer keineswegs an die Eigentlichkeit zu gelangen imstande. Trotzdem spricht Heidegger einerseits von der Möglichkeit einer Eigentlichkeit als der Seinsweise des Daseins, und merkwürdigerweise setzt er andererseits an vielen Stellen das dritte Moment "Sein-bei", das ursprünglich formal-indifferent das Sein des Daseins ausmachen soll, mit der uneigentlichen Seinsart des Verfallens gleich.

Um dieser widerspruchsvollen Situation zu entgehen, können wir nicht umhin, folgenderweise zu interpretieren, wie uns dünkt. 1) Angenommen, Heideggers Gleichsetzung des "Sein-bei" mit dem Verfallen sei in einem *strickten* Sinne gemeint, d. h. das dritte Moment sei *ausnahmslos immer* dem Verfallen gleich und weiter nichts, so würde sich das Dasein, solange es durch eine so strukturierte Sorge aufgebaut ist, gar nicht versprechen, eigentlich zu existieren. Das widerspricht aber offenbar dem

ganzen Gedankengang Heideggers in *Sein und Zeit*. Überdies würde
unter dieser Voraussetzung die Sorgestruktur ihres *formal-indifferenten*
Charakters verlustig gehen, was aber wiederum schwer hinzunehmen
wäre. So müßte angenommen werden, daß das "Sein-bei" zwar ein formal-
indifferentes Strukturmoment ist, daß es doch aus irgendwelchen
Gründen zunächst und zumeist zum größten Teil unter dem uneigentlichen Verfallen zum Vorschein kommt, daß es aber in einer eigentlichen
Gestalt möglicherweise auftreten kann. 2) Es ist in der Tat so. Das "Sein-
bei" kann als ein an sich neutrales Seinsmoment angesehen werden, das
aber zunächst und zumeist in der durchschnittlichen Alltäglichkeit in der
Weise des uneigentlichen Verfallens erscheint. Dieses Moment gibt nur
insofern den Anschein, als ob es völlig mit dem Verfallen gleichzusetzen
wäre, als man einzig und allein auf seine Erscheinungsweise in der Alltäglichkeit achtgibt und von diesem Aspekt her allein die Seinsart des Daseins
beurteilt. Und Heidegger selbst hat von diesem Gesichtspunkt Gebrauch
gemacht. Dieses Vorgehen war deshalb unumgänglich, weil er, um ohne
Vorurteil sachgerecht mit den Analysen anzufangen, von der sogenannten
indifferenten, doch in Wirklichkeit meistens im Modus der Uneigentlichkeit gefärbten, durchschnittlichen Alltäglichkeit ausgegangen ist.
Wenn aber die Analysen noch weiter gehen und endlich an die Zeitlichkeit
als den Seinssinn des Daseins gelangen, so zeigt sich, daß das "Sein-bei" an
sich formal-indifferent ist, daß das eigentliche "Sein-bei" hinsichtlich
seiner Zeitlichkeit als der "Augenblick" interpretiert wird, während das
uneigentliche "Sein-bei" als das "Gegenwärtigen" herausgestellt wird.
3) Diese einzig mögliche konsequente Interpretation läßt aber nicht
einfach vermuten, daß man jene bei Heidegger sehr oft vorkommende
Gleichsetzung des "Sein-bei" mit dem Verfallen nicht ernst zu nehmen
braucht. Über diese Gleichsetzung darf man sich nicht leichtsinnig
hinwegsetzen, weil das "Sein-bei" zunächst und zumeist doch noch zum
großen Teil in der Alltäglichkeit im Modus des uneigentlichen Verfallens
aufzutreten geneigt ist. Die formal-indifferente Sorgestruktur hat die
starke Tendenz, die Uneigentlichkeit in sich zu beschließen. Darauf deutet
in der Tat die Tatsache hin, daß jene Gleichsetzung häufig in *Sein und Zeit*
zum Ausdruck kommt. Gerade deswegen hat es den Anschein, als wäre die
Sorge mit der Uneigentlichkeit durchsetzt, ja sogar unter Umständen ein
Komplex von Eigentlichkeit und Uneigentlichkeit. Dieses Problem darf
nicht so formal konsequent leichtsinnig erledigt werden. Es ist im wesentlichen heikel, und es wird sich zeigen, daß es gerade hierin eine wesentliche
Beziehung zu unserer Fragestellung hat, die nach dem Verhältnis der

Eigentlichkeit zur Uneigentlichkeit fragt. Denn daß das "Sein-bei" doch endlich meistens in das uneigentliche Verfallen zu geraten geneigt ist, bedeutet, daß das eigentliche "Sein-bei" d. h. der Augenblick schwer verwirklicht wird und in das uneigentliche Gegenwärtigen oft verstrickt wird; was wiederum schon deutlich genug ein Problem des Verhältnisses von Eigentlichkeit und Uneigentlichkeit zueinander anzeigt.

Wenn dem so ist, werden sich folgende Fragen unvermeidlich erheben. (1) Was sind denn die formal-indifferenten Seinsstrukturen des Daseins? Denn wir müssen uns ihrer wieder versichern, weil die Uneigentlichkeit die hartnäckige Tendenz hat, in das Sein des Daseins leicht unversehens einzuschleichen. (2) Warum ist die ursprünglich formal-indifferent sein sollende Sorgestruktur sehr häufig so gekennzeichnet, daß sie die verfallende Uneigentlichkeit in sich enthält? Worin besteht der Grund dafür? (3) Wie wird dies alles aus der Zeitlichkeit her ausgelegt? Denn die Zeitlichkeit allein gibt, so wird erklärt, den ausschlaggebenden Schlüssel her. (4) Welche Bedeutungen hat dieses Problem für unsere Fragestellung nach dem Verhältnis der Eigentlichkeit zur Uneigentlichkeit?

Diese Fragen werden wir im folgenden weiter bedenken. Abgesehen davon, ist es jedenfalls ganz klar, daß an Heideggers Darlegungen in *Sein und Zeit* dunkle Schwierigkeiten haften. Man muß sich beispielsweise vor einer voreiligen Interpretation hüten, daß die indifferente Alltäglichkeit einfach mit der formal-indifferenten Struktur selbst identifiziert werden kann. Zwischen dieser und deren alltäglichem Modus der Uneigentlichkeit muß unterschieden werden. Ferner darf nicht leichtsinnig hingenommen werden, daß die Sorgestruktur ohne weiteres aus der Existenzialität, Faktizität und Verfallen besteht. Das enthält heikle Probleme, die noch mit Sorgfalt bedacht werden sollen. Jedenfalls hat alles, was Heidegger in Bezug auf Indifferenz, Alltäglichkeit, Uneigentlichkeit, Verfallen und Sorge usw. exponiert hat, so wenig eine formal klare Konsequenz, wie vielmehr Schwierigkeiten in sich, die sich vermutlich aus den Bemühungen, bisher unbeschrittene Gebiete Schritt für Schritt aufzuhellen, ergeben haben. Angesichts dieser Dunkelheiten wird es eine unumgängliche Aufgabe sein, sich eine sachgemäß konsequente Interpretation verschaffen zu suchen.

§5. Formal-indifferente Seinsstrukturen des Daseins

Wir wollen zuerst kurz das Problem, das sich auf die formal-indifferente Seinsstruktur des Daseins bezieht, betrachten. Wie gesagt, ist das Dasein

ein Seiendes, dem es in seinem Sein um dieses geht, und gerade dieses macht
den formalen Sinn der Existenz aus. Diese Existenz wird ferner aufgrund
des In-der-Welt-seins als ihrer Grundseinsverfassung exponiert. Diese
Explikationen gelangen schließlich, indem sie durch die tiefgreifenden
Analysen über *Weltlichkeit, Mitsein* und *In-sein* hindurchgehen, an
das Endergebnis, daß das Sein des Daseins in der *Sorge* besteht. Nun,
damit wir uns nochmals der formal-indifferenten Seinsstruktur des
Daseins vergewissern, wollen wir lakonisch die durchgeführten Analysen
durchgehen. Dabei braucht unser Rückblick nicht so eingehend zu sein;
uns gilt nur, die *fundamentalen* formal-indifferenten Seinsstrukturen
festzustellen.

(1) Nun, in der *ersten* Problematik der *Weltlichkeit* wurde gezeigt,
daß das Dasein als In-der-Welt-sein zunächst in einer Umwelt existiert,
daß es dabei stets in dem umsichtigen Besorgen mit dem innerweltlich
zuhandenen Seienden zu tun hat; dieses Phänomen so wie das Welt-
phänomen ist übrigens, so wird immer wieder gesagt, bisher völlig über-
sprungen worden (SZ 65 ff., 95, 98 ff., 200–212). Abgesehen davon, wird
als das Sein dieses Zuhandenen dann die "Bewandtnis" (SZ 84) herausgestellt; diese aber kann erst durch das ontologisch verstandene
"Bewendenlassen" von der Art zustande kommen, die das Zuhandene
vorgängig auf die Bewandtnis*ganzheit* hin freigibt. Diesem letzteren
aber liegt ferner, sagt Heidegger, eine Erschließung dessen voraus,
woraufhin die Freigabe des Zuhandenen im Ganzen erfolgt. Das bedeutet,
daß das Dasein letztlich aus einem Worumwillen her an den Bezugszusammenhang der Bewandtnisganzheit sich verweist. Aufgrund dieses
Sichverweisens, das seinerseits aus dem Worumwillen her möglich wird,
wird das Begegnenlassen von Seiendem erst ermöglicht. Dieses Phänomen
ist gerade das von der Welt, und das, was die Struktur der Welt ausmacht,
wird als die "Bedeutsamkeit" (SZ 87) bestimmt. Die "Bedeutsamkeit"
heißt das "Bezugsganze des Be-deutens" (SZ 87), in dem sich das Dasein
letztlich sein Sein und Seinkönnen in der Welt "zu verstehen gibt", d. h.
"bedeutet" (SZ 87) und auf diese Weise von seinem Worumwillen her den
Bezugszusammenhang von Um-zu "be-deutet". Nun, wichtig ist, daß es
dem Dasein in seinem Sein wesenhaft um dieses geht, daß es sich also stets
darauf entwerfen muß, worumwillen es selbst ist, und daß es
infolgedessen, solange es im wesentlichen *verstehend* existiert, gleichzeitig
damit immer mit *der Bedeutsamkeit zu tun haben* muß, die *die Weltlichkeit ausmacht.* "Im Verstehen des Worumwillen ist die darin gründende
Bedeutsamkeit *miterschlossen*" (SZ 143 – gesperrt vom Verfasser), so sagt

Heidegger selbst in der Tat. Also das Dasein muß *immer zugleich* mit seinem Entwurf auf sein Seinkönnen in seiner Vertrautheit mit der Bedeutsamkeit irgendwie *das innerweltlich zuhandene Seiende*, d. h. die sogenannte "Welt" mit Anführungszeichen, *in der Weise des umsichtigen Besorgens begegnen lassen.* Diese Seinsweise muß in diesem Sinne, wie wir schon früher angezeigt haben, als ein dem Dasein unentbehrliches Seins-moment, gleichgültig, ob es eigentlich oder uneigentlich existiert, angesehen werden.

(II) In der *zweiten* Problematik des *Mitseins* wird nun weiter gezeigt, daß "das Dasein wesenhaft an ihm selbst Mitsein ist" (SZ 120), was aber selbstverständlich "existenzial-ontologisch" gemeint ist (SZ 120). Heidegger sagt immer wieder: "Das Mitsein ist ein existenziales Kon-stituens des In-der-Welt-seins" (SZ 125); "das In-sein ist *Mitsein* mit Anderen" (SZ 118) usw. Nun, "das innerweltliche Ansichsein" dieser Anderen ist "Mitdasein" (SZ 118). Das Mitdasein ist die "eigene Seinsart von innerweltlich begegnendem Seienden", d. h. von den Anderen (SZ 125). Das Dasein ist in diesem Sinne Mitsein mit dem Mitdasein, und das eigene Dasein ist auch als für Andere begegnend Mitdasein. Mitsein ist sicher "ein *Sein zu* innerweltlich begegnendem Seienden" (SZ 121); dieses Seiende jedoch wird nicht "besorgt", sondern steht "in der Fürsorge" (SZ 121), die "geleitet wird durch die *Rücksicht* und *Nachsicht*" (SZ 123), so wie dem Besorgen die Umsicht gehört. Die Fürsorge kann natürlich verschiedene Formen aufnehmen; als deren extreme Möglich-keiten werden z. B. die "einspringend-beherrschende" und die "vor-springend-befreiende" Fürsorge genannt (SZ 122), wobei die letztere als eine eigentliche angesehen wird (SZ 122, 298); aber "zunächst und zumeist" hält sich die Fürsorge in den "defizienten oder zum mindesten indifferenten Modi", d. h. "in der Gleichgültigkeit des Aneinander-vorbeigehens" auf (SZ 124); diese "Modi der Defizienz und Indifferenz charakterisieren das alltägliche und durchschnittliche Miteinandersein" (SZ 121). Aber abgesehen von diesen mannigfaltigen Modifizierungen des Miteinanderseins, muß jedenfalls festgestellt werden: "sofern Dasein über-haupt *ist*, hat es die Seinsart des Miteinanderseins" (SZ 125). Zum Sein des Daseins gehört nämlich wesensmäßig das Mitsein mit Anderen, und diese Sachlage erfüllt auch die Bedeutsamkeit, d. h. die Weltlichkeit (SZ 123), so daß die Welt des Daseins "Mitwelt" (SZ 118) ist. Aus diesem Grund muß das Dasein, so muß man beurteilen, stets genau ebenso wie bei dem Zu-tun-haben mit dem innerweltlich Zuhandenen in dem umsichtigen Besorgen, nur als Mitsein mit den Anderen in der Weise der

Fürsorge existieren, gleichgültig, ob es eigentlich oder uneigentlich existiert. Im übrigen tritt in den Analysen des Mitseins der Begriff des "Man" (SZ 126 ff.) als "das 'realste Subjekt'" (SZ 128) des alltäglichen Miteinanderseins auf, aber es muß für den uneigentlichen Modus dieses Mitseins gehalten werden, obwohl es als wesentliches Existenzial, so wird gesagt, zur Verfassung des Daseins gehört. Der Begriff des Man hat vielmehr enge Beziehung zur erst später zu erklärenden Seinsart des "Verfallens" (SZ 166-180).

(III) Nun, in der *dritten* Problematik des *In-seins* wird die Erschlossenheit nach ihrer Konstitution gefragt. Die Erschlossenheit bedeutet, wie früher erwähnt, daß "das Dasein in der Weise *ist*, sein Da zu sein" (SZ 133). Das "Da" bedeutet, daß es als In-der-Welt-sein "gelichtet", selbst "die Lichtung *ist*" (SZ 133). "*Das Dasein ist seine Erschlossenheit*" (SZ 133). Dieses Da zu sein, ist gerade das Sein, darum es dem Dasein in seinem Sein geht, so sagt Heidegger (SZ 133). Also, die Erschlossenheit ist gleichbedeutend mit dem formalen Sinn der Existenz. Nun, als die "*primären*" (SZ 133) Konstitutionsmomente des Seins der Erschlossenheit werden dort drei Seinsweisen herausgestellt: nämlich, *Befindlichkeit, Verstehen* und *Rede*. Andererseits wird das "*alltägliche*" (SZ 133) Sein des Da als *Verfallen* bestimmt. Von diesen vier Begriffen aber können die ersten drei für die formal-indifferente Seinsstruktur gehalten werden, die das Dasein, ohne Ansehen von Eigentlichkeit und Uneigentlichkeit, wesensmäßig in sich beschließt, während der lezte Begriff des Verfallens allein als das konstitutive Moment der Uneigentlichkeit betrachtet werden kann. Wir wollen jetzt kurz diesem Punkt nachgehen.

(1) Die *Befindlichkeit* ist nach Heidegger der ontologische Titel dessen, was ontisch als Stimmung bekannt ist. Seiner Ansicht nach ist das Dasein in der Stimmung "als das Seiende erschlossen, dem das Dasein in seinem Sein überantwortet wurde als dem Sein, das es existierend zu sein hat" (SZ 134); in der Stimmung ist das Dasein gewissermaßen "vor sein Sein als Da gebracht", d. h. vor "das Sein des Daseins als nacktes 'Daß es ist und zu sein hat'" (SZ 134). Dieses 'Daß es ist' wird als die "Geworfenheit" des Daseins in sein Da bestimmt, die die "Faktizität der Überantwortung" andeutet (SZ 135). Die Befindlichkeit ist dem Dasein wesentlich, denn das Dasein befindet sich stets in seiner Geworfenheit. In diesem Sinne muß die Befindlichkeit, d. h. die darin erschlossene geworfene Faktizität als die dem Dasein wesenhafte Seinsstruktur angesehen werden. So wird z. B. von der "uneigentlichen Befindlichkeit" der Furcht und von der eigent-

lichen "Grundbefindlichkeit" der Angst die Rede sein können (SZ 140 ff., 184 ff., 339 ff.). Die befindliche Geworfenheit muß also als solche eine formal-indifferente Struktur sein.

(2) Nach Heidegger, gleichursprünglich mit der Befindlichkeit konstituiert das *Verstehen* das Sein des Daseins. Das Dasein ist "primär Möglichsein" (SZ 143), allerdings, "durch und durch *geworfene* Möglichkeit" (SZ 144), und diese Möglichkeit ist "die ursprünglichste und letzte positive ontologische Bestimmtheit des Daseins" (SZ 143 f.). Das bedeutet, daß das Dasein je das ist, was es sein kann, daß das Dasein "die Möglichkeit des Freiseins für das eigenste Seinkönnen ist" (SZ 144). Verstehen ist "das existenziale Sein des eigenen Seinkönnens des Daseins selbst" (SZ 144) und "entwirft das Sein des Daseins auf sein Worumwillen ebenso ursprünglich wie auf die Bedeutsamkeit als die Weltlichkeit seiner jeweiligen Welt" (SZ 145). Dieser verstehende Entwurf ist dem Dasein so wesentlich, daß er zweifelsohne als die formal-indifferente Grundart seines Seins angesehen werden muß. In der Tat differenziert sich diese Seinsart, der Art, daß das Verstehen "entweder eigentliches, aus dem eigenen Selbst als solchem entspringendes, oder uneigentliches" ist (SZ 146).

(3) Nun, "mit Befindlichkeit und Verstehen existenzial gleichursprünglich" ist ferner die *"Rede"* (SZ 161). Die Rede ist "ursprüngliches Existenzial der Erschlossenheit", deren "existenziale Verfassung", "konstitutiv" für die "Existenz" des Daseins, ja für "die Existenzialität der Existenz" (SZ 161). Das befindliche Verstehen birgt nämlich in sich die Möglichkeit der Auslegung, von der sich weiter die Aussage ableiten läßt, und diese hat den sprachlichen Ausdruck zur Folge. Nun, "die existenzial-ontologische Fundament der Sprache ist die Rede" (SZ 160). Das heißt: vor der Auslegung und Aussage *"spricht sich* die befindliche Verständlichkeit des In-der-Welt-seins als *Rede* aus" (SZ 161), die die "Artikulation der Verständlichkeit" ist und den "Sinn" artikuliert als das "Bedeutungsganze" (SZ 161). "Die Hinausgesprochenheit der Rede ist die Sprache" (SZ 161). Die Rede durchdringt den geworfenen Entwurf in voller Breite, "bestimmt gleichursprünglich Befindlichkeit und Verstehen" (SZ 133) so gründlich, daß durch sie allein erst das sinnhafte Bedeutungsganze des Seinverständnisses des Daseins als In-der-Welt-sein artikulierbar wird. Erst aufgrund dessen kommt die Sprache zustande. Sicher spricht die Rede zunächst und zumeist "in der Weise des besorgend-beredenden Ansprechens der 'Umwelt'" (SZ 349), doch vollzieht sich einzig und allein in der Rede, die vor allem in dem ausgezeichneten Modus

des "Hörens" und "Schweigens" auftreten kann (SZ 163 ff., 272 ff.), das
Seinsverständnis des Daseins in seinem eigenen In-der-Welt-sein so wie in
seinem Mitsein. Die Seinsweise der Rede gehört dem Sein des Daseins so
wesentlich, daß es nicht zu weit ginge, wenn man sagte, daß hierin einer der
wichtigsten Kerne des Heideggerschen Existensbegriffs versteckt liege.
Abgesehen davon, könnte man mindestens wegen dieses wesentlichen
Charakters der Rede für die Existenz annehmen, daß sie auch ein
entscheidendes formal-indifferentes Strukturmoment des Daseins
ausmacht, was allerdings von Heidegger selbst nicht so deutlich zum
Ausdruck gebracht ist.

(4) Im Unterschied zu den oben erwähnten drei Strukturmomenten der
Erschlossenheit ist das *Verfallen* als die Seinsweise, in der das Dasein
alltäglich seine Erschlossenheit ist, die durch Gerede, Neugier und
Zweideutigkeit charakterisiert wird, keine formal-indifferente Seinsart in
dem Sinne, in dem darunter eine sowohl der Eigentlichkeit als auch der
Uneigentlichkeit zugrunde liegende, neutrale Seinsverfassung verstanden
wird. Sondern das Verfallen ist, wie gesagt, das Charakteristikum der Uneigentlichkeit des Daseins. Das Verfallen bedeutet, daß *zum einen* das
Dasein "zunächst und zumeist bei der besorgten 'Welt' ist" und "an die
'Welt' verfallen ist", daß es aber *zum anderen* wegen dieses "Aufgehens
bei der 'Welt'", d. h. wegen der "Verfallenheit an die 'Welt'", "den
Charakter des Verlorenseins in die Öffentlichkeit des Man" hat, nämlich
"das Aufgehen im Miteinandersein, sofern dieses durch Gerede, Neugier
und Zweideutigkeit geführt wird" (SZ 175). Auf diese Zweiseitigkeit des
Verfallens werden wir aber später wieder zu sprechen kommen. Für den
Augenblick genügt uns die Feststellung, daß das Verfallen eigens die
Uneigentlichkeit bedeutet, und daß es, wie aus dem obigen Charakteristikum einleuchtet, eine enge Beziehung zur besorgend-fürsorgenden
Seinsweise in der Alltäglichkeit hat, die schon in den Analysen der Weltlichkeit und des Mitseins zum Vorschein gekommen ist.

(IV) Nun, erst nach diesen Analysen wird das Sein des Daseins als Sorge
aufgewiesen. Die Sorge faßt in sich, wie gesagt, die "*formal* existenziale
Ganzheit des ontologischen Strukturganzen des Daseins" (SZ 192 –
gesperrt vom Verfasser) und bedeutet: "Sich-vorweg-schon-sein-in-
(der-Welt-)als Sein-bei (innerweltlich begegnendem Seienden)" (SZ 192).
In Bezug auf diese Sorgestruktur müssen, soweit es für das vorstehende Problem nötig ist, folgende Bemerkungen vermerkt
werden.

(a) Wenn die Sorge wirklich das Sein des Daseins in seiner *formal-
indifferenten* Strukturganzheit erfüllt, so müßte sie die obengenannten

formal-indifferenten Seinsweisen in ihrer Gesamtheit in sich beschließen, weil sie sonst der "*formal* existenzialen *Ganzheit* des ontologischen Struktur*ganzen* des Daseins" nicht würdig sein könnte. In der Tat könnte man im Ungefähren sagen, daß alles, was unter der Problematik der *Weltlichkeit* und des *Mitseins* aufgezeigt worden ist, in das dritte Moment der Sorge, d. h. in das "Sein-bei" zusammengefaßt ist, und daß gerade die alltägliche Seinsweise dieses "Sein-bei" nichts anderes als das *Verfallen* ist, das in der dritten Problematik aufgewiesen worden ist. Aber aus den Stellen, wo die Erläuterung von der Sorgestruktur zur Sprache kommt (SZ 191 ff.), und nicht zuletzt aus den ganzen Darlegungen von *Sein und Zeit*, läßt sich leicht entnehmen, daß Heidegger unter dem "Sein-bei" vorwiegend das "Sein-beim *besorgten* innerweltlichen *Zuhandenen*" (SZ 192 – gesperrt vom Verfasser) versteht und nicht zusammen damit *deutlich* an *das Mitsein mit den Anderen* zu denken neigt. Freilich bemerkt Heidegger oft, daß das besorgende "Sein-bei" eine Tendenz hat, sich mit der "Öffentlichkeit des Man" (SZ 192) zu verbinden. Aber diese Tendenz beschränkt sich unter anderem nur auf das *alltägliche verfallende* "Sein-bei". Bei dem Verfallen geht das Dasein, wie gesagt, in der besorgten zuhandenen "Welt" auf, und dieses Aufgehen zieht unumgänglich die Verlorenheit in die Öffentlichkeit des Man nach sich, so denkt Heidegger. Aber wenn das an sich formal-indifferente "Sein-bei" auch in der *eigentlichen* Weise auftreten kann, wie wird sich dann das besorgende "Sein-bei" mit dem "Mitsein" mit den Anderen verknüpfen? Das bleibt aber bei Heidegger völlig dunkel. Wie früher erwähnt, spricht er von dem *eigentlichen* Besorgen gar nicht und nur bloß vorübergehend von dem *eigentlichen* Miteinandersein in der Form der vorspringend-befreienden Fürsorge (SZ 122, 298). Abgesehen davon, läßt der Ausdruck *"Bei"* in dem "Sein-bei (innerweltlich begegnendem Seienden)" als dem dritten Moment der Sorge durchblicken, daß dieses überwiegend an dem Sein-*bei* dem Zuhandenen orientiert ist, und nicht so sehr an dem Mitsein *mit* den Anderen. Gewiß ist das Mitdasein auch als ein "innerweltlich begegnendes Seiendes" bestimmt, und insofern könnte angenommen werden, daß das Mitdasein in das "innerweltlich begegnende Seiende", bei dem das dritte Moment des "Sein-bei" ist, miteinbegriffen wäre. Aber schon ist darauf hingewiesen, daß dieses "innerweltlich begegnende Seiende" nicht so, wie bei dem "innerweltlich begegnenden *zuhandenen* Seienden", "besorgt" wird, sondern in der "Fürsorge" steht. Da die Verhaltungsweise zu den beiden, wenn auch gemeinsam innerweltlich begegnenden Seienden doch verschiedenartig sein muß, könnte sich nicht ohne weiteres annehmen lassen, daß das *fürsorgende* Mitsein *mit* den Anderen im vornherein in das *besorgende* Sein-*bei* dem Zuhandenen völlig eingeschlossen wäre. Aus den ganzen Darstellungen von *Sein und Zeit* her ist es auch klar, daß Heidegger mehr Gewicht auf die Analyse des Besorgens legt, als auf die der Fürsorge. Dafür spricht am besten die Tatsache, daß er in der zeitlichen Interpretation der Alltäglichkeit nur die Zeitlichkeit des Besorgens (SZ 352 ff.) berücksichtigt, wobei allerdings, beiläufig bemerkt, wiederum bloß von dessen Konstitution durche die *uneigentliche* Zeitlichkeit allein gesprochen wird, und gar nicht von der durch die eigentliche, und daß er hingegen zur Zeitlichkeit der Fürsorge niemals die geringste deutliche Stellung genommen hat. Sicher kommt wohl diese Ungleichmäßigkeit daher, daß er von Anfang an nicht beabsichtigt hat, "eine vollständige Ontologie des Daseins" (SZ 17) zu geben. Jedoch spricht sie deutlich genug dafür, daß das Problem der Fürsorge nicht gründlich durchdacht ist. Auf alle Fälle, wenn man aus dem dritten Moment des Sein-bei das Mitsein mit den Anderen auslassen wollte, so darf es keineswegs hingenommen werden. Das dritte Moment "Sein-bei" soll ja doch immerhin die Seinsweise des Daseins bezeichnen, in der das Dasein nicht nur, indem es sich aus dem Worumwillen her an die Bezüge von Um-zu verweist, das Zuhandene im umsichtigen Besorgen begegnen läßt, sondern auch sich in der Fürsorge zu dem Mitdasein verhält. Diese Seinsweise soll als indifferente sowohl für die Eigentlichkeit als auch für die Uneigentlichkeit gelten, so müßte interpretiert werden.

(b) Nun, wie steht es mit den drei "primären" Konstitutionsmomenten des Seins der Erschlossenheit: Befindlichkeit, Verstehen und Rede? Sind sie völlig in die Sorgestruktur einbegriffen worden? Inwiefern ist die Sorgestruktur in dieser Hinsicht befähigt, sich als die *formal*-indifferente *Ganzheit* des Seins des Daseins auszugeben? Wenn wir mit dieser Frage der Sachlage nachsehen, stoßen wir wiederum noch auf eine andere Schwierigkeit. Es ist nämlich so, daß sich sicher Befindlichkeit und Verstehen in der Sorgestruktur deutlich und fest niedergelassen haben, während die Rede allein keine ausdrückliche Stellung darin gefunden hat. Denn das "Sich-vorweg" in der Sorgestruktur hat sich gerade daraus abgeleitet, daß sich das Dasein verstehend auf eine Möglichkeit entwirft und so immer sich vorweg, über sich hinaus, ist, und ebenso stammt das "Schon-sein-in-(der-Welt)" daher, daß das Dasein ihm selbst überantwortet, ja schon in eine Welt geworfen ist; was aber die Rede betrifft, so hat sie offensichtlich in der dreigliedrigen Sorgestruktur keinen Platz, obwohl sie, wie gesagt, als ein "ursprüngliches Existenzial", "gleichursprünglich" mit der Befindlichkeit und dem Verstehen konstitutiv für die Existenzialität der Existenz sein soll. Das ist eines der auffälligsten und merkwürdigsten in den Darlegungen von *Sein und Zeit*. Aber daran ist kein Zweifel, daß Heidegger an vielen Stellen in der Tat die Rede als ein ebenso wichtiges Strukturmoment des Seins des Daseins ansieht, wie die Befindlichkeit, das Verstehen und das Verfallen (vgl. SZ 181, 295 f., 269 f., 335). Trotzdem hat er die Sorge nicht so formuliert, daß sie, mit der Rede einbegriffen, eine viergliedrige Strukturganzheit bildete. Das liegt höchstwahrscheinlich daran, daß er im voraus die Sorgestruktur, gerade in Entsprechung zu der später zu erhellenden dreigliedrigen Zeitlichkeit, in ihrer Dreigliedrigkeit bestimmen mußte; die Zeitlichkeit, die sich in der ekstatischen Einheit von drei Ekstasen zeitigt, muß ja doch als der Sinn der Sorge eben in der genauen Entsprechung zu dieser Sorge herausgearbeitet werden. Aber abgesehen davon, verwurzelt die Rede, wie erwähnt, so gründlich in der Existenz des Daseins, daß man sich über dieses Strukturmoment nicht leichtsinnig hinwegsetzen darf. In einem gewissen Sinne könnte auch möglicherweise angenommen werden, daß sie gerade deshalb in der Sorgestruktur keinen selbständigen Platz gefunden hätte, weil sie so innig Befindlichkeit und Verstehen bestimmt und beherrscht, weil sie sogar das Sein des Daseins, das sich in einem sinnhaften Bedeutungsganzen seines Seinsverständnisses in hörend-schweigender Weise ausspricht, von seinem Grunde aus durchdringt, so daß die Rede fast der Existenz als solcher völlig einverleibt worden wäre. Andererseits könnte aber auch vermutet werden, daß Heidegger zu der Zeit, da er *Sein und Zeit* schrieb, noch nicht auf den klaren Gedanken über die Tragweite der Rede gekommen wäre. In der Tat ist das Problem der Rede und Sprache in ihrem Zusammenhang mit der Zeitlichkeit und dem Sein hinsichtlich seiner entschiedenen Lösungen meistens in *Sein und Zeit* offen gelassen (vgl. SZ 160, 349, und 349 Anm. von der Auflage von 1927). Wenn man aber Heideggers Behandlungsarten der Sprache in seinen verschiedenen Perioden zusammenhält, wie etwa die erste Phase in seiner Dissertation und Habilitationsschrift, die zweite in *Sein und Zeit* und die dritte in dem späteren Heidegger (z. B. "Das Haus des Seins ist die Sprache."), so leuchtet sofort ein, daß das Problem der Rede in *Sein und Zeit* für die ganzen Gedankenentwicklungen Heideggers von entscheidender Bedeutung ist. Seine späteren Gedanken sind, so könnte man sogar interpretieren, in der Weise entstanden, daß Heidegger das Problem der Rede und Sprache, über das er sich selbst in *Sein und Zeit* noch nicht gänzlich klar war, nachher weitergedacht hat und seiner unermeßlichen Tiefe und Tragweite innegeworden ist; wir können uns aber leider unter der Einschränkung unserer jetzigen Abhandlung auf dieses Problem nicht näher einlassen. Wir müssen uns augenblicklich damit begnügen, darauf hinzuweisen, daß ein tiefes Verständnis für die Tragweite der Rede nicht nur für die Interpretation von *Sein und Zeit*, sondern auch von den ganzen Gedanken Heideggers von ausschlaggebender Bedeutung ist, daß die Rede, wie gesagt, in *Sein und Zeit* trotz ihrer Wichtigkeit merkwürdigerweise keinen ausdrücklichen Platz in der Sorgestruktur gefunden hat und daß infolgedessen in Bezug auf das Befugnis der Sorgestruktur, die *formal* existenziale *Ganzheit* des ontologischen Strukturganzen des Daseins zu sein, einige Fragwürdigkeiten bleiben müßten.

(c) Nun, wir möchten schließlich auf das Wort "als" in der Bestimmung der Sorge aufmerksam machen. In der Definition der Sorge ist das "Sich-vorweg-schon-sein-in-(der-Welt)", d. h. der geworfene Entwurf, mit dem "Sein-bei (innerweltlich begegnendem Seienden)" durch ein kleines Wort "als" verbunden. Dieses "als" soll vorsichtig genommen werden. Das heißt, daß das erste und zweite Moment, das gerade "überhaupt und indifferent" (SZ 192) ein *geworfenes* In-der-Welt-sein-*können*" (SZ 192 - gesperrt vom Verfasser), nämlich den "geworfenen Entwurf" (SZ 223) bedeutet, sich eben "als" das "Sein-bei" konkretisiert und "als" diese Form offenbar macht. Die Befindlichkeit und das Verstehen sind "*primäre* Konstitution des Seins der Erschlossenheit" (SZ 133, gesperrt vom Verfasser), d. h. des In-seins. Nun, "das 'Sein bei' der Welt ist ein im In-sein fundiertes Existenzial" (SZ 54). Das bedeutet: das "Sein-bei" in der Sorgestruktur ist gewissermaßen ein *sekundäres* Moment, das in den *primären* Konstitutionsmomenten des *In-seins*, d. h. in dem geworfenen Entwurf als dem "Sich-vorweg-schon-sein-in-(der-Welt)" *fundiert* ist. Gerade deswegen ist das "Sich-vorweg-schon-sein-in-(der-Welt)" nicht direkt mit dem "Sein-bei" verknüpft, sondern vermittels des Wörtchen "als", und das deutet an, daß sich das *primäre* Moment gerade "als" das *sekundäre* "Sein-bei" in Wirklichkeit offenbar macht. Und nicht nur das, sondern darüber hinaus tritt das "Sein-bei" *zunächst und zumeist*, obwohl es in dem geworfenen Entwurf als dem primären Moment gründet, doch in der *durchschnittlichen Alltäglichkeit* gerade, wie gesagt, in der Seinsart des *Verfallens* auf. Diese Seinsart des verfallenden "Sein-bei" aber ist in dem primären "In-Sein", d. h. dem geworfenen Entwurf, fundiert und kommt erst auf dessen Grunde zum Vorschein. Aber wie? Gerade hierin liegt der Grund dafür, daß das Dasein alltäglich verfallend existiert, daß es eine starke Tendenz hat, meistens in dem uneigentlichen Sein-bei zu existieren, daß infolgedessen das dritte Moment der ursprünglich formal-indifferent sein sollenden Sorgestruktur trotzdem zum großen Teil mit der verfallenden Uneigentlichkeit gleichgesetzt zu werden geneigt ist. Es ist nämlich folgendes: "Das Aufgehen im Man und bei der besorgten 'Welt' offenbart so etwas wie eine *Flucht* des Daseins vor ihm selbst als eigentlichem Selbst-sein-können" (SZ 184); "das Verfallen des Daseins an das Man und die besorgte 'Welt'" ist "'Flucht' vor ihm selbst" (SZ 185). Doch, in der fliehenden Abkehr vor ihm selbst ist das Wovor der Flucht schon erschlossen, und dies kommt in der Grundbefindlichkeit der Angst an den Tag. Die verfallende Flucht gründet nämlich in der Angst. In der Angst wird die Bewandtnisganzheit, mit der vertraut das Dasein mannigfaltig zerstreut beschäftigt war, völlig unbedeutsam und belanglos, sinkt sich zusammen; "die alltägliche Vertrautheit bricht in sich zusammen" (SZ 189) und das Dasein ist beinahe vor dem "Nichts" (SZ 186 f., 276 f.) gebracht. Aber gerade hierin stellen sich das Wovor und das Worum der Angst als Nichts von innerweltlich Seiendem heraus, sondern eben als die Welt als solche, besser gesagt, als das In-der-Welt-sein des Daseins selbst. In der Angst wird dem Dasein sein vereinzeltes eigentliches In-der-Welt-sein-können als geworfener Entwurf als solcher erst sichtbar. In der Angst wird ihm selbst sein Dasein in seiner "Unheimlichkeit", d. h. das ursprüngliche geworfene In-der-Welt-sein als "Un-zuhause", erst durchsichtig (SZ 188 f., 276 f.). "Das In-sein kommt in den existenzialen 'Modus' des *Un-zuhause*" (SZ 189). Gerade deshalb, weil sich das Dasein in seinem Grunde des In-der-Welt-seins in der Angst befindet, so sucht es "in der Weise der ausweichenden Abkehr" (SZ 136) vor ihm selbst in das Man und in die besorgte "Welt" die Flucht. "Die verfallende Flucht *in* das Zuhause der Öffentlichkeit ist Flucht *vor* dem Unzuhause, d. h. der Unheimlichkeit, die im Dasein als geworfenen, ihm selbst in seinem Sein überantworteten In-der-Welt-Sein liegt. Diese Unheimlichkeit setzt dem Dasein ständig nach und bedroht, wenngleich unausdrücklich, seine alltägliche Verlorenheit in das Man" (SZ 189). Aber "das beruhigt-vertraute In-der-Welt-sein ist ein Modus der Unheimlichkeit des Daseins, nicht umgekehrt. *Das Un-zuhause muß existenzial-ontologisch als das ursprünglichere Phänomen begriffen werden*" (SZ 189). Dieses ursprünglichere Sein des Daseins wird noch erst durch die Aufweisung des eigentlichen Ganzseinkönnens des Daseins, d. h. durch die Analysen von Tod und Gewissen, in seiner vollen Breite und Tiefe ans Licht gebracht, worauf wir aber hier näher einzugehen verzichten müssen. Jedenfalls hat sich hoffentlich damit gezeigt, daß das Dasein, weil es sich als unheimliches In-der-Welt-sein vor und um sich selbst ängstigt, alltäglich vor ihm selbst fliehend, sich "als" das uneigentliche verfallende "Sein-bei" offenbar macht und zerstreut bei dem Man und der besorgten "Welt" die Beruhigung sucht. Das ist ja der Grund dafür, daß das "Sein-bei" zunächst und zumeist mit dem uneigentlichen Verfallen gleichgesetzt zu werden neigt, daß die Sorgestruktur als aus Existenzialität, Faktizität und Verfallen bestehend erklärt wird. Dennoch, solange das "Sein-bei", wie gesagt, ursprünglich formal-indifferent gemeint sein sollte, muß es noch auch eigentlicherweise auftreten können. Dafür spricht in der Tat der Begriff der "Situation", die das in der eigentlichen Entschlossenheit erschlossene Da ist, als welches das Dasein existiert (SZ 299 f., 382 ff.). Darauf können wir später zu sprechen kommen.

§6. *Zeitlichkeit*

Jetzt soll die formal-indifferente Struktur des Seins des Daseins von der Seite der Zeitlichkeit her kurz betrachtet werden. Wir haben hier nicht die Absicht, das *Ganze* der reichhaltigen Analysen über die Zeitlichkeit in Betracht zu ziehen; unser Vorhaben ist bescheidener und darauf beschränkt, die Problematik der Zeitlichkeit nur im Zusammenhang mit der formal-indifferenten Seinsstruktur zu behandeln, ihren fundamentalen Charakter in dieser Hinsicht zu betrachten und den dabei wiederum entstehenden Schwierigkeiten zu begegnen. Für die Erörterung der formal-indifferenten Seinsstruktur des Daseins ist die Heranziehung der Zeitlichkeit doch unumgänglich, weil sie der Seinssinn des Daseins ist. Der Sinn ist das, was ermöglicht. Die Zeitlichkeit ist daher das, was das Sein des Daseins im Ganzen "*aus seinem Grunde*" (SZ 436) ermöglicht. Alle analysierten Strukturen sollen endlich "in die ursprüngliche Struktur der Seinsganzheit des Daseins", d. h. "die Zeitlichkeit", "zurückgenommen" werden und darin "ihre 'Begründung' erhalten" (SZ 436). Infolgedessen ist die Zeitlichkeit auch die "ursprüngliche Bedingung der Möglichkeit der *Sorge*" (SZ 372). Alle Seinsweisen des Daseins sind nach Heidegger letztlich erst aus der Zeitigung der Zeitlichkeit als ihre verschiedenen Modi zu begreifen. Nun, wenn wir das Wesen dieser Zeitlichkeit feststellen wollen, begegnen uns einige Schwierigkeiten. Die Zeitlichkeit wird nämlich in *Sein und Zeit* erst nach der Herausarbeitung des eigentlichen Ganzseinkönnens des Daseins behandelt. *Von der Seite dieser Behandlungsweise her gesehen*, in der die Zeitlichkeit als der Seinssinn des Daseins erst in Zusammenhang mit der vorlaufenden Entschlossenheit herausgearbeitet und abgehoben wird, hat es den Anschein, als ob die Zeitlichkeit, die doch alle Seinsweisen des Daseins begründen soll, indessen durch das eigentliche Ganzseinkönnen des Daseins bedingt wäre. Die Zeitlichkeit wird in der Tat "*phänomenal ursprünglich*", so wird gesagt, "*am eigentlichen Ganzsein des Daseins, am Phänomen der vorlaufenden Entschlossenheit erfahren*" (SZ 304), "in Hinblick auf das eigentliche Ganzseinkönnen des Daseins herausgestellt" (SZ 372), und zeigt sich, "zuerst an der vorlaufenden Entschlossenheit" (SZ 331). Das Kapitel, in dem die Zeitlichkeit zuerst an den Tag gebracht wird, folgt tatsächlich dem Kapitel, in dem die vorlaufende Entschlossenheit aufgezeigt worden ist. Aber *von der Seite dessen her gesehen, was den Grund des Seins des Daseins ausmacht*, so liegt die Zeitlichkeit umgekehrt, wenngleich sie "sich ursprünglich" an der vorlaufenden Entschlossenheit "bekundet" (SZ 304), doch als die "Bedingung" dieser eigentlichen Seinsweise ebenso ihr selbst zugrunde, wie auch allen anderen Seinsweisen des Daseins. "Die Zeitlichkeit der vorlaufenden Entschlossenheit" ist "ein *ausgezeichneter* Modus ihrer selbst" (SZ 304, gesperrt vom Verfasser). "Zeitlichkeit kann sich in verschiedenen Möglichkeiten und verschiedenen Weisen *zeitigen*. Die Grundmöglichkeiten der Existenz, Eigentlichkeit und Uneigentlichkeit des Daseins, gründen ontologisch in möglichen Zeitigungen der Zeitlichkeit" (SZ 304).

Wenn dem so ist, scheint es, daß *diese* Zeitlichkeit, *die* als der letzte Grund und Sinn des Daseins jedem Verhalten des Daseins zugrunde liegt, die also in diesem Sinne sowohl die Eigentlichkeit als auch die Uneigentlichkeit ermöglicht, vorausgesetzt, daß es eine solche wirklich gebe –, wohl *die formal-indifferente Zeitlichkeit* als solche sei, die sich je nachdem als ein "ausgezeichneter" Modus der eigentlichen Zeitlichkeit wie auch als eine uneigentliche Zeitlichkeit zeitigen kann. Gewiß könnte man, wenn man Darlegungen Heideggers sorgfältig nachgeht und sie konsequenterweise interpretieren will, eine solche vermutlich auch von Heidegger selbst im heimlichen gedachte *formal-indifferente Zeitlichkeit* herausfinden. Es ist nämlich folgendes. Das Dasein "*kann*" "als seiendes" "*überhaupt*" "schon immer" "auf sich zukommen", "die Möglichkeit in diesem Sich-auf-sich-zukommenlassen als Möglichkeit" aushalten, und ist so "in seinem Sein überhaupt zukünftig" (SZ 325). Das Dasein "*ist* überhaupt" ferner immer "als ich *bin*-gewesen" und kann "sein eigenstes 'wie es je schon war', d. h. sein 'Gewesen' sein" (SZ 325 f.). Das Dasein aber kann weiter überhaupt stets nur als "das handelnde Begegnenlassen des umweltlich *Anwesenden*", d. h. "in einem Gegenwärtigen dieses Seienden" und in der Seinsweise der "*Gegenwart* im Sinne des Gegenwärtigens" existieren (SZ 326). "Diese phänomenalen Charaktere des 'Auf-sich-zu', des 'Zurück auf', des 'Begegenlassen *von*'" sind Kennzeichen von "Zukunft, Gewesenheit, Gegenwart" (SZ 328). Diese "ekstatische" Einheit der Zeitigungen von Zukunft, Gewesenheit und Gegenwart macht die formal-indifferente Zeitlichkeit aus, so könnte man annehmen.

Aber Heidegger geht nicht so vor, daß er allererst solch eine formal-indif-ferente Zeitlichkeit abhebt und dann ihre verschiedenen Zeitigungsmodi entwickelt. Sein Verfahren ist nicht so formal, daß er zuerst ein oberstes Prinzip aufstellte und dann seine Besonderungen entfaltete, wie es oft bei einem deduktiv aufgebauten System der Fall wäre. So ist denn auch die Sorgestruktur selbst, wie erwähnt, meistens nicht in ihrer formal-indif-ferenten Verfassung, sondern in ihrer engen Verbindung mit dem uneigent-lichen Verfallen erläutert. Heideggers phänomenologische Analysen gehen nicht so formal-deduktiv vonstatten, sondern sie beginnen mit der durchschnittlichen Alltäglichkeit, um dann Schritt für Schritt in die ursprüngliche Interpretation vertieft zu werden. Diesem seinen Vorgehen liegt sein eigener Gedanke zugrunde, daß sich in den alltäglichen Phäno-menen in ihrem Grunde das unheimliche In-der-Welt-sein als solches versteckt, daß "Uneigentlichkeit mögliche Eigentlichkeit zum Grunde hat" (SZ 259). Als dieses tiefste eigentliche Ganzsein des Daseins wird die vorlaufende Entschlossenheit herausgearbeitet, an der erst die Zeitlichkeit als der Seinssinn des Daseins ans Licht gebracht wird. So wird gesagt: "Zukünftig auf sich zurückkommend, bringt sich die *Entschlossenheit* (gesperrt vom Verfasser) gegenwärtigend in die Situation. Die Gewesen-heit entspringt der Zukunft, so zwar, daß die gewesene (besser gewesende) Zukunft die Gegenwart aus sich entläßt. Dies dergestalt als gewesend-gegenwärtigende Zukunft einheitliche Phänomen nennen wir die *Zeitlichkeit*" (SZ 326). Also der "*Gehalt*" des "*Sinnes*" von Zeit-lichkeit wird "phänomenal" "aus der Seinsverfassung der vorlaufenden Entschlossenheit *geschöpft*" und dieser Gehalt "*erfüllt* die Bedeutung des Terminus Zeitlichkeit" (SZ 326, gesperrt vom Verfasser). Damit ist vor allem gemeint, daß zu der Sinnbestimmung der Zeitlichkeit einzig und allein die Erfahrung der vorlaufenden Entschlossenheit Befugnis hat, daß man sie dabei mit Hilfe des vulgären Zeitbegriffs nicht verstehen darf. Die Zeitlichkeit, die an der vorlaufenden Entschlossenheit "erfahren" und daraus hinsichtlich ihres Sinngehaltes "geschöpft" wird, zeigt das "ursprüngliche" Phänomen der Zeit. Beispielsweise kann "das ursprüng-liche Phänomen der *Zukunft*" eigens nur darin gefunden werden, daß das Dasein in seiner vorlaufenden Entschlossenheit, die "ausgezeichnete

Möglichkeit aushaltend", "sich auf sich zukommen läßt" (SZ 325). In diesem Sinne wird die Zeitlichkeit, die an der vorlaufenden Entschlossenheit erfahren wird, die *ursprüngliche Zeitlichkeit*" und somit auch "die *ursprüngliche Zeit*" genannt (SZ 326 f., 328 ff.).

Aber wenn dem so ist, bestimmt und bedingt die vorlaufende Entschlossenheit nicht die Zeitlichkeit, die doch als der Seinssinn sowohl die Eigentlichkeit als auch die Uneigentlichkeit begründen soll? Auf diese Frage muß man einerseits freilich deshalb bejahend antworten, weil in der Tat, wie oben gezeigt, die vorlaufende Entschlossenheit allein berechtigt ist, die *sinnhafte Bedeutung* der ursprünglichen Zeitlichkeit zu bestimmen. Jedoch muß man andererseits, wenn man unter dem Begriff der "*Bestimmung* und *Bedingung*" das versteht, was einen *sachlichen Grund* eines Sachverhaltes ausmacht, vielmehr sagen, daß die vorlaufende Entschlossenheit keineswegs die Zeitlichkeit *bestimmt* und *bedingt*, sondern daß im Gegenteil die Zeitlichkeit als solche eben die vorlaufende Entschlossenheit "ermöglicht", weil sie gerade der Seinssinn des Daseins ist und der Sinn das ist, was "ermöglicht". So wird in der Tat gesagt: "Nur sofern das Dasein als Zeitlichkeit bestimmt ist, ermöglicht es ihm selbst das gekennzeichnete eigentliche Ganzseinkönnen der vorlaufenden Entschlossenheit" (SZ 326). In diesem Sinne scheint hier eine Zirkelstruktur vorhanden zu sein. Denn die Zeitlichkeit bestimmt und ermöglicht *sachlich* die vorlaufende Entschlossenheit, aber umgekehrt, die vorlaufende Entschlossenheit bestimmt und ermöglicht die unsprüngliche *sinnhafte Bedeutung* der Zeitlichkeit. Zum Beispiel kommt das Dasein "überhaupt" immer auf sich und ist in diesem Sinne "zukünftig"; erst aufgrund dessen kann es denn auch eigentlich vorlaufend-entschlossen zukünftig existieren; aber das ursprüngliche Phänomen der "Zukunft", oder der sinnhafte Gehalt des "Auf-sich-zu-kommens" läßt sich in seinem "ausgezeichneten" Modus erst allein aus der Erfahrung der vorlaufenden Entschlossenheit schöpfen. Aber abgesehen davon, muß die "*ursprüngliche*" Zeitlichkeit, die *sinnhaft* aus der vorlaufenden Entschlossenheit herausgenommen wird, aus dem obengenannten Grunde gleichzeitig *sachlich* die *eigentliche* Existenz ermöglichen, und in diesem Sinne auch die "*eigentliche*" Zeitlichkeit genannt werden (SZ 326 f., 329 ff.). "Das im Hinblick auf die Entschlossenheit gewonnene Phänomen selbst", d. h. "die Zeitlichkeit" stellt "nur eine Modalität der Zeitlichkeit" selbst dar (SZ 327), nämlich die *eigentliche* Modalität der Zeitlichkeit. "Die Zeitlichkeit der vorlaufenden Entschlossenheit" ist "ein ausgezeichneter Modus ihrer selbst" (SZ 304). Das bedeutet, daß diese

Zeitlichkeit nicht nur die *ursprüngliche* Zeitlichkeit, sondern auch die *eigentliche* Zeitlichkeit ist, die die eigentliche Seinsweise des Daseins in ihrem Sinn ausmacht. Deswegen wird die Zeitlichkeit, die *sinnhaft* an der vorlaufenden Entschlossenheit "erfahren" und daraus geschöpft wird, die aber zugleich *sachlich* diese ermöglicht, die "*ursprüngliche und eigentliche Zeitlichkeit*" genannt (SZ 326 f., 329 ff., gesperrt vom Verfasser). Diese Zeitlichkeit wird deshalb "*ursprünglich*" genannt, weil jedes Zeitverständnis hierin seinen *Ursprung* finden muß. So ist sie die "ursprüngliche Zeit" (SZ 329). Das vulgäre Zeitverständnis, das die Zeit als "eine pure, anfang- und endlose Jetzt-folge" ansieht (SZ 329), kommt erst dadurch zustande, daß die ursprüngliche und eigentliche Zeitlichkeit sich zu einer uneigentlichen Zeitlichkeit modifiziert, die dann ihrerseits die "Jetzt-Zeit" zeitigt (SZ 326, 329); diese echte Genesis des vulgären Zeitbegriffs zu verfolgen (SZ 404–428), muß hier aber verzichtet werden. Die an der vorlaufenden Entschlossenheit erfahrene ursprüngliche Zeit heißt ferner deshalb "*eigentlich*", weil sie den Seinssinn der Eigentlichkeit des Daseins ausmacht. Diese "*ursprüngliche und eigentliche*" Zeitlichkeit zeitigt sich stets in der ekstatischen Einheit ihrer Ekstasen und primär aus der Zukunft, hat aber den entscheidenden Charakter der Endlichkeit (SZ 328 ff.). Dieser Charakter ist für das Verständnis der Zeitlichkeit und somit des Seins des Daseins von entscheidender Bedeutung, jedoch können wir darauf nicht eingehen.

Nun, Heidegger setzt an einer Stelle (SZ 330) die "ursprüngliche und eigentliche" Zeitigung der Zeitlichkeit fast mit der "Zeitlichkeit" überhaupt gleich. Das bedeutet, daß er nicht so, wie vorher angedeutet, in Wirklichkeit zuerst die formal-indifferente Zeitlichkeit als Prinzip aufstellt und dann zu ihren eigentlichen und uneigentlichen Besonderungen fortschreitet, sondern daß er denkt, das *ursprüngliche* Wesen der Zeitlichkeit solle in der "ursprünglichen und eigentlichen" Zeitlichkeit gesehen werden. Wenn man allerdings seinen Gedankengang nach *der sachlichen Ordnung der Begründung* hinnehmen dürfte, könnte man sagen, daß die Zeitlichkeit *überhaupt*, die alle Seinsweisen des Daseins begründen soll, in der ekstatischen Einheit der Zeitigungen von Zukunft, Gewesenheit und Gegenwart besteht, die je "auf-sich-zu", "Zurück auf" und "Begegnenlassen von" bedeuten, daß sich diese Zeitlichkeit an der vorlaufenden Entschlossenheit "*ursprüngliche und eigentlich*" zeitigt, und daß in diesem Modus alles Zeitverständnis seinen Ursprung finden soll. Aber natürlich entsteht auch bei dieser Interpretation wiederum ein Problem, woher der sinnhafte Gehalt der Zeitlichkeit

überhaupt geschöpft werden soll, und diese Frage läßt sich nicht
anders beantworten, als daß die Zeitlichkeit endlich aus der vorlaufenden
Entschlossenheit her verstanden werden muß; aber warum muß sie aus
dieser her verstanden werden? Wie ist die aus dieser entnommene Zeitlichkeit *ursprünglich*? Diese Frage läßt sich nicht mehr bloß von der
Betrachtung über das Wesen der Zeitlichkeit entscheiden, sondern nur aus
den Grundgedanken Heideggers her, daß das Dasein in der vorlaufenden
Entschlossenheit allein ursprünglich und eigentlich existieren kann. In
diesem Sinne wird die Zeitlichkeit auf alle Fälle hinsichtlich ihrer sinnhaften ursprünglichen Bedeutung doch endlich durch die vorlaufende Entschlossenheit bedingt, aber hinsichtlich der sachlichen Begründungs-
ordnung macht sie umgekeht den letzten Grund aus, in dem alle Seinsweisen des Daseins verwurzeln sollen. Die Zeitlichkeit ist gewissermaßen
"ratio essendi", während "ratio cognoscendi" davon die vorlaufende Entschlossenheit ist; den Zirkel dieser Art muß man sich irgendwie gefallen
lassen, wie es scheint, wenigstens in dem philosophischen Denken.

Nun, inwiefern ist ferner die Zeitlichkeit die Bedingung der Möglichkeit
der Sorge? Das Sich-vorweg von der Sorgestruktur gründet, wie leicht
einzusehen ist, in der Zukunft, und das Schon-sein-in in der Gewesenheit.
Wie steht es aber mit dem Sein-bei? Wenn man auf eine formal-indifferente Weise denkt, muß wohl das Sein-bei, das seinerseits selbst für ein
formal-indifferentes Seinsmoment gehalten werden mußte, auch in der
formal-indifferenten Zeitigung der Zeitlichkeit gründen, d. h. in dem
"Begegnenlassen von", in dem "handelnden Begegnenlassen des umwelt-
lichen Anwesenden", in dem "Gegenwärtigen dieses Seienden", in der
"Gegenwart im Sinne des Gegenwärtigens" (SZ 326). In der Tat sagt
Heidegger selbst: "Formal verstanden ist jede Gegenwart gegenwärtigend"
(SZ 338). Also, "formal verstanden" ist die dritte Ekstase der Zeitlichkeit
die "gegenwärtigende Gegenwart", und das Sein-bei von der
Sorgestruktur muß ja in dieser gründen. Aber an der Stelle, in der die Zeitlichkeit der Sorge erwähnt ist (SZ 325 f.), ist *einmal* das dritte Moment
der Sorge wiederum, wie oft geschieht, als das "verfallende Sein-bei"
(SZ 328) genommen, und dazu kommt *alsdann* noch hinzu, daß der
zeitliche Sinn dieses Moments im Gegensatz zu der ursprünglichen und
eigentlichen Zeitlichkeit, die an der vorlaufenden Entschlossenheit an den
Tag kommt, interpretiert und so in der Form, die die Uneigentlichkeit in
sich enthält, betrachtet wird; was die Sachlage kompliziert. Abgesehen
davon, wird dort am Ende gezeigt, daß sich die formal-indifferent
verstandene dritte Ekstase der "gegenwärtigenden Gegenwart" derart

modifizieren kann, daß die eigentliche Gegenwart nichts anderes als der "Augenblick" ist, während sich die uneigentliche Gegenwart als das "uneigentliche, augenblicklos-unentschlossene Gegenwärtigen" (SZ 338), "in dem das *Verfallen* an das besorgte Zuhandene und Vorhandene *primär* gründet" (SZ 328), zeitigt. Worin aber der Unterschied zwischen der augenblicklich-gegenwärtigenden Gegenwart und der verfallend-gegenwärtigenden Gegenwart besteht, bleibt eine Frage, worauf wir später nochmal zu sprechen kommen, weil dieses Problem auf unser Thema eng bezogen ist.

Nun, die ursprüngliche und eigentliche Zeitlichkeit läßt sich weiter, indem der Charakter ihrer Eigentlichkeit betont und herausgehoben wird, mit anderer Terminologie ausdrücken. Dies wird dadurch ermöglicht, daß die Zeitlichkeit des Verstehens exponiert wird. Das Verstehen besagt *"entwerfend-sein zu einem Seinkönnen, worumwillen je das Dasein existiert"* (SZ 336) und dieser Charakter bildet, wie gesagt, die "Existenzialität" des Daseins, dem es in seinem Sein um dieses geht. So hängt die Eigentlichkeit und Uneigentlichkeit des Daseins, so könnte man sagen, von der Seinsweise des Verstehens ab. Da es sich dem Dasein im entwerfenden Sichverstehen um das Auf-sich-zukommen aus einer Möglichkeit handelt, so gründet das Verstehen primär in der Zukunft. Dem Verstehen liegt also die "formal-indifferente" (SZ 337) Zukunft, d. h. das Auf-sich-zukommen, das Sich-vorweg zugrunde. Wenn sich diese Zukunft aber eigentlich zeitigt, dann wird sie mit dem Ausdruck "Vorlaufen" (SZ 336) bezeichnet, weil ja sich das Dasein in der *vorlaufenden* Entschlossenheit allein eigentlich verstehend entwerfen kann. Dagegen hat die uneigentliche Zukunft den Charakter des "Gewärtigens" (SZ 337), weil das Dasein in dem uneigentlichen Verstehen aus dem Besorgten her seiner *gewärtig* ist. Wenngleich das Verstehen primär in der Zukunft gründet, ist doch diese gleichursprünglich immer durch Gewesenheit und Gegenwart bestimmt, weil sich die Zeitlichkeit stets in der ekstatischen Einheit ihrer drei Ekstasen zeitigt. Die eigentliche Gegenwart, die der eigentlichen Zukunft entspringt, heißt, wie erwähnt, der "*Augenblick*" (SZ 328, 338), weil das eigentliche Dasein im "'Augenblick' auf die erschlossene Situation 'da' zu sein" (SZ 328) sucht. Hingegen wird die uneigentliche Gegenwart das "*Gegenwärtigen*" genannt, das das Verfallen an die besorgte "Welt" ermöglicht (SZ 328, 338). Ferner läßt sich die eigentliche Gewesenheit mit dem Terminus "*Wiederholung*" bezeichnen, weil sich das Dasein im Vorlaufen *wieder holt* in das eigenste Seinkönnen *vor*" (SZ 339). Demgegenüber ist die uneigentliche Gewesenheit die "*Vergessenheit*", weil "sich das Dasein in seinem eigensten geworfenen Seinkönnen *vergessen* hat" (SZ 339). "Und nur auf dem Grunde dieses Vergessens kann das besorgende, gewärtigende Gegenwärtigen *behalten* und zwar das nichtdaseinsmäßige, umweltlich begegnende Seiende" (SZ 339).

Im allgemeinen könnte also zusammengefaßt werden, daß die Zeitlichkeit überhaupt in der ekstatischen Einheit der Zeitigungen von "Auf-sich-zu", "Zurück auf" und "Begegnenlassen von", d. h. Zukunft, Gewesenheit und Gegenwart, besteht, daß sie sich aber an der vorlaufenden Entschlossenheit als die *ursprüngliche und eigentliche Zeitlichkeit* zeitigt, daß doch hierin eine zirkelnde Struktur von der sachlichen Begründung und der sinnhaften Bedeutung zu finden, ja sogar unvermeidlich ist, daß sich diese ursprüngliche und eigentliche Zeitlichkeit, indem ihr Charakter der Eigentlichkeit betont wird, als die eigentliche Zeitlichkeit herausstellt, die ihrerseits mit eigener Terminologie bestimmt wird: nämlich, die *eigentliche Zeitlichkeit* zeitigt sich in der ekstatischen Einheit von "*Vorlaufen-Augenblick-Wiederholung*", während sich die *uneigentliche Zeitlichkeit* in der ekstatischen Einheit von "*Gewärtigen-Gegenwärtigen-Vergessen (oder Behalten)*" zeitigt. Die eigentliche Zeitlichkeit ermöglicht die Eigentlichkeit der Existenz, und die uneigentliche Zeitlichkeit deren Uneigentlichkeit. Diese Zeitlichkeiten können aber natürlich weitere Modifizierungen aufnehmen; überdies entspringt vor allem der uneigentlichen Zeitlichkeit wesensmäßig die "Weltzeit", die die "Innerzeitigkeit" ausmacht, in der das innerweltliche Seiende begegnet; wenn die ekstatischen Charaktere der Weltzeit nivelliert und besonders ihr Charakter der Öffentlichkeit betont wird, kommt echterweise der vulgäre Zeitbegriff als "Jetzt-Zeit" zustande (vgl. SZ 404–428). Aber diese Probleme können wir hier leider nicht berücksichtigen. Diese Einschränkung kann sich indes dadurch rechtfertigen, daß unser Ziel gerade darin besteht, aufgrund der oben durchgeführten allgemeinen Feststellungen das Problem des Verhältnisses der Eigentlichkeit zur Uneigentlichkeit jetzt erneut zu bedenken und zu Schlußbetrachtungen zu gelangen.

§7. *Dialektisches Verhältnis von Eigentlichkeit und Uneigentlichkeit*

Unser Thema war das Verhältnis der Eigentlichkeit und Uneigentlichkeit zueinander, vor allem aber das Verhältnis *der* Eigentlichkeit *zur* Uneigentlichkeit. Unser Interesse galt der Seinsweise, in der sich die Eigentlichkeit, die Uneigentlichkeit *modifizierend* und *verändernd*, als konkretes

In-der-Welt-sein verwirklichen läßt. Ob eine solche ontologische Möglichkeit vorhanden ist oder nicht, war der Ausgangspunkt unserer Frage.
Denn das eigentliche Dasein kann doch als In-der-Welt-sein nicht isoliert
existieren, sondern muß irgendwie sich zu der "Welt" und den Anderen
verhalten, eventuell diese Momente, die zunächst und zumeist das Dasein
in das uneigentliche Verfallen zu geraten veranlassen, eigentlicherweise
überwinden und so ein völlig eigentliches In-der-Welt-sein verwirklichen
können. Sonst wäre das eigentliche Dasein eines konkreten In-der-Welt-
seins verlustig gehen. Aber schon am Anfang unserer Diskussion wurde
darauf hingedeutet, daß dieses Problem nicht leichtsinnig erledigt werden
kann, sondern einige Schwierigkeiten und heikle Aspekte in sich enthält.
Beispielsweise wird das Besorgen, in dem sich auch das eigentliche Dasein
halten soll, doch in Wirklichkeit bei Heidegger einseitig durch die uneigentliche Zeitlichkeit konstituiert. Und das Sein-bei der Sorgestruktur wird
äußerst oft mit dem uneigentlichen Verfallen gleichgesetzt. Wenn dem so
ist, müßte doch angenommen werden, daß nach Heideggers Ansicht selbst
das eigentliche Dasein schließlich nicht völlig eigentlich zu existieren
imstande wäre, sondern mit der Uneigentlichkeit wesensmäßig zu tun
haben müßte, bzw. unter Umständen wieder in die Uneigentlichkeit
abstürzen müßte. Das bedeutet, daß es für das Dasein sehr schwierig ist,
sich als vollständig eigentliches In-der-Welt-sein durchzusetzen. So erhebt
sich notwendig eine Frage, ob denn eine vollkommene Umänderung der
Uneigentlichkeit durch die Eigentlichkeit und zwar zum Zweck der
Verwirklichung eines völlig eigentlichen In-der-Welt-seins möglich ist,
und wie überhaupt denn darüber von Heidegger nachgedacht ist.

Um diesem Problem nachzugehen, haben wir einen recht großen
Umweg gemacht. Unsere Exposition hat bei der Überprüfung einer
angeblich "dritten" Existenzmöglichkeit der Indifferenz angesetzt, die
noch neben der Eigentlichkeit und Uneigentlichkeit bestehen soll. Unsere
Erörterung aber hat herausgestellt, daß die "indifferente Alltäglichkeit",
die in der Regel als jene dritte Möglichkeit angesehen zu werden geneigt
ist, nur ein bloß uneigentlicher Modus einer *genuinen formal-
indifferenten Seinsweise* des Daseins ist. Diese letztere mußte vielmehr
dafür gehalten werden, was unter der sogenannten Indifferenz verstanden
werden soll. Unsere darauf folgenden längeren Betrachtungen haben
dann, mit mehreren Schwierigkeiten kämpfend, der Feststellung des
Inhalts dieser formal-indifferenten Seinsstrukturen gegolten; sie schienen
vielleicht von unserem eigentlichen Thema einigermaßen abzukommen,
doch haben sie dazu beigetragen, verworrene Verhältnisse in Ordnung zu

bringen. Das Dasein hat nämlich formal indifferent die Existenzstruktur, gemäß der es ihm als In-der-Welt-sein in seinem Sein um dieses geht, und diese Seinsstruktur wird als Sorge gefaßt. Obwohl das dritte Moment der Sorge häufigst mit dem uneigentlichen Verfallen gleichgesetzt ist, doch muß es unserer Meinung nach ursprünglich für ein formal-indifferentes Seinsmoment gelten, so zwar, daß darunter sowohl das umsichtige Besorgen mit dem Zuhandenen als auch das fürsorgende Mitsein mit den Anderen verstanden werden müssen. Diese formal-indifferenten Seins-strukturen der Sorge werden ferner hinsichtlich ihres Grundes in der Weise aufgewiesen, daß sie alle in der Zeitigung der Zeitlichkeit gründen, obgleich es in Bezug auf die Feststellung der ursprünglichen und eigentlichen Zeitlichkeit einige Schwierigkeiten gab, die einem Zirkel ähnelten.

Nun, wichtig ist, daß das Dasein "primär" geworfener Entwurf ist, d. h. "Sich-vorweg-schon-sein-in-(der-Welt)" ist. Mit dem Ausdruck der Zeitlichkeit gewendet, kommt das Dasein "zukünftig" auf sich zu, und ist immer schon "gewesen", oder besser gesagt, "gewesend". Befindlichkeit und Verstehen, Faktizität und Existenzialität, Geworfenheit und Entwurf, sind "primäre" Konstitutionsmomente des Daseins. Darin ist auch das "Sein-bei" "fundiert". Dieses entsteht nämlich erst "als" eine konkrete Erscheinungsform des geworfenen Entwurfs. Die Zukunft und Gewesen-heit entlassen aus sich die "Gegenwart". Aber dabei wird das alltägliche Dasein hinsichtlich seines geworfenen In-der-Welt-sein-könnens, wenn auch unausdrücklich, doch schon in der Angst dessen inne, daß sein eigenes In-der-Welt-sein selbst eine "Unheimlichkeit" und ein "Unzuhause" in sich birgt. Gerade deshalb weicht das alltägliche Dasein in der Abkehr vor ihm selbst aus, flieht vor seinem eigenen Sein, und verfällt der besorgten "Welt" und dem Man. Dadurch kommt das "Sein-bei" alltäglich im uneigentlichen Modus zum Vorschein. Besonders unheimlich ist dem alltäglichen Dasein der Tod als "die eigenste, unbezügliche, gewisse und als solche unbestimmte, unüberholbare Möglichkeit" seines eigenen Daseins (SZ 258). Ferner ist ihm die "Nichtigkeit", die den geworfenen Entwurf durchdringt, auch unerträglich (SZ 285). Umso mehr ist "das alltägliche Sein zum Tode als verfallendes eine ständige *Flucht vor ihm*" (SZ 254), und umso mehr stürzt das Dasein in die "Nichtigkeit des *un*eigentlichen Daseins im Verfallen" (SZ 285) ab. Abgesehen von diesen wichtigen Aspekten des Daseins von Tod und Gewissen, in die wir leider nicht näher eingehen können, ist jedenfalls das unheimliche In-der-Welt-sein als solches der Anlaß dazu, daß das "Sein-bei" oder die

"Gegenwart" zunächst und zumeist im Modus des uneigentlichen Verfallens auftritt. Indessen muß für ausgeschlossen gehalten werden, daß das "Sein-bei" stets uneigentlich wäre, daß es kein eigentliches "Sein-bei" geben könnte. Denn unsere früheren Darlegungen haben genug gezeigt, daß das Begegnenlassen vom Zuhandenen im umsichtigen Besorgen dem Dasein ebenso unentbehrlich und wesensnotwendig ist, wie das Miteinandersein mit den Anderen in der Fürsorge. Überdies hat die Betrachtung über die Zeitlichkeit ergeben, daß in Hinblick auf das "Sein-bei" sich die eigentliche Zeitlichkeit in der Tat als der vorlaufend-wiederholende "*Augenblick*" zeitigen kann, während sich die uneigentliche Zeitlichkeit als das gewärtigend-vergessende (behaltende) "*Gegenwärtigen*" zeitigt. Es besteht also jetzt kein Zweifel daran, daß es ein eigentliches "Sein-bei" geben kann. Sonst würde in der Tat kein wirklich eigentliches In-der-Welt-sein möglich sein. Aber wie kann sich die Eigentlichkeit, die die Uneigentlichkeit überwindet, als konkretes In-der-Welt-sein verwirklichen?

Damit wir uns darüber klar werden, wollen wir das *verfallende Sein-bei* mit der *Situation* vergleichen, in der das Dasein entschlossen sein Da und so auch die jeweilige faktische Bewandtnis der Umstände eigentlicherweise erschließt. Das *Verfallen*, das natürlich etwa mit "Nachtansicht" oder mit der "ontischen Aussage über die 'Verderbnis der menschlichen Natur'" (SZ 179) gar nichts zu tun hat, sondern durchaus "ein ontologischer Bewegungsbegriff" (SZ 180) ist, bedeutet, daß das Dasein "aus ihm selbst" "in die Bodenlosigkeit der Nichtigkeit der uneigentlichen Alltäglichkeit" "stürzt" (SZ 178) und in die Uneigentlichkeit "hineingewirbelt" ist (SZ 179). Dieses Verfallen enthält, wie früher schon angedeutet, zwei Seiten in sich; nämlich, es ist "ein ausgezeichnetes In-der-Welt-sein", "das von der 'Welt' *und* dem Mitdasein anderer im Man völlig benommen ist" (SZ 176, gesperrt vom Verfasser). Das Verfallen zeigt also an, daß das Dasein *einmal* "bei der besorgten 'Welt'" "aufgeht" und *dann* gleichzeitig damit "in die Öffentlichkeit des Man" "verloren" ist (SZ 175). Nun, Müller-Lauter hat einmal sehr scharf darauf hingewiesen, daß "das 'Verfallen an das Man' und das 'Aufgehen in der Welt' durchaus nicht wesenhaft zusammengehören", daß "hier nicht von einem, sondern von zwei Phänomenen die Rede sein muß".[6] Sicher scheint es, zwischen dem besorgenden Aufgehen in der wirklichen "Welt" und dem Schweben in der aufenthaltslosen Möglichkeit des Man einen scharfen Unterschied ziehen zu müssen. Jedoch haben diese zwei verschiedenen Phänomene trotzdem die eine Gemeinsamkeit. Sie besteht nämlich darin, daß das Dasein im Verfallen durchaus von dem *Un-eigenen benommen* ist, zu sich selbst als dem, was sich *zueigen* ist, was sich selbst *eigensterweise* gehört, nicht zurückkommt, und sozusagen sich selbst in das *Uneigene* bzw. *Uneigentliche* "entäußert" hat. Aufgrund der Sichselbstvergessenheit und Sichselbstverlorenheit könnte man doch erst in die "Welt" und in das Mitsein im Man eingehen. Wenigstens, wie uns scheint, liegt der Gedanke dieser Art der Bestimmung der verfallenden Uneigentlichkeit Heideggers zugrunde; sonst wäre doch die Verknüpfung beider Momente in ein und demselben Verfallen nicht denkbar. Natürlich erhebt sich hier eine Grundfrage, ob der eigenste Bereich des Daseins so strukturiert ist, daß er, eindeutig von dem uneigensten Bereich unterschieden, für sich allein als etwas Selbstständiges und Festes bestehen kann, oder ob es vielmehr dem Dasein nicht unvermeidlich sein müßte, stets im Uneigenen verstrickt zu werden, gewissermaßen "sichanders" zu werden, und so erst durch die Überwindung der "Entfremdung" zu dem wahren Selbst zurückkommen zu können. Doch, wir müssen vorläufig diese Frage beseite schieben und begnügen uns mit der Annahme, daß der Definition Heideggers des Verfallens ein Grundgedanke zugrunde liegen muß, der darauf hinausläuft, daß das *Eigene*, bzw. *Eigentliche* gerade in dem scharfen *Unterschied* und *Gegensatz* zu dem *Un-eigenen*, bzw. *Un-eigentlichen* zu suchen sein muß und demzufolge in der strengen *Absetzung* und *Distanzierung* von dem letzteren besteht.

In der Tat wird unter dieser Annahme allein der Sinn des folgenden Satzes Heideggers allererst verständlich: in der Entschlossenheit, die die dem Man wesenhaft verschlossene *Situation* erst erschließt, ist "das verstehende besorgende Sein zum Zuhandenen und das fürsorgende Mitsein mit den Anderen jetzt *aus deren eigensten Selbstseinkönnen heraus* bestimmt" (SZ 298, gesperrt vom Verfasser). In Unterschied zum Verfallen wird dem Dasein erst in der *Entschlossenheit* das *Eigene*, bzw. *Eigentliche* durchsichtig; das Dasein wird sich dessen jetzt voll selbstbewußt, und *aus diesem fest bewußt erwählten eigentlichen Selbstseinkönnen heraus* will es sich von nun an zu der "Welt" und den Anderen verhalten. Die Eigentlichkeit ist, wie schon erwähnt, die Seinsart, in der sich das Dasein aus seinem *eigensten* Seinkönnen her versteht, entwirft und so in der Welt zu existieren entschlossen ist. Dagegen meint die Uneigentlichkeit die Seinsart, in der es sich, von dem *Uneigenen benommen*, aus der "Welt" und dem Man versteht. Die Grenze zwischen der Eigentlichkeit und Uneigentlichkeit liegt einzig und allein darin, ob das Dasein aus dem selbstgewählten eigenen Selbstseinkönnen heraus existieren will oder nicht.

Aber das Problem ist, auf welche Weise dann das entschlossene Dasein gerade aus diesem eigentlichen Selbstseinkönnen heraus sich in der Situation zu der "Welt" und den Anderen verhält und es so zur vollkommenen Eigentlichkeit bringen kann. Denn: "Die Entschlossenheit bringt das Dasein gerade in das jeweilige besorgende Sein bei Zuhandenem und stößt es in das fürsorgende Mitsein mit den Anderen" (SZ 298); sie modifiziert "gleichursprünglich" die "Entdecktheit der 'Welt'" und "die Erschlossenheit des Mitdaseins der Anderen" (SZ 298). Aber wie? Heidegger antwortet: "*Die zuhandene 'Welt' wird nicht 'inhaltlich' eine andere, der Kreis der Anderen wird nicht ausgewechselt, und doch ist das verstehende besorgende Sein zum Zuhandenen und das fürsorgende Mitsein mit den Anderen jetzt aus deren eigenstem Selbstsein- können heraus bestimmt*" (SZ 297 f., gesperrt vom Verfasser). Alles kommt es darauf an, sich als besorgendes Sein-bei und fürsorgendes Mitsein "*primär auf sein eigenes Seinkönnen*" zu entwerfen (SZ 263, gesperrt vom Verfasser). Die "*Inhalte*" des Besorgens und der Fürsorge werden nicht "*anders*" und "*gewechselt*", sondern einzig und allein verwirk- licht sich die Eigentlichkeit nur dadurch, daß sich das Dasein "auf sein eigenstes Seinkönnen" hin und heraus "entwirft" und "bestimmt". Mit anderen Worten, nicht, daß die *Inhalte* des Entwurfs eigentlich werden, aber die *Weise* des Verhaltens, in der sich jetzt das entschlossene Dasein zu der "Welt" und den Anderen verhält, hat die *Modalität der Eigentlichkeit* gewonnen. Das heißt, daß jetzt zu der "Welt" und den Anderen gerade *aus dem selbstbewußt gewählten eigensten Selbstseinkönnens heraus* Stellung genommen wird, und nicht mehr aus "der Möglichkeit des Man-selbst" (SZ 263). Die Eigentlichkeit der Existenz besteht darin, daß die *Weise* des Entwurfs, *nicht seine Inhalte*, so geändert wird, daß er sich *ausschließlich aus dem eigensten Selbstseinkönnen her* bestimmt. Aber das Problem ist, auf welche Weise das Dasein nun *aufgrund dieses eigensten Selbstsein- könnens* weiter zu dem Zuhandenen und den Anderen Stellung nimmt; was für *neue eigentliche* möglichkeiten des Verhaltens zu diesen geöffnet wird, welcher *ontologischem* Weg der *eigentlichen* Verhaltens*weise* zu der "Welt" und den Anderen gebahnt wird. Ohne daß der *ontologische* Weg der *eigentlichen* Verhaltens*weise* zu ihnen zu finden wäre, würde doch die noch so feste Entschlossenheit, sich aus sich selbst heraus zu entwerfen, zu dem konkreten In-der-Welt-sein, das das Sein-bei und Mitsein in sich beschließt, gar nicht taugen.

Aber gerade in diesem Punkt bleiben Heideggers Darlegungen völlig in Dunkelheiten. Denn das besorgende Sein-bei dem Zuhandenen wird am Ende, wie oft hingewiesen ist, ausschließlich als durch die *uneigent-*

liche Zeitlichkeit konstituiert aufgewiesen, und sogar ist von der Zeitlichkeit der Fürsorge, die allein die ontologische Verhaltensweise zu den Anderen ermöglichen könnte, gar nicht das Geringste von Heidegger gesprochen worden. Sicher erwähnt er als den Grund der Möglichkeit des umsichtigen Besorgens so wie des theoretischen Entdeckens von *innerweltlichem Seiendem* die "horizontale Verfassung der ekstatischen Einheit der Zeitlichkeit" (SZ 365), auf deren Grunde die *Welt* erschlossen werden kann. Die Zeitlichkeit als ekstatische Einheit hat nämlich "so etwas wie einen Horizont"; zu jeden Ekstasen gehört ein "Wohin" der Entrückung, das "das horizontale Schema" genannt wird (SZ 365). Demgemäß wird zum Beispiel "das horizontale Schema der *Gegenwart*" durch das "*Um-zu*" bestimmt. Also ensprechend diesem horizontalen Charakter der ekstatischen Zeitlichkeit ist für das Dasein, solange sein Seinssinn in einer solchen Zeitlichkeit liegt, grundsätzlich die ontologische Möglichkeit garantiert, sich mit der durch Bedeutsamkeit ermöglichten Bewandtnisganzheit von Um-zu zu befassen. Aber dieser Horizont-charakter der Zeitlichkeit ist formal-indifferent. Wie das Dasein auf diesem horizontalen Schema eigentlicherweise das Besorgen und nicht zuletzt die Fürsorge gestalten soll, bleibt bei Heidegger absolut im Dunkel. Wenn dem so ist, müßte das Dasein, noch so eigentlich zu existieren entschlossen, doch, wenn es konkret mit dem Besorgen befaßt ist, wieder der uneigentlichen, Zeitlichkeit verfallen, und hinsichtlich der Fürsorge keine Ahnung haben können, wie es sich ontologisch-zeitlich verhalten soll. Das bedeutet, daß bei Heidegger schließlich *die neue ontologische Möglichkeit der eigentlichen Seinsweisen* des Daseins, in denens es sich zu der "Welt" und den Anderen verhält, keineswegs ernstlich bedacht und vollkommen dunkel dahingestellt gelassen worden ist.

Die gleichen Resultate lassen sich ferner aus der Betrachtung über die Zeitigung der Zeitlichkeit von "Sein-bei" folgern. Das "Sein-bei" ist in seinem zeitlichen Sinne, wie gesagt, formal-indifferent verstanden, die "Gegenwart" oder das "Begegnenlassen von" (SZ 326, 328). Die "Gegenwart" meint das "Begegnenlassen", sie soll aus dem "Gegenwärtigen" "des umweltlich Anwesenden" her verstanden werden (SZ 326). "Formal verstanden ist jede Gegenwart gegenwärtigend" (SZ 338). Nun, diese Gegenwart zeitigt sich in der vorlaufenden Entschlossenheit als "das *unverstellte* Begegnenlassen dessen, was sie handelnd ergreift" (SZ 326, gesperrt vom Verfasser). Dieses ist gerade die *eigentliche Gegenwart*. Diese aber wird genauer der "*Augenblick*" genannt (SZ 328, 338), weil sich das entschlossene Dasein gerade "aus dem

Verfallen zurückgeholt" hat, "um desto eigentlicher im 'Augen*blick*' auf die erschlossene Situation 'da' zu sein" (SZ 328). Das entschlossene Dasein *blickt* sozusagen *auf* die Situation mit den unverfälschten Augen und will *unverstellterweise* das umweltlich Seiende begegnenlassen. Die "*eigentliche Gegenwart*" "nennen ·wir *Augenblick*", sagt Heidegger (SZ 338). Nun, "im Unterschied von Augenblick als eigentlicher Gegenwart nennen wir die uneigentliche das *Gegenwärtigen*. Formal verstanden ist jede Gegenwart gegenwärtigend, aber nicht jede 'augenblicklich'. Wenn wir den Ausdruck Gegenwärtigen ohne Zusatz gebrauchen, ist immer das uneigentliche, augenblicklos-unentschlossene gemeint" (SZ 338). Dieses uneigentliche Gegenwärtigen ist "das *Gegenwärtigen*, in dem das *Verfallen* an das besorgte Zuhandene und Vorhandene *primär* gründet" (SZ 328). Das Verfallen "hat" in dem Gegenwärtigen "seinen existenzialen Sinn" (SZ 338). Im Gegensatz zu diesem verfallend-uneigentlichen Gegenwärtigen ist die eigentliche Zeitlichkeit des Augenblicks gerade "eine *Entgegenwärtigung* des Heute" (SZ 391, 397).

Übrigens fällt uns hier auf, daß das Wort "Gegenwärtigen" im doppelten Sinne gebraucht wird: einmal zur *formalen* Bezeichnung der Gegenwart, zum anderen als die Terminologie für die *uneigentliche* Gegenwart. Die erstere Sachlage ist wahrscheinlich dadurch verursacht, daß Heidegger bei seinem Versuch der Aufweisung der Zeitlichkeit an der vorlaufenden Entschlossenheit (SZ 326) schon das Wort "*Gegenwärtigen*" im formal-indifferenten Sinne von "Begegnenlassen von" gebrauchen muß, um nachher die dritte Ekstase der Zeitlichkeit überhaupt als die "*Gegenwart*" zu bezeichnen; er bezweckt im vornherein, die "Gegenwart" im Wortklang mit dem "Gegenwärtigen" abzuleiten. Aber diese an der vorlaufenden Entschlossenheit erfahrene *ursprüngliche und eigentliche Gegen*wart wird gleich darauf (SZ 328) wieder der *Augenblick* genannt. Warum er aber schon bei seiner ersten Aufzeigung der ursprünglichen und eigentlichen Zeitlichkeit an der vorlaufenden Entschlossenheit den Ausdruck "Augenblick" nicht gebraucht hat, läßt sich nur dadurch erklären, daß er beabsichtigt, die dritte Ekstase der "Gegenwart" dort im voraus in Anknüpfung an dem "Gegenwärtigen" klingen zu lassen. Abgesehen davon hat aber andererseits sein Vorgehen, den Ausdruck "Gegenwärtigen" auch als eine Bezeichnung für die *uneigentliche* Gegenwart zu benutzen, schwerwiegende Folgen nach sich gezogen. Denn dadurch ist der Anschein entstanden, als ob sich das formal-indifferente Gegenwärtigen dem Wortklang nach leicht mit dem uneigentlichen Gegenwärtigen zu verbinden neigte, als ob das Gegenwärtigen, d. h. Begegnenlassen eigens in Anlehnung an das uneigentliche Gegenwärtigen gedacht würde, als ob der Augenblick mit dem formal-indifferenten Gegenwärtigen kaum zu tun hätte. Aber scheint es nur so zu sein; in Wirklichkeit soll sowohl dem Augenblick wie auch dem uneigentlichen Gegenwärtigen das formal-indifferente Gegenwärtigen zugrunde liegen. Indessen ist die klare Abgrenzung dieser drei Begriffe gegeneinander dadurch erschwert; sogar verknüpft sich dieser Sprachgebrauch mit der Tatsache, daß das Sein-bei meistens bei Heidegger mit dem uneigentlichen Verfallen gleichgesetzt ist, und gibt dem Verdacht den Ansporn, daß bei ihm doch endlich der ontologischen Möglichkeit des augenblicklichen, d. h. eigentlichen Gegenwärtigens nicht so eine tiefgehende Rechnung getragen wurde.

Nach Heidegger ist die Gegenwart in der Entschlossenheit, d. h. in dem *Augenblick*, "aus der Zerstreuung in das nächste Besorgte" "zurückgeholt" (SZ 338), und in diesem Sinne "eine Entgegenwärtigung" (SZ 391, 397). Im Unterschied zu dem uneigentlichen Verstehen, das sich aus dem uneigentlichen Gegenwärtigen zeitigt, zerbricht sich im Augenblick die verfallende Angewiesenheit auf die "Welt" und das Man, und so zeitigt sich der Augenblick "aus der eigentlichen Zukunft" (SZ 338). Die Gegenwart ist aber darüber hinaus "in der Zukunft und Gewesenheit *gehalten*", "in der eigentlichen Zeitlichkeit gehalten", "in der Entschlossenheit *gehalten*" (SZ 338). Also, "das Gegenwärtigen, in dem das *Verfallen* an das besorgte Zuhandene und Vorhandene *primär* gründet, bleibt im Modus der ursprünglichen Zeitlichkeit *eingeschlossen* in Zukunft und Gewesenheit" (SZ 328). Das verfallende Gegenwärtigen nimmt daher noch nicht Überhand. Obgleich so *gehalten* und *eingeschlossen*, "*blickt*" doch das *augenblickliche* Dasein mit den unverfälschten "*Augen*" auf die Situation (SZ 328). So ist der Augenblick "die entschlossene, aber in der Entschlossenheit *gehaltene* Entrückung des Daseins an das, was in der Situation an besorgbaren Möglichkeiten, Umständen begegnet" (SZ 338). Auf diese Weise will das augenblickliche Dasein "das unverstellte Begegnenlassen dessen, was sie (d. h. die Entschlossenheit-ergänzt vom Verfasser) handelnd ergreift" (SZ 326), erzielen. Aber "die Ekstasen sind nicht einfach Entrückungen zu . . . Vielmehr gehört zur Ekstase ein 'Wohin' der Entrückung" (SZ 365). Das Wohin der Gegenwart ist der Bereich des "Um-zu" (SZ 365). Also, das entschlossene Dasein, das, um das Zuhandene in Bezügen von Um-zu "unverstellt" begegnen zu lassen, die Entrückung dazu durchführt, muß sich nun jetzt wirklich in die Um-zu-Bezüge hineinwagen. Was folgt aber darauf? Das Dasein muß dann wirklich das umweltlich Seiende "besorgen" und für die Anderen "fürsorgen". Indessen wird das Besorgen merkwürdigerweise durchaus durch die uneigentliche Zeitlichkeit (Gewärtigen-Behalten-Gegenwärtigen) konstituiert, und von der Zeitlichkeit der Fürsorge ist, wie gesagt, gar keine Rede. So bleibt doch immer noch völlig dunkel, wie das Dasein wirklich eigentlich mit der "Welt" und den Anderen befaßt sein könnte.

Was bedeutet denn überhaupt dieses merkwürdige Verhältnis zwischen Eigentlichkeit und Uneigentlichkeit? Das kann *entweder* bedeuten, daß Heideggers Ansicht nach das noch so eigentliche Dasein trotzdem, wenn es sich wirklich in das Besorgen und die Fürsorge einläßt, doch am letzten Ende stets etwa der uneigentlichen Zeitlichkeit verfallen müßte und also kein *vollkommen eigentliches* In-der-Welt-sein zu erzielen imstande wäre; dann aber würden seine wiederholten Versicherungen, das Dasein könne doch eigentlich existieren, unverständlich, sogar widersprächen sie dem Vorstehenden; in der Tat wird doch das ganze Werk der Herausstellung der Möglichkeit des eigentlichen Daseins gewidmet. *Oder aber* das bedeutet, daß das Dasein im Gegenteil irgendwie einst zu einem sich eigentlich zeitigenden Besorgen wie einer eigentlichen Fürsorge zu gelangen und die vollkommene Eigentlichkeit zu erreichen vermöchte; dann aber würde die Tatsache, daß Heidegger von dieser eigentlichen Zeitigung der Zeitlichkeit von dem vollkommen eigentlichen In-der-Welt-sein keineswegs spricht, total unverständlich. Beide Möglichkeiten der Interpretationen müssen also auf die unverständliche Inkonsequenz der Heideggerschen Darlegungen stoßen und daran scheitern. Aber wie wäre es, wenn seine Absicht dahingehend interpretiert werden könnte, daß das Dasein noch so eigentlich zu existieren entschlossen, doch immer wieder in die Uneigentlichkeit verstrickt werden müßte, daß es ständig erst durch dieses wiederholte Verfallen in die Uneigentlichkeit und durch die immer erneute Bemühung um die Wiederherstellung seiner Eigentlichkeit aus der Uneigentlichkeit das Ideal des eigentlichen Existierens erreichen könnte? Nicht, daß sich also die Eigentlichkeit unvermittelt direkt gewissermaßen mit einem Schlag, die Uneigentlichkeit völlig überwindend, verwirklichen läßt, noch, daß die Eigentlichkeit und Uneigentlichkeit bloß parallel zueinander im Gegensatz stehen, sondern wenn die wahre Existenz gerade darin bestünde, daß sich die Eigentlichkeit, die Uneigentlichkeit in sich enthaltend, in einem ständigen Fortgang befindet, von niedriger Eigentlichkeit zur höheren zu steigen, so daß das Dasein in dem stetigen dynamischen, ja sogar dialektischen Prozeß von Uneigentlichkeit zur Eigentlichkeit erst wirklich existieren könne?

In der Tat denkt Heidegger auch so. Die Eigentlichkeit ist, wie früher gesagt, "die Wahrheit der Existenz" (SZ 221); die Entschlossenheit ist "die ursprünglichste, weil *eigentliche* Wahrheit des Daseins" (SZ 297). Doch, *"das Dasein ist, weil wesenhaft verfallend, seiner Seinverfassung nach in der 'Unwahrheit'"* (SZ 222). Heidegger sagt daher: das Dasein "hält sich gleichursprünglich in der Wahrheit und Unwahrheit. Das gilt 'eigentlich' gerade von der Entschlossenheit als der eigentlichen Wahrheit. Sie eignet sich die Unwahrheit eigentlich zu. Das Dasein ist je schon und demnächst vielleicht wieder in der Unentschlossenheit" (SZ 298 f.). "Die Unentschlossenheit bleibt gleichwohl in Herrschaft" (SZ 299); "auch der Entschluß bleibt auf das Man und seine Welt angewiesen" (SZ 299). Wenn dem so ist, kann das Dasein mit seiner noch so festen Entschlossenheit doch nicht unmittelbar sein volles eigentliches In-der-Welt-sein fertig bringen, noch können die Eigentlichkeit und Uneigentlichkeit bloß gegensätzlich parallel nebeneinander stehen, sondern das Dasein existiert *gleichursprünglich, gleichermaßen* in der Eigentlichkeit und Uneigentlichkeit, d. h. in der Wahrheit und Unwahrheit der Existenz, derart zwar, daß dieses gleichursprüngliche Gleichgewicht nie so bewegungslos stillsteht, wie vielmehr sich in der dynamischen, ja dialektischen Bewegung befindet. Und gerade hierin wird der Heideggerschen Daseinsanalytik gegenüber, die diesen dialektischen Prozeß der Existenz, wenn auch unausdrücklich schon in den Blick behaltend, doch nie zu einer tiefgreifenden Problematik gedeihen ließ, eine mögliche immanente Kritik sich geltend machen können.

IV. SCHLUß

Das Dasein kann sicher eigentlich existieren; es kann sich gewiß aus dem Verfallen zurückholen und jetzt aus seinem eigensten Selbstseinkönnen heraus auf das eigentliche In-der-Welt-sein entwerfen. Aber die Eigentlichkeit der Existenz besteht keineswegs einfach darin, daß das Dasein auf diese Weise, sich zurückholend, gerade mit unverfälschten *Augen auf* die Situation *blick* und sich *entschließt*, sich eigentlich zu der "Welt" und den Anderen zu verhalten. Denn sobald es sich dann wirklich in das konkrete In-der-Welt-sein einläßt, ist es unvermeidlicherweise gezwungen, sich mit der *uneigentlichen* Zeitlichkeit des Besorgens zu

befassen und trotz der Möglichkeit einer eigentlichen vorspringend-
befreienden Fürsorge immer wieder in das ihm wesensmäßige
uneigentliche Verfallen abzustürzen. Daß Heidegger, obwohl sein Begriff
des Sein-bei ursprünglich als ein formal-indifferentes Seinsmoment
angesehen werden mag, doch niemals die *ontologische Möglichkeit der
eigentlichen Modifizierung dieses Momentes* deutlich angezeigt hat,
könnte wahrscheinlich als Anzeichen dafür genommen werden, daß nach
seiner Ansicht das Dasein doch im Grunde genommen ein vollkommen
eigentliches In-der-Welt-sein nicht einfach, nicht so ohne weiteres
unmittelbar, zu verwirklichen vermag, sondern daß es dem Dasein
wesentlich wäre, in die uneigentliche Seinsweise ständig verflochten und
darin verstrickt zu werden. Dasein existiert ja doch in der Wahrheit und
zugleich in der Unwahrheit; die Eigentlichkeit und Uneigentlichkeit sind
ihm gleichermaßen wesenhaft. Wenn dem aber so wäre, müßte das
eigentliche Wesen der Existenz, d. h. die Wahrheit der Existenz nicht
einfach darin gesucht werden können, daß sich das Dasein aus dem
Verfallen zurückholt und entschlossen auf die Situation *blickt*, sondern
gerade darin, daß dieser Versuch, eigentlich zu existieren, immer wieder in
die Uneigentlichkeit wesensmäßig hineingezogen wird, daß das Dasein
alsdann in diese Uneigentlichkeit gewissermaßen sich "entäußert" und
"entfremdet" und gerade durch dieses notwendige "Sichanderswerden"
hindurch sich als ein höheres eigentliches Selbst "wiederherzustellen"
bemüht. Die Eigentlichkeit muß vielmehr also darin bestehen, daß sie die
Uneigentlichkeit in sich enthält, durch sie hindurchgeht, und so in diesem
dialektischen Prozeß von Eigentlichkeit und Uneigentlichkeit erst das
werden kann, was sie ursprünglich war. Nach Heideggers Ansicht aber ist
die Eigentlichkeit so strukturiert, daß sie mehr oder weniger erst dadurch
entsteht, daß sich das Dasein das, was sich zueigen ist, eben, sich von dem
Un-eigenen absetzend, zuzueignen bemüht ist. Die Eigentlichkeit ist, wie
gesagt, die Seinsart, in der sich das Dasein einzig und allein aus seinem
eigensten Selbstseinkönnen heraus und darauf hin versteht und entwirft.
Aber wenn die Eigentlichkeit des Selbst als solche im wesentlichen gerade
nicht dadurch entstehen könnte, daß sie sich von dem Uneigenen
distanzierte, sondern dadurch, daß sie das Uneigene in sich übernähme
und die daraus notwendig ergbende Entzweiung des Selbst auf ein höheres
Selbst hin zu synthetisieren bemüht wäre? "Denn die notwendige Entzweiung ist ein Faktor des Lebens, das ewig entgegensetzend sich bildet,
und die Totalität ist in der höchsten Lebendigkeit nur durch
Wiederherstellung aus der höchsten Trennung möglich" (Hegel).[7] Die

"wahre" *eigentliche Eigentlichkeit* müßte dann in jenem "Ganzen" gefunden werden, in dem die Eigentlichkeit, durch die Uneigentlichkeit hindurchgehend, sich als das Ganze vollendet. Das Dasein kann nicht so formal-indifferent, *neben* der verfallenden uneigentlichen Seinsweise, eine *andere* danebenstehende Eigentlichkeit gerade in einem einfältigen Gegensatz zu der Uneigentlichkeit haben, sondern die Eigentlichkeit liegt eben darin, daß sie mit der Uneigentlichkeit verflochten ist, sich in einem diese Uneigentlichkeit überwindenden fortdauernden Entwicklungsprozeß befindet und durch diese dialektische Zickzackbewegung hindurch erst zu sich selbst als das wahre Selbst *werden* kann. Der Eigentlichkeit ist es daher in diesem Sinne unvermeidlich, sich gewissermaßen der Negation der *Un*eigentlichkeit zu unterziehen und sich als ein wahres Ganzes erst durch die Negation dieser Negation erreichen zu können. Das Dasein müßte also stets bereit sein, die negative Kraft der Uneigentlichkeit der "Welt" und der Anderen zu erfahren, und erst durch die Wiederherstellung daraus sich als ein höheres Selbst zu verwirklichen.

Heidegger hat gerade diesen dialektischen Werdegang des eigentlichen Selbst, wie uns scheint, leider nicht zu einem tiefbohrenden Problemthema gemacht, obwohl sich diese Thematik, wie wir bisher gesehen haben, aus seinen mit mehreren Schwierigkeiten und Inkonsequenzen einbegriffenen Darlegungen *notwendigerweise* als die wichtigste Problematik, die weitergedacht werden müßte, ergeben hat. Immerhin scheinen uns alle Schierigkeiten in seinen Darlegungen auf diesen Punkt zurückzuführen, und die Richtung, die wir in Bezug darauf angedeutet haben, würde wahrscheinlich den einzigen Weg bahnen, den Verwicklungen zu entgehen und den Grundgedanken Heideggers am konsequentesten zu interpretieren und weiter zu entwickeln. Das bedeutet aber natürlich nicht, daß Heideggers Versuch, vor allem das eigenste Selbstsein als geworfenes In-der-Welt-sein-können in seiner unermeßlichen Tiefe der Unheimlichkeit, ja sogar in seinem abgründigen Grunde ans Licht zu bringen, einfach außer Acht gelassen und unberücksichtigt werden darf. Ganz im Gegenteil, liegt sein hervorragendstes Verdienst gerade darin, daß er die brutale Faktizität des endlichen In-der-Welt-seins, seinen angstvollen nichtigen Abgrund, sein Sein zum Tode und nicht zuletzt das durchaus durch Nichtigkeit durchdrungene Schuldigsein des Daseins auf die unerhört eindringliche Weise zergliedert und herausgearbeitet hat. Trotzdem existiert das Dasein stets in seinem unheimlichen geworfenen Entwurf doch zugleich ständig als das Sein-bei und das Mitsein. Aufgrund seines abgründigen endlichen Da-seins existiert doch das Dasein, sich der

Negation des Uneigentlichen in der Welt unterziehend und sich daraus wiederherzustellen trachtend, im Mitsein und Sein-bei; hierin allein und dergestalt allein entfaltet sich die "Totalität" des "Lebens". Diese Totalität ist natürlich in ihrem Grunde dem Tod und der Nichtigkeit ausgesetzt. Aus diesem unheimlichen Grunde heraus aber existiert das Dasein, und zwar konkreterweise stets in der Dimension des Sein-bei und des Mitseins. Dieses "Ganze" des "Lebens", das sich auf dem Grunde der Nichtigkeit abspielt, das sich doch angesichts der Anderen und der "Welt" "ewig entgegensetzend" bildet, macht die Wahrheit der Erschlossenheit des menschlichen Daseins aus. Dieses Ganze erlebend in seinem Wesen einzusehen und aufgrund dieser Einsicht sein eigenes endliches Leben durchzuleben und zu runden, ist die Aufgabe des Menschen und der Philosophie.

Heidegger aber, wie uns scheint, liegt weniger daran, dieses Ganze gerade angesichts der Anderen und der "Welt" mit Einsicht durchdenkend und erlebend als ein lebendiges philosophisches Ganzes zu vollbringen, als vielmehr, besonders seit seiner Kehre, *auf der aus dem Verfallen gänzlich zurückgeholten eigentlichen Existenz stehend*, der verfallenen ganzen Welt gegenüber, mit der gerade in *Sein und Zeit* nicht eingehend verfolgten, doch wichtigen Problematik der "Rede" und "Sprache" vertiefterweise gerüstet, sein Seinsdenken zu radikalisieren. Dieses Denken hat eine ungeheure Tragweite und eine unerschöpfliche Tiefe. Aber abgesehen davon, scheint uns doch auf alle Fälle nicht verleugnet zu werden, daß schon in *Sein und Zeit* sich eine starke Tendenz unverkennbar durchblicken läßt, die eigentliche Existenz mehr oder weniger in jenem Punkt zu sehen, in dem *sie sich entschieden von der verfallenden Uneigentlichkeit zurückholt*, obwohl sie doch immer wieder als etwas Mitsein und Sein-bei in sich Enthaltendes erklärt ist. Diese Tendenz verbindet sich andererseits selbstverständlich mit Heideggers Grundposition in *Sein und Zeit*, daß das Dasein ausschließlich nicht ontisch, sondern ontologisch, nicht existenziell, sondern existenzial, und zwar im *Gegenzug zur verfallenden Auslegungstendenz der bisherigen philosophischen Kategorien,* in Hinblick auf die Existenzialien, vor allem hinsichtlich seines Seins und dessen Sinnes herausgestellt werden soll. Wegen dieser an der *Eigentlichkeit des Daseins* orientierten, im *Gegenzug zur verfallenden Tendenz* durchgeführten, phänomenologisch-hermeneutischen Analysen wurden sicher Existenzphänomene des Daseins auf eine unerhört eindringliche Weise herausgestellt. Wenn man sich aber zu sehr von der *verfallenden Auslegungstendenz* absetzen wollte, so daß das Wesen der eigentlichen *Existenz* eigens nur in seiner *scharfen Abgrenzung gegen* sowohl das zuhandene bzw. vorhandene innerweltlich Seiende wie auch das innerweltlich begegnende Mitdasein gesucht würde, würde es andererseits schwerwiegende ernste Folgen nach sich ziehen müssen. Einige Beispiele dafür sollen jetzt schließlich ganz kurz betrachtet werden.

Die Weltlichkeit der Welt, "'worin' ein faktisches Dasein als dieses 'lebt'" (SZ 65), ist gewiß in einer deutlichen *Abhebung gegen das* innerweltliche Seiende, das ihm erst in dieser existenten Welt begegnen kann, als Existenzial und die Bedingung der Möglichkeit der Entdecktheit von dem letzteren herausgearbeitet worden. Aber trotzdem ist das Dasein auf der anderen Seite "in seinem 'Geschick' verhaftet mit dem Sein" des innerweltlichen Seienden (SZ 256), "auf eine begegnende 'Welt' angewiesen" (SZ 87); "geschichtliche Welt ist faktisch nur als Welt des innerweltlichen Seienden" (SZ 389); das Sein dieses Seienden "steht nicht im Belieben des Dasein" (SZ 366); so kann "die konkrete Ausarbeitung der Weltstruktur überhaupt und ihrer möglichen Abwandlungen" erst nach der "Ontologie des möglichen innerweltlichen Seienden" in Angriff genommen werden (SZ 366). Wenn dem so ist, ist in die Weltlichkeit der Welt als "ein Existenzial" (SZ 64), das das Sein des innerweltlichen Seienden bedingt und ermöglicht, umgekehrt das Sein des innerweltlichen Seienden, besser gesagt, ihre Gebundenheit an den innerweltliche Seiende tief eingedrungen. Das bedeutet, daß es einseitig wäre, wenn die eigentliche Existenzweise *bloß in der Entgegensetzung gegen* das innerweltlich Zuhandene bzw. Vorhandene betrachtet würde. Vielmehr muß bei der Bestimmung der Existenz ihr Verhaftetsein mit dem innerweltlichen Seienden noch eingehender berücksichtigt werden.

Ferner bestimmt sich das Dasein, wie gesagt, als Mitsein mit den Anderen. Aber "Dasein kann nur dann *eigentlich es selbst* sein, wenn es sich von ihm selbst her dazu ermöglicht" (SZ 263). Freilich geht das eigentliche Dasein dadurch seines Mitseins keineswegs verlustig, aber das eigentliche Selbst kommt bei Heidegger einzig und allein dadurch zustande, daß das Dasein sich "*von ihm selbst her*" ermöglicht. Jedoch, gegen diesen Gedanken sind von vielen Philosophen eine Menge Einsprüche erhoben. Sie arbeiten gemeinsam alles in allem darauf hin, hervorzuheben, daß sich das Dasein nicht "*von sich selbst her*" allein zur Eigentlichkeit zu gelangen entschließen kann, sondern daß der Anruf auf die Eigentlichkeit in Wirklichkeit erst von den Anderen kommt, und daß das eigentliche Selbst in einem wesensmäßigen Bezug zu dem anderen Selbst steht, das "Du" heißt. Ohne "Du" kann das "Ich" nicht sein, heißt es.[8] Dieses Problem erreicht überdies seinen Höhepunkt in der Diskussion um das Gewissensphänomen. Nach Heidegger wird alle vulgäre Gewissensauslegung abgewiesen; der Gewissensruf kommt von dem Dasein selbst auf das Dasein selbst, und demgemäß entschließt sich das Dasein "*von sich selbst her*" zu seinem eigensten Schuldigseinkönnen. Aber es gibt dagegen eine Kritik, die behauptet, daß der Gewissensruf ohne Mitsein mit dem Anderen unmöglich sei.[9] Wir wollen aber hier nicht darauf eingehen, ob die Kritik dieser Sorte zutrifft oder nicht. Jedenfalls kommt die Tatsache, daß um das Problem des Mitseins eine Menge Widersprüche erhoben wurden, vermutlich daher, daß das eigentliche Selbst zu sehr in seiner Entgegensetzung gegen die "Anderen" gedacht ist, so daß eine Schwierigkeit empfunden worden ist, ein wirkliches Mitsein wiederherzustellen.

Mit dem Vorstehenden hängt weiter ein anderes Problem zusammen, daß von Heidegger zwar die Genesis der Historie aus der Geschichtlichkeit mit Recht ausgezeichneterweise von ihrem Grunde aus aufgezeigt wurde, daß aber auf das umgekehrte Verhältnis gar keine Rücksicht genommen ist. Gewiß ist die Historie erst aufgrund der Geschichtlichkeit möglich. Aber wenn Heidegger sagt, daß die eigentliche Geschichtlichkeit "nicht notwendig der Historie bedarf" (SZ 396), d. h. daß sie ohne historische Kenntnisse vor allem über die "Welt-Geschichte" (SZ 387 ff.) bestehen kann, dann erhebt sich Fragwürdigkeit. Die eigentliche Geschichtlichkeit, die der historischen Erkenntnisse entbehrte, könnte sich doch unter Umständen einer Gefahr verschreiben, sich blind zu benehmen. Wissen und Handeln müssen einander immer in Anspruch nehmen; Geschichtlichkeit und Historie müssen sich miteinander ergänzen. Dem oberen Heideggerschen Gedanken liegt schließlich seine Grundposition zugrunde, daß das Wesen des eigentlichen Selbst nachdrücklich in dem Punkt gesucht wird, in dem sich das Dasein aus dem Verfallen an das Man und die "Welt" zurückholt; denn Historie kann durch Heranziehung des Gesichtspunktes von Mitsein und "Welt" allererst in ihrer Breite und Tiefe zustande gebracht werden. Der Gedanke, daß die eigentliche Geschichtlichkeit nicht immer der Historie bedarf, hängt also eng damit zusammen, daß die Eigentlichkeit im Grunde genommen dadurch zustande kommt, daß sich das Dasein "*von sich selbst her*" zu sich selbst entschließt und dabei die Rücksicht auf die "Welt" und die Anderen nicht so eine große Rolle zu spielen vermag.

Aber wenn das Dasein nur vermittels des Eindringens in die "Welt" und die Anderen, und durch die Wiederherstellung aus der dort sich ergebenden Entfremdung, erst eigentlich es selbst werden könnte? Wenn die Wahrheit und die Eigentlichkeit in ihrem dialektischem Verflochtensein mit der Unwahrheit und der Uneigentlichkeit erst zu dem *werden* könnten, was sie ursprünglich sind? Sicher gähnen im Grunde des menschlichen Daseins unheimliche Abgründe von Tod und Nichtigkeit; die Endlichkeit des Daseins sollte nicht verdeckt werden. Trotzdem spielt sich die "Totalität des Lebens" als das wahre Ganze auf diesem abgründigen Grunde ab. Die Aufgabe der Philosophie besteht darin, durch dieses Ganze hindurchgehend den wahren Ort der "wahrhaften Unendlichkeit" ergreifen zu suchen.

Tokyo

ANMERKUNGEN

Ursprünglich wollte der Verfasser hier seine ausführlichen Anmerkungen sowie seine gründlichen Auseinandersetzungen mit wichtiger Heidegger-Literatur beifügen; aber er hat darauf verzichtet, weil er befürchtet hat, daß diese Abhandlung damit zu umfangreich sein würde.

[1] G. Misch, *Lebensphilosophie und Phänomenologie*, 2. Aufl., 1931, S.1.
[2] Die Abkürzung "SZ" bedeutet: M. Heidegger, *Sein und Zeit*, 7. Aufl., 1953. Die darauf folgende Zahl zeigt die Seitenzahl dieser Ausgabe, wo die Zitate zu finden sind.
[3] E. Husserl, *Ideen zu einer reinen Phänomenologie und phänomenologischen Philosophie*, Erstes Buch, in: Husserliana, Bd. III, 1950.
[4] E. Husserl, op. cit., S. 336.
[5] E. Husserl, op. cit., S. 335.
[6] Müller-Lauter, *Möglichkeit und Wirklichkeit bei Martin Heidegger*, 1960, S. 74.
[7] Hegel, *Differenz des Fichte'schen und Schelling'schen Systems der Philosophie*, Philosophische Bibliothek, Bd. 62a, 1962, S. 14.
[8] Vgl. M. Buber, *Die Schriften über das dialogische Prinzip*, 1954.
[9] Vgl. H. Kuhn, *Begegnung mit dem Sein*, 1954.

PART II

PHENOMENOLOGY
IN THE JAPANESE INHERITANCE

TADASHI OGAWA

THE KYOTO SCHOOL OF PHILOSOPHY AND PHENOMENOLOGY*

INTRODUCTION

The philosophers of the so-called "Kyota School" can be compared with a solar system, the center of which is Kitarō Nishida (1870–1945). Nishida was called to a professorship at the Imperial University at Kyoto in 1911. Through his teaching and his philosophical writings he played such a preeminent role in Japan's academic philosophy that many talented philosophers gathered around him. Nishida is almost the only philosopher in Japan who since the Meiji Restoration (1868) built up an independent and original philosophy. His philosophy influenced his students, e.g., H. Tanabe, Baron Kuki, G. Miyake, K. Nishitani, etc.[1] Therefore this essay treats primarily Nishida's philosophy and its relationship to Husserl's phenomenology.

1. HUSSERL AND NISHIDA

Nishida, in the Far East, steeped himself, with his strong desire for knowledge, in the European and American philosophy of his time. In Japan he was the first to recognize the important meaning of the philosophers of that time, e.g., Bergson, Husserl, Meinong, Brentano and James. His sympathy with James' philosophy is to be found already in his diaries before the emergence of his first work. He retained a strong affinity to Bergson all his life.

Nishida was born in 1870 in Kanazawa. He experienced the rise of the Japanese empire and died in 1945 at the same time as the fall of the empire. Nishida is eleven years younger than Husserl (1859–1938). Husserl's first work, *Logische Untersuchungen*, appeared at the turn of the century (1900–1901). Nishida's first major work appeared in the year 1911, entitled *A Study of Good*. This work already evinces his characteristic thought. Nishida's main work therefore appeared eleven years later. Nishida was the first one in Japan to present Husserl's phenomenology in its connection with Neo-Kantianism and with the Brentano school.[2] Another similarity between Husserl and Nishida is their life-long

Nitta/Tatematsu (eds.), Analecta Husserliana, Vol. VIII, 207–221. All Rights Reserved.
Copyright © 1978 by D. Reidel Publishing Company, Dordrecht, Holland.

interest in logic and the bases of mathematics. It is well known that Husserl began his studies and his academic career as a mathematician and that his dissertation and his thesis for habilitation dealt with mathematics, specifically the foundations of mathematics. On the other hand, Nishida had some hesitation when he had to choose his professional field. He chose philosophy because the problem of life was of greatest interest to him. But he retained his interest in the bases of mathematics throughout his life, as the countless letters from Nishida to Professor Shimomura show. Shimomura was a student of Nishida who dedicated himself to the bases of mathematics.

There is also a similarity in their mode of thinking. Professor Yamauchi, his eminent student, called Nishida "a man who thinks as he writes."[3] On the other hand it was once said of Husserl: "Husserl: ne pensait qu'en écrivant."[4] Both philosophers dedicated their lives to the analysis of "things themselves." It is no exaggeration to say that they died thinking and writing. Nishida related himself to Husserl's phenomenology in a somewhat different manner than the philosophers of the Kyoto School did, such as Tanabe, Yamauchi, Miyake, Nishitani and others. Because Nishida was the founder of the school and a contemporary of Husserl, Bergson and James, he did consider phenomenology to be valuable, but he maintained a critical distance from it. In addition it should be mentioned that Nishida practiced Zen meditation in the temple. He built up his system of philosophy on the experience of Zen meditation. Thus Professor Dr. Keiji Nishitani says, rightly: "Nishida's philosophy, owes its originality to the East Asian spiritual tradition, that is, to Buddhism."[5]

II. THE QUESTION AS TO THE ESSENCE OF CONSCIOUSNESS

1. *Pure Experience and True Being (Jitsuzai)*

On February 18, 1927, Nishida wrote to his student Risaku Mutai, who was then studying with Husserl in Freiburg in Breisgau: "Phenomenology does not yet think through the basic problem: what consciousness is. The principle according to which variations of consciousness are to be distinguished and determined has not been clearly and distinctly worked out."[6] Here Nishida poses the question of how consciousness is to be determined to be consciousness, that is, the question as to the essence of consciousness. He wrote in a letter to Professor Miyake:

... the descriptive way of dealing with reality, as phenomenologists do it, is inadequate. Experience in phenomenology does not overcome subjectivity. ... Phenomenology begins with consciousness, basing itself on the modern subjectivism since Descartes. We, however, must first of all think through the Self, that is, consciousness as such.[7]

Nishida was dissatisfied with phenomenology on the following points:

a. Its subjectivism, that is, the circumstance that consciousness is regarded only as an individual existing entity. In other words, experience always belongs to an individual person.

b. The difference between the subject and the object is presupposed as already self-evident.

c. Comprehension, feeling, and will are considered to be only different levels.[8]

In actuality the question is put in such a way that phenomenology does define the essence of consciousness as consciousness of something, that is, as directedness toward something, but that it does not go on to inquire into the essence of intentionality. It seemed to Nishida that insofar as the essence, that is the basis, of consciousness is not discussed, phenomenology is not adequately established.

Nishida's first major work was entitled *A Study of Good* because the problem of life, namely practical philosophy, was for him the most important. He said in the foreword: "Although the first half of the book is devoted to purely theoretical investigations, the problem of life is the center of the final purpose."[9] In the diary he kept before the publication of the book he expressed what was his intention in his system of philosophy.[10] He meant to make his decision for research into life. On this point he radicalized the question as to the essence of consciousness and found a way out of solipsism.

The *A Study of Good* consists of four studies: (1) on pure experience, (2) on real being, (3) the good, and (4) religion. The concept of "pure experience" also appears in William James. But with Nishida this concept is independent of the historical connection.

Pure experience is immediate consciousness, thus a psychological concept insofar as it is a kind of consciousness. Yet it is never a subjectivistic concept.

If one directs one's attention to the sequence in which this work originated, the second part, "Real Being" (Jitsuzai) originated first. Therefore the first section of the second part is entitled "Starting Point of the Investigation."[11]

According to Nishida's conception philosophy is investigation into truth. There is only one truth. In this conception of philosophy Nishida is linked with Occidental philosophy. Investigation into true being begins with doubting everything that one can doubt. Philosophy must begin with the original immediate consciousness that one cannot doubt. By means of doubting the traditional two theories Nishida dismisses the dualism that is based on the difference between the mind and the thing. One can call into question both realism and idealism. Realism is questionable because it thinks that things in the external world are independent and separated from consciousness. Idealism is questionable because it thinks that behind consciousness there is a spirit/mind that accomplishes acts. Both theories correspond merely to the assumption that common sense claims "Immediate consciousness, which one cannot doubt" – that is the fact of intuitive experience, namely knowledge of the conscious phenomenon. But the conscious phenomenon is an essence *before* the difference between subject and object, thus before the realm of the subject-object relation. Therefore, Nishida says: "The consciousness appearance-phenomenon of what is present and the becoming-consciousness of what is present are simply the same. It is not possible, therein, to distinguish between subject and object."[12] According to Nishida direct consciousness itself, without separation of the seer seeing and the seen, is the unconditionally indubitable thing with which philosophy must begin. Common sense thinks that the essence of things that have a certain nature is independent and transcendent to and outside of consciousness. But it is impossible to phantasize about the nature of the thing outside of or separated from consciousness. He says: "It is impossible to consider the thing itself to be independent of our consciousness. The same applies also to the soul. What we can know is only the act of knowing, of feeling and of will, *never* the soul itself" (*al. loc.*). The so-called external thing, that is the thing-phenomenon, is not transcendent and independent of our consciousness. The immediate primordial fact is the consciousness-phenomenon, never the thing-phenomenon. Our body is also a part of our consciousness-phenomenon. Consciousness is not in the body, but vice versa, the body is in the consciousness-phenomenon. Nishida criticizes the dualistic way of thinking that wants to consider the consciousness-phenomenon and the thing-phenomenon as equivalent. Experience teaches us, on that subject, that there is only the consciousness-phenomenon. The thing-phenomenon is only that aspect of the consciousness-phenomenon that is common to everyone and unchanged. According to Nishida the conscousness-phenomenon does not consist only in the soul which is separated from the thing.

He says: "True being, strictly observed, cannot be called either consciousness-phenomenon or thing-phenomenon."[13] Compared to the principle of
Berkeley's philosophy, direct and real being is not passive, but an
independent activity. *Esse est actio.* Direct and real being is an activity, an
action itself. But then is not Nishida's philosophy a kind of phenomenalism? By no means. Because phenomenalism maintains that objects
of experience and comprehension are only the appearance of things in
contrast to the things themselves. But Nishida maintains that only the consciousness-phenomenon is true being and that the consciousness-phenomenon develops itself as true being. Regarded in rough outline, for
Nishida, Berkeley's and Hegel's philosophy are united.

2. *Pure Experience*

The first part of Nishida's first major work is entitled "pure experience."
The first part was written later than the second and third parts. After
Nishida discusses real being and the good, he has to express his basic
experience thoroughly and clearly. The first part is meant to lead
altogether simply into Nishida's thought.

Why did European philosophy basically want to transcend the realm of
experience, although it proceeds from experience? For the concept of
experience in Europe includes the following three assumptions:

a. Knowing, feeling and will are distinguished from the beginning from
each other in their classification.

b. The difference between subject and object is firmly retained.

c. The bearer of experience is the individual person. Experience is
always made by the individual person.

Nishitani asserts: "It is thought that experience belongs to
consciousness and that consciousness consists of the relation between the
thing and the soul, that is, between the object and the subject. The subject
of experience, it is said, cannot transcend its frame as an individual being
entity."[14] Contrary to the European way of thinking, Nishida thinks that
one need not leave the realm of experience in order to construct a
metaphysics. Vice versa, one can deepen experience.[15] In willing, living
and comprehending, we achieve it by means of experience.

What, then, is pure experience? According to Nishida, experience
means to know the fact, after becoming aware of it and putting prejudices
aside. "Pure" means the state of experience *per se*, without adding thought
or judgment. Thus pure experience is prepredicative experience. Nishida
says:

Pure experience and immediate consciousness are the same. If the state of consciousness is experienced directly, *per se*, there does not yet arise a differentiation between the subject and the object. Knowing is united with its object.[16]

When pure consciousness is viewed in the dimension of time, it is called present-consciousness. Consciousness of otherness cannot be experienced; said in another way, consciousness of otherness does not become a self-datum. Whether it is consciousness of self, or recollection of the past, that is presentifying of the past: when one makes a judgment about it, then that which is produced by the judgment is no longer pure experience. At the moment of judging, the unity of pure consciousness is interrupted. On the other hand, the interruption of this unity is the self-differentiating evolution of pure experience. The differentiation consists in the higher unifying, which is also produced by the opposition. Pure experience is present-consciousness *per se*. In thought about what has been, that is, in presentifying what has been, it is present-consciousness which presentifies what has been. In judgment, the unity of present-consciousness, that is, of pure experience, is destroyed. It splits into presentifying consciousness and the presentification of what has been.

Pure experience is prior to the relation of consciousness to its object. According to Nishida it is viewed as something prior to directedness-toward-something. It is the original experience *per se*, on which intentionality is grounded.

Nishida carries out the reflection[17] on the nature of pure experience. Although pure experience is implicitly complicated, in the moment of perception it is always a simple fact. When past consciousness gains a new meaning in present consciousness, on the one hand the experience of the past consciousness is not the same as past consciousness, even when it is only a repetition of past consciousness. That which is analyzed by present consciousness is still not the same as present consciousness.

The synthesis of present consciousness extends itself as far as attention reaches. The state of pure experience is by no means a single attentiveness. For example, when a mountain climber scales a cliff or a pianist plays his favorite piece with masterly perfection, a series of movements of attention are shown. In each case the attention is always turned to the thing. In the character of attention there is no gap through which thinking could enter. The sequence of attentions shows a unifying progression prior to the difference of subject and object.

Nishida grounds the purity and immediacy of pure experience on the fact of the unity of concrete consciousness. He says: "Consciousness actually forms a system."[18] Consciousness, regarded *per se*, is a chaotic unity out of which various manifold states of consciousness differentiate themselves. Let us once more call to mind an image. For example, when I perceive the whole of a given image with a movement of my eyes, the visual perception is not one-pointed. I see the image in the course of the attentive visual perceptions. The whole image appears implicitly already at the beginning. The whole image becomes explicit by means of the whole sequence of attentive seeing. Nishida interprets this as a systematic evolution of consciousness. He says, "Actual consciousness is a systematic development. While consciousness develops itself from itself with strict unity, we do not lose pure experience."[19]

Even in the case of external perception, attention is not directed by the external thing, but by an unconscious unifying power of consciousness. Attention, turned toward the external thing, is actually directed by the unity of consciousness. By means of the concept "unity of consciousness" (unifying power of consciousness), Nishida designates consciousness as a system that develops itself out of itself. Thus thinking overcomes psychologism, or element psychology. For, with Nishida, consciousness is, as a unified system, true being. For Nishida consciousness is an ontological concept. There is neither inward nor outward experience.

Only pure experience is true being. As already stated, the purity of pure experience consists in the unity of consciousness, that is, of experience. When a moment of pure experience distances itself from the unity of the present and relates itself to something else, it is no longer experience of the present, but a meaning (sense). Insofar as the unity rules over experience, the whole experience is real being. The act of uniting is actually that of willing. Nishida says: "The essence of the will is . . . the act of the present in the present."[20] Pure experience is the contemplation of the fact, therefore meaningless. Various meanings and judgments are produced by the self-differentiation of experience itself. The self-differentiation of experience is accomplished by the process in which pure experience develops itself into a meaning (sense), that is, a judgment, but we cannot be distanced from pure experience. Therefore meaning-formation, as judgment, is based on relating present-consciousness to past-consciousness. But how is it possible? By the present-consciousness' becoming integrated into the large system of consciousness and united with it.[21]

First I should like to summarize what has been presented above. Pure experience shows itself as present-conscousness. In the present it is the unity of consciousness, which, as an attentive gaze, is itself the focal point of consciousness. Meaning (sense) or judgment results from present-conscousness' distancing itself from the unity of itself. Thereby present-consciousness becomes a past consciousness, that is, a represented consciousness. Thence a difference between subject and object comes to expression. Consciousness estranged (alienated) from itself is united in the greater consciousness by means of now-consciousness (pure experience shows itself as present-consciousness. In the present it is the unifying. Therefore meaning (sense) or judgment is based on the interruption of the unity of consciousness. Yet the interruption is produced by the opposition between present-consciousness and past-consciousness. Nishida says:

Pure experience and its meaning (sense) or judgment express the bilaterality of the same consciousness. Which side appears, depends on the position taken toward the same consciousness. On the one hand consciousness must be unified; yet on the other hand it must differentiate itself. Consciousness relates itself implicitly to the Other. The present must always be viewed as a part of the greater system.[22]

3. The Self-differentiation of True Being

Thinking is first of all an act that brings a relationship to unification. In judgment the fact of the original pure experience is analyzed and developed into the subject and the predicate. Judgment is in the deepest sense the self-division of the primordial (originary), of true being, namely. Behind every judgment lies the experience of that which it unifies. According to Nishida, in the background of pure logic judgment there is intuitive pure experience. For example: the judgment: "The horse is galloping" is based on the pure experience in which the galloping horse as such appears. In Nishida's philosophy a basic experience is "that only when we have surrendered ourselves to the object of thought and have lost our self in it, do we see thinking accomplished."[23] He continues, "that thinking develops itself out of itself" (al. loc.). Therefore it is Nishida's basic view of truth when he says: "Truth (= the true law according to Buddhism) will emerge by means of our always losing our subjective self and becoming something objective."[24]

The self-differentiation of true being is formulated as follows:

The whole appears at first implicitly. Its content differentiates itself and presents itself. When this self-differentiating of true being is accomplished, the whole of true being is actualized and perfected (completed). In short, the One develops itself out of itself.[25]

For Nishida, being is effecting (realizing, causing to happen). The movement of self-development of true being is grasped dialectically. The meaning (sense) or judgment of pure experience estranges itself from pure experience. In this estrangement pure experience relates itself to itself as if to an Other. The relationship of pure experience to the Other unites itself in the third thing, namely in the greater consciousness. The dialectically grasped development if real being is constituted by Nishida in the following regions: (1) nature, (2) spirit/mind, (3) God.

The unifying act, or "the unifying self," is found in nature.[26] The act is the unifying act of our consciousness itself.[27] Pure being is the unity of subject and object. Nature is a concrete fact of the unity of subject and object.[28]

Spirit/mind is the unifying act of pure being and of its workings. Our Self is the inner unity in us. The Self is precisely the unifying act of pure being.[29]

The distinguishing characteristic of Nishida's thought consists of the following points: (1) the basic experience, namely pure experience, is grasped metaphysically as true being. (2) The development of true being (= pure experience) is regarded dialectically. (3) With Nishida, the ground (basis) of the world and my ground are the same, that is, the Self. Phenomenologists may assert, in objection to Nishida's thought, that he was metaphysically oriented from the beginning. Many objections may be raised against the presupposed dialectical method and his assertion about the Self in us and in the world.

But if one chooses to evaluate Nishida's posing of the question from a positive orientation, one must remember that Husserl viewed the intentionality of consciousness as unarguable. In the intentional-analyses the basic idea: "consciousness of something" is presupposed as self-understood. As has already been mentioned, Nishida found Husserl's subjectivism inadequate. In this connection, what Professor Rombach (Würzburg) said about intentionality is very understandable. He formulates intentionality as follows: "No object is thinkable without its subject; to every subject there belongs a corresponding object. Both are strictly related to each other."[30] The relationship of the two is shown as intentionality (directedness-to). But is it not necessary to inquire into the scope of the relation? That is just what Nishida has done. The question as to the essence of consciousness is concerned with the scope of the relation. For the essence of consciousness is what rules over consciousness. The scope of the relation is precisely that upon which the intentionality of consciousness is based. With Nishida the question as to the essence of consciousness is – as a question as to the scope of the relation – what is placed between the subject and the object. He inquires into and seeks the basis of the relation of the two as such, which precisely as a relation is neither subject nor object. With Nishida the relation as such is grasped as pure experience which is the immediate consciousness of the unity of subject and object. Nishida intended to constitute the various modes of being of pure experience in the areas: nature, spirit/mind, and God, as the unity of true being. When Rombach separates the basic intention of Heidegger's phenomenology from that of Husserl, then Nishida's question as to the essence of intentionality has something in common with Heidegger's criticism of Husserl. Rombach writes: "If intentionality is that relationship in which a subject as 'cognition-I' confronts an object as 'cognition-thing,' then intentionality itself is thereby a 'founded mode of Being-in-the-World'."[31] What the two have in common objectively is the question as to the mode of being of directedness to the object, thus of the relation between the subject and the object.[32] The question as to the mode of being of intentionality was carried out, with Nishida, as the self-differentiation of pure experience (of true being), and with Heidegger as fundamental-ontology.

III. THE PROBLEMATICS OF INTERSUBJECTIVITY

As is known, the problem of intersubjectivity and of the constitution of the Other is today still a gap in Husserl's phenomenology. Husserl's sketch of the constitution of the Other attempts to constitute as a psychophysical unity an alter ego out of the real corporeal appearance (phenomenon). With Husserl of course it depends on the egological attitude.[33] It seems to me that Husserl's egological posing of the question[34] made it very difficult to constitute, or understand, the Other-consciousness. Also the concept of empathy does not seem well grounded.[35] I have no room here to present thoroughly Husserl's theory of the constitution of the Other. What is the basis for the fact that the question as to Other-consciousness is not as sharply posed with Nishida as with Husserl, or perhaps is the problem very skillfully avoided?

Pure experience is immediate consciousness, the purity of which consists of the condition of unity of the subject and the object. Precisely that is present-consciousness, and experience of progression which is connected with it. If the present-consciousness is related to the Other, the unity of pure experience will be destroyed and at the same time the meaning (sense) or judgement arises.

The Other vis-à-vis pure experience is, according to Nishida, either past consciousness or consciousness of the foreign (the other).[36] For (1) only the consciousness-appearance (-phenomenon) is for Nishida real being. Therefore behind the consciousness-appearance one cannot assume a substratum of consciousness, namely the I or something similar. What we can know is only our act of knowing, feeling and willing. No substantial soul may be erected as a hypothesis behind the consciousness-appearance. The act is accessible to us only by means of the factual function.[37] (2) The fact that consciousness must belong to someone, indicates nothing more than the fact that consciousness must have a unity. Nishida says, about this: "Although in an individual consciousness the consciousness of Today and that of Yesterday are dependent on one another, because of their belonging to the same system they are the same consciousness. One's own consciousness is related to Other-consciousness as the consciousness of Today to that of Yesterday."[38] The difference between the consciousness of Today and that of Yesterday arises from the fact that, alienated from pure experience, one wants to assume that consciousness appears under the form of time. But already prior to the form of time there is the unifying act of consciousness, because the difference between the consciousness of Today and that of Yesterday is based on the unity underlying them.[39]

The act of consciousness, which produces the unity between the two consciousnesses, is not conditioned by time; rather, vice versa, time comes into being by means of the act of its unity-producing.[40] Both the consciousness of Yesterday and Other-consciousness are united with present-consciousness or pure experience and become a true being because they belong to the same system. The act of consciousness that produces unity is accomplished by means of the development of true being. But what is the uniting entity in the production of unity? The Logos.[41] Nishida writes: "The Logos is the unifying power of all things. It is at the same time the unifying power of consciousness. It brings things and consciousness into being."[42] Therefore the Logos is that which unifies the consciousness of Today and that of Yesterday, one's own consciousness and Other-consciousness, into one great system of consciousness. At the basis of our consciousness there is something General, by means of which we can understand each other. Nishida was able to overcome solipsism also by his interpretation of the General in our consciousness as Logos. Pure experience (true being) is, as already stated, prior to the individual I. That which differentiates itself from pure experience carries the meaning (sense) of past consciousness, or Other-consciousness. Other-consciousness and present-consciousness belong together in the Logos, namely the General. In the theory of the Self Nishida carries out the grounding of inter-subjectivity. "Our Self, which unites spirit/mind, is originally the unifying act of real being."[43] The true Self is the unifying act of real being and, at the same time, of our consciousness, namely, of immediate experience. This Self is the focal point of consciousness and of the world. "The great Self in us includes in itself the other person and me."[44] In short, the unity of true being (of pure experience) is grounded in God. God is the Ground of the world. Consciousness-appearance is the only true-being, which in the focal point of our individual consciousness-appearances is the same as the unity of true being produced by God. The great Self is the same as the great Consciousness. That is the Self, which is found in the Ground of nature and of the mind/spirit.[45]

With Nishida the objectivity of the external world is grounded on *intersubjectivity*.[46] He says: "The so-called objective world outside us is also nothing else than a consciousness-appearance produced by a unifying act. The universality of appearances, which forms a unity among the individual consciousness, manifests itself as the objective world, independent of us. If a lamp here shines only for me, it is certainly a subjective illusion. By means of the fact that everyone here confirms the lamp, it becomes an objective fact."[47] Nishida actually had no subjective egological attitude, such as is found with Husserl. Pure experience is neither subjective nor objective; it is a basic experience prior to the object and the subject. In the basic experience there is no difference between "outer" and "inner." "Outer" means here transcendent to consciousness, independent of it. "Inner" means within the individual consciousness, immanent to it. The immediate, pure present-consciousness is joined together with the other consciousness (whether it be Other-consciousness or whether it be past consciousness) to make up the great consciousness. The relation of present-consciousness to the other is produced by the uniting act of the great consciousness, which itself grounds this relation. The other consciousness is, to be sure, a meaning (sense) alienated from pure experience, but the meaning (sense) is constituted by the great consciousness. But with Husserl, it has always remained in the dark how the other consciousness as meaning (sense) of pure experience is constituted. The How is not grounded sufficiently with the concept of the unifying act of true being. The question as to the How of the unifying of true being must be posed.

From the beginning, Nishida did not see in intentionality the grounding essence of consciousness. Pure experience and its meaning, or judgment, are the two sides of consciousness. The meaning of the other consciousness results, according to Nishida, from the development of pure experience. Pure experience exists as true being prior to the subject-object relation.

The analysis, or self-differentiation of true being is grasped as thought, or as meaning- (sense-) formation. Other-consciousness, as a part of true being, or as the meaning of pure experience, is united with the great consciousness.

IV. CONCLUSION

The difference between the two thinkers can first of all be summarized in the following points: (1) Nishida lacks the consciously formed methodology, such as is found in Husserl's phenomenological reduction. Therefore pure experience is assumed from the beginning as a metaphysical essence.[48] (2) Intentionality (consciousness of something) is for Husserl the only interpretation of consciousness. With Husserl the principle of intentionality is assumed as indisputable. Reflection on the act of consciousness with Husserl, therefore, cannot escape the cleavage of the I between the accomplishing-I of reflection and the reflected I. With Nishida, on the other hand, consciousness as such is true being. Attention, perception and thought are explained by the unity of true being. For Nishida, what is more important than intentionality (consciousness of something) is consciousness of the Self. What consciousness is, would not be grounded, if there were not true being.

NOTES

*Translated from the German by Barbara Haupt Mohr.

[1] In the works of Kuki and Nishitani one finds more of the influences of Heidegger's hermeneutic phenomenology than that of Husserl. Cf. Heidegger: *Unterwegs Zur Sprache* [On the Way to Language], p. 85, fourth ed. 1971, Neske Pfullingen. (Heidegger was mistaken there. Kuki was not a count, but a baron.) Baron Kuki, already in 1933, elucidated Heidegger's philosophy with admirable exactness. He also explicated 'Die Struktur von "Iki"' (1930) and 'Das Problem der Zufälligkeit' (1935). Professor Nishitani develops a philosophy of religion. In his main work, *Was ist die Religion?* (1961) he attempts to confront Heidegger's philosophy in the process of elucidating the essence of religion.

[2] 'On the Position of the Purely Logical School in Epistemology' (1911); 'The Idealistic Philosophy of the Present' (1917).

[3] *Nishida Kitarō - Documents of Contempories*, Iwanami (Tokyo), 1974, p. 5. Professor Yamauchi's main works: *Introduction to Phenomenology* (1929), *Metaphysics of Meaning* (1965).

[4] Robberechts *Husserl*, Editions Universitaires, Paris, 1964, p. 121.

[5] Introduction to the book *Nishida Kitarō* (Gendai Nihon Shiso Taikei, Vol. 22, publ. Chikuma-Shobo, Tokyo, 1968, which is edited and introduced by Professor Nishitani. P. 33.

I quote Nishida's first work from the text edited by Professor Nishitani. Other writings are quoted according to the second edition of the complete works. (Iwanami, Tokyo, 1965–66). In the following essay I often quote Nishida's text, out of consideration for readers in other countries.

[6] *Complete Works,* Vol. 18, p. 324.

[7] 5.2.1935, *Complete Works,* Vol. 18, p. 512.

[8] Cf. *Ideas I,* Paragraph 116.

[9] *A Study of Good,* ed. Professor Nishitani, p. 120.

[10] Nishida wrote in his journal (19. 7. 1905): "I am not a psychologist, nor a sociologist. I am becoming a researcher of life." (Quoted by Shimomura in his book: *Wakaki Nishida Sensei –* translation: My Teacher Nishida In His Youth, Tokyo, 1947, p. 54.)

[11] Cf. the foreword of the author, p. 119. The first part, 'Pure Experience,' is, it seems to me, the introduction to his whole philosophy.

[12] Ibid., p. 149.

[13] Ibid., p. 153.

[14] Ibid., Introduction, p. 21.

[15] Readers of the late Schelling will think of his metaphysical empiricism. cf. Schelling, *Complete Works,* Vol. XIII, pp. 112—132.

[16] Ibid., p. 122.

[17] Nishida does not take into consideration the justification of reflection.

[18] Ibid., p. 124.

[19] Ibid., p. 124.

[20] Ibid., p 125. "Thought is also a form of apperception, like will. Its unity, however, is subjective. Will is, rather, unity of subject and object." Ibid., p. 126.

[21] "Meaning (sense) or judgment designates the relation of present-consciousness to the other, namely the position of present-consciousness in the system of great consciousness." Ibid., pp. 126-7.

[22] Ibid., p. 127. Here one finds the same idea as in Husserl's concept 'horizon.' cf. *Husserliana,* Vol. I., p. 82: "Horizons are pre-indicated potentialities."

[23] Ibid., p. 130.

[24] Ibid., p. 179.

[25] Ibid., p. 159.

[26] Ibid., p. 174.

[27] Cf. *A Study of Good,* p. 175.

[28] "In short, the subjective unity in us and the objective unity in nature are the same. When this Sameness is observed objectively, it is the unifying power of nature. When it is viewed subjectively, it is the unity of knowing, feeling, and willing in our Self." Ibid., p. 175.

[29] "There is no unifying act without bound entities. No subjective spirit without objective nature." Ibid., p. 179.

[30] Rombach, H., *Gegenwart der Philosophie,* Alber, Freiburg in Breisgau, 1962, p. 69.

[31] Rombach, H., op. cit., p. 17.

[32] "Intentionality indicates the directedness of the subject to an object and the being-directed of the object for a subject. But not every position that the subject can take or reject is called intentionality, but rather, only those that make up the essence of the subject itself. At the same time, however, it constitutes the essence of the object, for this is ... in each case a manner of comprehensibility of that which is. Intentionality is therefore that Between, that arena, in which subject and object can first move apart from each other." (Rombach, op. cit.,

p. 69.) Rombach's interpretation of intentionality is quite close to Nishida's pure experience, when Rombach continues: "And they always have to appear together, for the Subject is nothing other than the structured locus of the Capacity-for-becoming-present of a particular kind of objects: and the object is nothing other than a manner of possible being-grasped by a subject." (al. loc.)

33 Cf. *Husserliana*, Vol. I, Paragraph 50, 51, and 54. *Husserliana*, Vol. VIII, pp. 56–63.

34 That is meant to designate the egological reduction to the solipsism of my ego. (*Husserliana*, Vol. I., Paragraph 41.)

35 Cf. *Husserliana*, Vol. I, Paragraph 54.

36 Ibid., p. 127 and p. 125. Other-consciousness and past consciousness are, with Nishida and Husserl, equivalent, Cf. *Husserliana*, Vol. VI, pp. 158-9. This interpretation is, however, problematic. "Similarly, within my Ownness, specifically in its living present-sphere, my past is given only through memory, and characterized in memory as past presentness, that is, as an intentional modification. The experiential preservation of it (= memory) as a modification is carried out, then, necessarily in unanimous syntheses of re-call; only thus is the past preserved as such. Just as my remembered past transcends my living present as its modification, so, similarly, does the presentified other (foreign) being do to one's own (in the present pure and lowest meaning (sense) of the primordially unified)." *Husserliana*, Vol. I, pp. 144–5.

37 "The not-functioning mind/spirit is unrecognizable." (Nishida, ibid., p. 149.)

38 Ibid., p. 153.

39 In order to be able to distinguish one thing from something else, there must be a unity underlying them. (Nishida, *Complete Works*, Vol. III, 158—9.)

40 Cf. Nishida, ibid., p. 166.

41 The word "Logos" is the translation of the Japanese word "*Ri*" (law or Logos). In Sanskrit it is *dharma* or *tathatā*.

42 Cf. Nishida, ibid., p 166.

43 Nishida, ibid., p. 178.

44 Nishida, ibid., p. 184.

45 The assertion that the great consciousness system coincides with the present-consciousness of my Self, or, that "the Self of the world have the same Ground, no, they are the same" (ibid., p. 227), derives from Buddhism.

46 Cf. Nishida, ibid., p. 161. Nishida had a great sympathy for P. W. Bridgman's Operationism, Cf. *Complete Works*, Vol. 9, pp. 223 ff. Here one finds the similar idea to the thought of Husserl's intersubjectivity—because pure experience is preflexive Being-sphere which one calls intersubjective. Cf. Editor's Preface in *Husserliana* XV, p. lxix.

47 Nishida, ibid., p. 161.

48 What Professor Miyake, as a phenomenologist, said against Nishida's absolute reflection, is very important. It is precisely with the absolute reflection that Nishida could insist that pure experience is the essence of consciousness. He bases it on the phenomenological theory of reflection. Reflection on our experience confirms that there is no reflection that – illuminating itself as a source of light – at the same time performs the act of reflection. In other words, with reflection there is always something dark left over that is not illuminated by reflection. Therefore no philosophical reflection, insofar as it is based on experience-reflection, can escape naiveté. According to Miyake, Nishida's absolute reflection is condemned to be impossible. (Cf. Miyake, *Heidegger's Philosophy*, Tokyo, 1975, I. and II. Chapter.)

49 The author now knows that the phenomenology of H. Schmitz, especially his concept of *schlichte Wahrnehmung* is very similar to that of Nishida's 'pure experience'. Cf. H. Schmitz, *Subjektivität*, Bonn 1968, pp. 36 and 46.

AFFECTIVE FEELING*

Emotional feeling is a mental phenomenon which differs in kind from sensation. That is its essence. Every mental phenomenon is a structure of meaning-qua-reality, or reality based on the *Aktualitätsbegriff.* Sensation, of course, also has this structure. The difference between sensation and mere representation lies in this point. However, the discriminatory relation on which sensation depends is a minimal content of a conscious unity, and is thus only a first stage in the establishment of a mental phenomenon. If it develops to the next stage of representation or thinking, the unified content becomes more positive and active; that is, the mental state more and more manifests its essence.

In this frame of reference, affective feeling can be described as a unity underlying various intellectual forces. It can be considered as an a priori of a priori. When Rimbaud writes "*A noir, E blanc, I rouge, U vert, O bleu,*" or when Baudelaire claims "*les parfums, les couleurs et les sons se respondent,*" the fusion of sense qualities to which they refer must be accomplished on the basis of affective feeling. Feeling is often described as passive, but I do not subscribe to this view. I cite the example of Goethe who, by transforming his own life into poetry, surmounted its pain. In a similar way we become free as we embrace and transcend the intellect on the feeling level.

It is a common idea that feeling differs from knowledge, and that its content is less clear. To this I reply that the affective feeling of a sensitive artist is not necessarily less clear to him than the special knowledge of a scientist. The alleged unclarity of feeling means nothing more than that it cannot be expressed in conceptual knowledge. It is not that consciousness in feeling is unclear, but rather that feeling is a more subtle and delicate form of consciousness than conceptual knowledge.

What then is the content of affective feeling as a form of consciousness? The special characteristic of every mental phenomenon consists in the immanence of its object. This is true of perception and of thought, and it is no less true of the intentional structure of feeling. Perhaps some one will say that purely subjective feeling has no intentional structure whatsoever, and that such is even the essence of feeling. I do not agree. There is always

Nitta/Tatematsu (eds.), Analecta Husserliana, Vol. VIII, 223—247. All Rights Reserved.
Copyright © 1978 by D. Reidel Publishing Company, Dordrecht, Holland.

some content which can be discovered in feeling when it is discriminated adequately. Its intentional structure may differ in kind from that of knowledge, but that does not mean feeling does not have its own kind of immanent object.

Of course, if we follow the lead of most experimental psychologists and say that feeling is simply a compound whose constituent factors all pertain to the differences of intellectual elements, then we could conclude that there are only two kinds of qualities of feeling – namely, the qualities of pleasure and pain. Here again I demur. A phenomenon of consciousness cannot be a mere compound; it must be a unity. The reality of a phenomenon of consciousness lies not in its elements but in its unity. This is true especially of feeling, as Wundt also holds. That which accompanies a representation which has been constructed is also a simple feeling. For example, the feeling of harmony (*Harmoniegefühl*) is not a mere combination of feelings, but constitutes one feeling in itself. Feeling is the fundamental unity, in which we discriminate an indefinite number of qualitative differences.

To know a thing is to determine it. Through determining what kind of a thing it is we can know it, and judgment is the form of such determination. But determination has various meanings, and is not a univocal term. That which cannot be determined in a certain standpoint may be determinable in a higher standpoint. Take, for example, the case of finite numbers. If we cannot intuitively determine a certain very large finite number, this does not mean it is unknowable. Or again, we cannot positively grasp an infinite number; that is, it is a number about which we have to think in negative terms. But by transcending the standpoint of discrete elements, and by perceiving the act of discrimination itself – in other words, by transferring the object of cognition from objectivity to subjectivity – we can treat infinite numbers according to the same laws that govern finite numbers. Cantor's "set theory" arose in this way.

To cite Meinong's example, the consciousness of "non-red" is not on the same level as the representation of "red." It is a consciousness of a higher order. In the standpoint of mere representation, "non-red" might be thought of as an undeterminable and contentless mode of consciousness. But it can become a clearly determined content of consciousness in a higher standpoint. "Non-red" might remain unclear in the domain of thought, but it does not follow that it is without content or unclear. Even a phrase like "round triangle" is not without content. For it clearly has its own content; the very fact that it can be thought to be irrational indicates that there is some content. Every determination of a certain thing implies the standpoint of its counter-determination. The latter is not on the same level as the former. For it involves relationship with a higher standpoint which is its concrete basis. This is also the reason that the negative judgment can be considered to be of an even higher order than the affirmative judgment. Not being (μὴ ὄν) is not on the same level as being (τὸ ὄν); it is the stage to a higher standpoint. In the higher standpoint, that which has been thought of negatively as counter-determination acquires positive content.

Now, I have argued above that feeling remains after all content of consciousness has been intellectually objectified. From the intellectual standpoint, it might be considered to be without content and indeterminate. But in the higher standpoint of feeling itself it possesses its own clear and determined content. The content of each phenomenon of consciousness is an act. The highest a priori which unifies these acts is personal unity, the content of which is precisely that of feeling. The living personality is not a mere abstract concept. It is a dynamic unity of acts, a dynamic relation of various a priori. Feeling is the content of consciousness appearing in this kind of dynamic unity. This is also the reason the empirical psychologist uses the *Ausdrucksmethode* in doing research on feeling, and distinguishes the qualities of emotional states according to modes of expression.

Perhaps we can say that without an expressive act in the broad sense there is no clear feeling. The expressive act clarifies the content of feeling. Artistic expression, in particular, gives clear shape to our feelings. A Raphael without hands could probably have nothing but unclear feelings. Feeling without expression probably indicates an ambivalent conflict between pleasant and painful states. The artist creates in order to clarify his own feelings. In his *Dichterische Einbildungskraft und Wahrsinn* Dilthey argues that what we normally term psychological laws are abstractions; in actual concrete spiritual life a representation is a *Leben* and a *Vorgang*. Representations arise, develop, and disappear in the flux of life. Feeling is the mode of consciousness which internalizes the forms of life, and gives form to the inner spirit. It is the source of mythology, philosophy, and particularly of poetry. It is a living force that is inscrutable, but it possesses its own laws. The transformations and permutations of human life are reflected in such types as Faust, Richard II, Hamlet, and Don Quixote. The varieties of artistic expression reflect the rich content of our own affective life.

Pure affective feeling is not comprised of static states of pleasure or pain. It is something deeply active within the living heart. Pleasure and pain are merely the result upon which we have reflected. We ascribe truth or falsity to the content of meanings included in the act of judgment; we similarly make value judgments of pleasure or pain concerning the content of personal unity that is the result of personal activity. The content of emotional states is no less rich in its own sphere than that of truth and falsity in judgment. Pleasure and pain are only abstract general terms. It may be that pleasurable or painful feelings have no true content of their own, but derive their content from the intellectual objects accompanying them. But just as the content of judgment is not exhausted by any mere combination of representations, so too the content of feeling is not exhausted by any mere combination of intellectual objects. The content of feeling is always a certain creative whole beyond the mere combination of its elements.

Feeling may be identified with the act of imagination in the broad sense in that it combines other acts - or in the sense that it functions as an a priori of a priori. For example, what Rickert terms *Homogenes medium* in consideration of the judgment "A is A" involves Kant's *Einbildungskraft*. It is the concrete standpoint that combines one act, the *Thesis*, with another act, its *Antithesis*. The content of the combination of this kind of formal acts constitutes the content of logical feeling. Even in the case of sensory feeling which seems to differ radically from it, there is a creative thing behind it. It is the living personality, of which feeling is its expression. Pure affective feeling that has not been adulterated by concepts in any way whatsoever is my term for this kind of creative thing. In this standpoint everything has aesthetic value; and that we must despise physical pleasure results from our having conceptualized it. When we view these feelings from the abstract standpoint of the self, as simple pleasure and pain, they lose their own integrity.

In the eye of the artist, lines are an interpenetration (*Durchdringung*) of straight lines and curves, and color is a combination of various tendencies (*Tendenz*). Generalizing this, we can say that the same sensations are combinations of acts in the creative standpoint. Artistic emotion combines these acts. It is the content of the act of imagination. Of course, we might not usually make this identification, but upon reflection we can conclude that imagination is always a part of such creative activity. Just as various truth values appear in the content of judgment, so too various aesthetic values arise in the content of the imagination. Of the two standpoints, imagination is a more fundamental, concrete one than judgment. Take the case of music. Music expresses a certain extremely pure creative thing, apart from external forms, that is, apart from the determination of intellectual objects. Music is a term for pure feeling. Schopenhauer goes so far as to say it expresses the thing-in-itself. My position is that pure feeling is transcendental in respect of knowledge. Accordingly we must recognize a transcendental quality of feeling, such as Kant spoke of concerning formal beauty, in the area of aesthetic content as well.

In this frame of reference, the content of feeling is a fundamental act of consciousness grounded in something creative at the heart of our spiritual life – that is, grounded in something dynamic. We should no more seek the content of feeling within its intellectual elements than seek the content of judgment within representations which are its elements. The content of the affective life is adequately expressible only through art. It appears when the content of our consciousness is entirely dynamic, and when it is entirely focused into one activity – when the self is one with its world.

The essence of consciousness is the internal act – the act without a subject. Its elements may all be analyzable into intellectual contents; but just as aesthetic symmetry is something more than a mere combination of lines, so too consciousness can be understood only in terms of some dynamic function. Spinoza's *Ethics* retains its character of being a profound, internal explanation of the emotional life. But I have to disagree with him when he states that "Love is pleasure accompanied by the idea of an external cause," or that "Joy is pleasure accompanied by the idea of a past thing which surpassed our hope in its event." In my view, love and joy are *forces* that well up from the depths of the self; they are active *powers*. Rather than love being a feeling of pleasure accompanied by the idea of an external cause, we should say that the idea of an external cause is established by love in and for itself.

Truly to know a thing one must experience a sympathetic union (*Mitfühlen*) with it. At the foundation of all knowledge there is what Lipps termed empathy. This is evident in the case of inter-personal communication; but an analogous process takes place in the experience of color and sound. An artist perceives color as a continuum of various *Dimensionen* with which he empathizes and moves in unison. A person watching a tightrope walker moves together with him, becomes one with the activity, through sympathetic coalescence in feeling. We know our own past by uniting with our own past selves – that is, by uniting with our past acts. We know an external cause by the present self uniting with the thinking self, and uniting with the act of thinking. In other words, the self expands, a larger and deeper self must move. If a person were to be reborn a king in the next life, what pleasure could it afford him if he had no memory of his previous life? Conversely, Dante states that in times of unhappiness there is no greater sadness than to think of better days. Joy and sadness are in this way tied up with the larger self. Love and hate create the objectifications of self and non-self. These emotions are the resonances of the larger and deeper self which is grounded in the union of acts. Pure music well expresses this spiritual resonance. St. Augustine rings a change on the same idea when he writes that "The spirit which in the urgency of love seeks self-knowledge already has knowledge of itself" (my source here is Bindemann). This kind of self-knowledge is of course not conceptual knowledge. For the form of true reality is "the form of the formless," its voice is "the voice of the voiceless." It is knowledge grounded in what Plotinus terms *schweigende Verstehen*.

I argue above that the emotional life is not comprised of the simple feelings of pleasure and pain combined in clusters of sensations and representations. Concrete reality is a more profound consciousness than cognitive knowledge; it is expressive of a more profound reality than the objects of cognition. The psychologistic view that feeling consists merely in the qualities of pleasure and pain, after the various intellectual elements have been eliminated, is tantamount to the stereotyped division of the human race into male and female. Contra Wundt, Titchener argues that such emotional states as excitement and depression, tension and relaxation are not simple elements, but are comprised of organic sensations that possess intellectual elements of a higher order than the feelings of pleasure and pain, which can be considered to accompany lower sensations. But we can classify these mood sets, in their status as independent concrete feelings, as simple on the same level as the feelings of pleasure and pain. In my view, a distinction must be made between the feelings of pleasure and pain which accompany the so-called sensations of a lower order, and pleasure and pain as generic qualities of the emotional life, that is, pleasure and pain as abstract concepts. Concretely given, every feeling is simple; abstractly considered, every feeling is pleasurable or painful, just as every judgment is true or false.

If the essence of feeling lies in the union of acts in the single center of the self, then it is appropriate to divide the basic qualities of feelings into those which combine and those which do not, and that we consider pleasure and pain to express the essence of feeling. Confusion results when empirical psychology does not clearly distinguish the above two senses of pleasure and pain. When we are pleased or displeased – for example, while looking at a painting or listening to music – these experiences are accompanied by a kind of complex organic sensation; and at the same time they are pleasurable or not. However, in this analysis, we must not forget we have shifted our reference point. When absorbed in appreciating a famous painting, I am not simultaneously aware of an organic feeling of pleasure. When I turn my feeling from the painting to note my aesthetic response, the original feeling has already disappeared. As Nietzsche states, *Leib bin ich ganz und gar*. As I give myself to an ideal, my whole body stirs, accompanied by physical changes. Changes of pulse and respiration similar to simple feelings may even occur. From a purely objective standpoint, this may look like an adulteration of organic states of pleasure or pain. However, a pure feeling is not a mosaic; a concrete feeling must be a unity. From the perspective of natural science, the elements in both cases may be unchanging. But just as a representation within judgment differs from a simple representation in itself, so too pleasure and pain as constituent elements of aesthetic feeling are different in kind from simple feelings of pleasure or pain. In the intentionality of a mental phenomenon the differences of meaning content must be viewed as the differences of reality.

In saying this I am not necessarily defending Wundt's classification of simple emotions. In Wundt's psychology there is often a confusion of basic standpoints. Wundt attempts to confirm his own theory by means of the "method of expression" as an experimental psychologist. But as Titchener argues concerning Wundt's methodology, the introduction of a distinction between the "method of expression" and the "method of impression" serves to invalidate Wundt's approach (*A Textbook of Psychology*, §72). Wundt cannot escape the criticism of having confused the standpoint of experimental psychology with that of introspective psychology.

For the above reasons, I grant there are innumerable simple elements in feeling. We can classify the qualities of feelings just as we can classify the qualities of sensations. Pleasure and pain are the generic terms for this classification. Wundt's "tri-directional" distinction is another way of thinking of the matter. Nevertheless, we cannot say that these are simple elements in the same sense as a specific perception of a color or sound. Moreover, the intensity of feeling, like the intensity of sensation, involves a relationship to an even greater unity. If the will is the highest unity of the self, then intensity of feeling involves a relationship with the will. Feeling which functions as motivation exhibits this kind of intensity.

Section 4 of the same chapter

Things in their concrete immediacy transcend all categories and classifications of speech and thought. The more we try to grasp concrete immediacy, the more it eludes us. To be conscious is already to have assumed some relative, mediated standpoint. We cannot fathom how the relative arises out of the absolute, nor what it means to say that we find ourselves in some standpoint. There is no consciousness prior to consciousness, and there is no knowledge prior to knowledge. At this point the authority of knowledge is eclipsed. If so, is the absolute unknowable? Or outside of the domain of consciousness? But to be able to make even these determinations of the absolute is to bring it within the sphere of the relative. True reality in its immediacy does not pertain to either knowing or not-knowing. The absolute is no mere hypothesis. It is an inexhaustible force.

In the realm of consciousness, what we term knowledge is a process of purification of experimental contents within a certain standpoint. The more these contents are purified, the more knowledge acquires objectivity. Error arises from a confusion of standpoints – and we term this confusion being subjective. In the standpoint of knowledge, the subjective is antinormative and this involves error. The domain of feeling, though itself a subjective state, is nevertheless a horizon of consciousness which clarifies its own content by objectifying its own standpoint. Knowledge cannot reflect upon its own standpoint. Philosophy itself does not have a life apart from the standpoint of thought. Philosophy is a reflection upon the totality of relationships, but it does not escape being abstractly two-dimensional. It is still the content of an act; it is not the free act in itself. However, consciousness does have a ground which transcends the act, and is concretely three-dimensional. In a material phenomenon, the act and the consciousness of the act are considered to be different; in consciousness these two must be one, and there must be consciousness of the act itself. (From this perspective, philosophy is the content of only one kind of act, at the ground of which there is something personal. In this sense I am interested in Dilthey's concept of *Weltanschauungslehre*.) It is pure subjective feeling that is the underlying unifying consciousness of the activity of philosophy.

A confusion of standpoints becomes error in the domain of knowledge, but it can be the very truth in feeling. Error can express a profound humanity, as in Rodin's statement that the uglier a being is in nature, the more beautiful it is in art (*plus un être est laid dans la nature, plus il est beau dans l'art*). Anything that falls into contradiction in one standpoint can become potential content in a higher standpoint. What is impossible in the standpoint of representation can constitute potential content in the standpoint of thought. What appears as contradiction in the standpoint of cognition can constitute positive content in the standpoint of feeling. In short, feeling is the content of the will, which is the act underlying all acts. If imagination (*phantasie*) occupies a middle ground between the will and knowledge, we can say that feeling is the content of the imagination. As the act underlying all acts, the will is the ultimate standpoint. Consciousness attains its ultimate consummation in the will. In the will we touch the thing-in-itself. However, at the point of thoroughgoing realization of the will we advance, as it were, one step beyond the tip of a pinnacle. We enter the realm of true reality where the distinction between knowing and not-knowing is transcended. Salomom Maimon interprets Kant's thing-in-itself as a "limit concept," and he combines it with Leibniz' notion of *petites perceptions*. But I hold that in attaining to this ultimate point there must be a further leap. The world of *petites perceptions*, if interpreted as the thing-in-itself, would not be the infinitesimal unit of knowledge, but the clear world of affectivity and the will.

The above consideration provides a framework in which feeling and cognitive knowledge can be properly discussed. Knowledge becomes clearer the more it reflects upon its own process, whereas feeling disappears the more we turn our attention to it. Feeling is the underlying condition of subjectivity. It is the consciousness of acts, consciousness of the a priori. It naturally evaporates as we turn it into an object of attention. Feeling rather deepens and becomes purer the more it is absorbed in its own object.

If we follow Brentano and distinguish phenomena of consciousness according to their respective intentionalities, then feeling can be described as a non-determinable determination. After the analogy with infinite numbers, it can be described as involving a relationship with a negative object. In what way, then, can we be conscious of our own affective states? Consciousness of feeling already entails knowledge, and thus can no longer be said to be pure feeling. The answer is that we can objectify feeling and determine its content in the standpoint of the will, the act underlying all acts. Again after the analogy with numbers: though wholly negative in the standpoint of finite numbers, infinite numbers can be considered to be positive in a higher standpoint. This is true, for example, in set theory. Although feeling remains undeterminable in the standpoint of knowledge, it has its own positive content in the standpoint of the will. The purer the latter standpoint, the clearer the content of feeling becomes. Aesthetic feeling well exemplifies this. Plato's world of Ideas was, to Plotinus, the realm of the Beautiful. It is through reflection and conceptualization that we lose the quality of pure aesthetic feeling, and fall back, as it were, into the unclear, stereotyped world of emotions described in terms of pleasure and pain.

Feeling of which we are conscious is no longer feeling. Feeling is always a mode of present consciousness. Concerning the other consciousness-phenomenon, it cannot be said that the same phenomenon appears in consciousness. It appears especially true of feeling because of the subjectivity of feeling, in contradistinction to the objectivity of knowledge. Our subjective states are forever changing. But conversely, if consciousness is the self-realization of the universal, then at the ground of all consciousness there is an eternal a priori; and feeling, the a priori of a priori, is the eternal present. Pure feeling has its own intentional structure. We can become conscious of feeling by intellectualizing it and translating the intellectual content into dynamic terms. This is because feeling is the unity of various acts. It is thus the content of the I as the union point of various a priori. It is accompanied by the act of expression. Consequently pure feeling is dynamic, and this also is spontaneously accompanied by the activity of the body. In this light, the expressive act is a process of symbolization, of which artistic forms are an outstanding example.

There is no such thing as universal validity in feeling such as postulated to obtain in the realm of perception. As Kant says in reference to the judgment of taste (*Geschmacksurteil*), in the realm of aesthetic sensibility it is impossible to set up universal laws in the same sense as can be postulated in the sphere of intellectual judgment. Knowledge stands on the a priori of the understanding, whereas feeling is the content of the free personal unity that is the a priori of all a priori. Its unity is not intellectually determinable. It always discovers and creates its own unity, by a process of turning inward upon itself. Knowledge, in its own standpoint, unifies all empirical content, but feeling unifies the a priori themselves by returning to itself and discovering unity within itself. Therefore there are no universal criteria in the realm of affectivity. To posit universal criteria in any sense is already to fall back into the standpoint of conceptual knowledge. Kant's notion of a *sensus communis* does not entirely avoid criticism in this light. The a priori of feeling is the a priori of a free and pure act prior to concepts. Therefore each feeling must be creative, and uniquely individual. There is no room to insert universality in the sense of a common sense. We can only recognize universality as an organic unity of transcendental personality.

The purer feeling becomes, the more beautiful it becomes. It achieves its own purity apart from the adulteration of concepts. The more removed from conceptual determination, the more feeling becomes transempirical and beautiful. Lipp's concept of empathy must also indicate a transcendental unity of acts in such a sense. While we *empathize* with the movement of the tightrope walker, we do not *think* we are the tightrope walker. We become one with his activity in the transcendental realm. This kind of sympathetic coalescence is the foundation of artistic intuition. It functions at the ground of conceptual understanding as well. When intellectual content becomes a pure act transcending the standpoint of knowledge, it too possesses aesthetic significance. Moreover, normally unaesthetic emotions can become beautiful ones in the standpoint of pure humanity.

Kant states that aesthetic feeling accompanies the act of reflective judgment. Reflective judgment subsumes the universal under the particular. That is, it determines the formal teleological quality of nature. The reflective judgment is thus the reverse of the intellectual act of determinative judgment, which subsumes the particular under the universal. Reflective judgment views the a priori itself from the standpoint of the will, the a priori of a priori, and thus views the act itself as object. Kant also states that sensory things are pleasant (*angenehm*) and are accompanied by interest (*Interesse*), but sensory experiences also have internal development that is continuous in itself, and they have their own a priori. Every act which can be reflected upon in the standpoint of pure personality is accompanied by aesthetic emotion. In the standpoint of pure personality, every sensory experience is beautiful. It is only when adulterated by concepts that the emotions become unaesthetic.

(End of chapter)

From Chapter 6, 'Various Continuities of Empirical Content', Sect. 1, NKZ III, p. 99 ff.

Consider a series of red colors constituting one continuum. Let us say the transition from each shade of red to the next is extremely gradual – that is, the degree of difference between adjacent shades is infinitely small. Mathematically speaking, any part of this continuum is infinitely divisible. That is, not only is it *überalldicht*; every point constitutes the "limit point," or the "differential coefficient," of the series. When every such limit point is included within it without exception, it is described as "self-dense" (*insichdicht*) and "closed" (*abgeschlossen*). The series then becomes a perfect set, which Cantor refers to as "*Ordnungstypus theta*" (G. Cantor, *Beiträge zur Begründung der transfiniten Mengenlehre*, Sect. 10). Now a continuous series is comparable to a series of real numbers; whereas a series of discrete elements is comparable to a series of rational numbers. What is termed a limit point is a point that cannot be attained by division. A set involves the notion of totality, as in Dedekind's definition of a "section" (*Schnitt*).

Here we can also cite Leibniz, who writes: *chacune de ses substances contient dans sa nature legem continuationis seriei suarum operationum.* Any real must be a complex whole, characterized by the continuity of its parts. A set of discrete elements not exhibiting continuity within itself cannot be said to be real. In a continuum of red colors constituting a set of limit points, the totality is no longer a material unity. It becomes a mental unity – that is, an autonomous act with its own integrity in itself. Some psychologists talk about discrete sensations as elements of mental phenomena. But discrete sensations are the product of intellectual analysis; they are not living sensations in themselves. Colors such as red or blue can be viewed as mental or material phenomena. When directly combined in themselves, without any hypothesis of external associations, we can describe them as mental phenomena, the transformations of which are internal to the given set. Mental phenomena are self-supporting, and exhibit their own internality and immediacy.

Of course, we may not usually think of something continuous as a mental act. For example, a physical force, as an independent activity, is something continuous in itself. But mental activity is not only continuous in itself; it is also self-generative and self-developmental, and contains the laws of change within itself. In Leibniz' phrase, it contains *legem continuationis seriei suarum operationum et tout ce qui lui est arrivé et arrivera.* To do so it must contain its own end within itself; it must have a teleological structure. This is also true of biological phenomena. But in mental phenomena purpose is built into the very intentionality of consciousness. Only in this instance is purpose truly immanent, and self-operative. A material phenomenon is on the contrary merely governed by the force on inertia. Consequently the connection of a material force is merely accidental. Biological phenomena do exhibit teleological processes immanent to themselves. Nevertheless, biological processes are not completely autonomous, and thus can be subsumed within a mechanistic explanation of life. In the terms of Kant's third *Critique*, the teleology of nature is merely a regulative principle (*regulative Grundsätze*) and not a constitutive principle (*constitutive Grundsätze*). Only in mental phenomena does purpose become the constitutive principle. There are no mental phenomena in which purpose does not operate immanently – that is, in which the unity is not internal. This view is shared widely by experimental psychologists.

Let us now return to the question of continuity. A set must have both elements and a form. We can call the latter the a priori of the set. Cantor distinguishes in a set (*Menge*) between that which has been ordered (*geordnet*) (what he terms *Ordinalzahl*) and that which is not. But in an ordered set (*geordnete Menge*) there must be a "form of order" (*Ordnungstypus*). The latter established the order (*Ordnung*). In finite numbers, the cardinal numbers (*Kardinalzahl*) and ordinal numbers (*Ordinalzahl*) function similarly. But in infinite numbers they do not. Originally cardinal and ordinal numbers are different concepts. But in infinite numbers the "form of order" brings out their autonomy. When a type of order requires special treatment – that is, when it manifests its autonomy – the ideal elements of a set become real. The operations (*Operationen*) of ordinal numbers express the relationships among ideal elements. We can say that the transition from finite numbers to the idea of transfinite numbers involves a transition from the real to the ideal. This entails a transition to a higher reality. A series of finite numbers corresponds to a system of determined entities; and no matter how it approaches to an infinite series, it can never transcend the scope of finitude. This is analogous to the fact that an entity determined in time and space can never transcend the forms of time and space. In limit numbers (*Limitzahl*) we transcend the realm of empirical facts determined by time and space, and we enter into the realm of pure thinking. A limit number is no longer an object of perception. It is entirely an ideal entity. In contrast to it, a thing which can be perceived is always finite.

Now in regard to self-consciousness, the self which has been reflected upon is the self after the analogy of finite numbers. But the true self cannot be attained by reflection. Rather, we can think of it after the analogy of limit numbers, which may also be described in terms of the act of reflection itself. In finite numbers each element is thought of as real; but in transfinite numbers the *form* of their order is more real than the elements. The unifying form is a higher integrity than the elements unified. In other words, the act itself becomes autonomous. Every point of a set constitutes a limit point – that is, is self-dense, and thus establishes a set of the above kind of ideal elements. Each of its elements is a differential coefficient of the whole. A set exhibits a system of elements of a higher order of reality.

Dedekind's definition of a "section" (*Schnitt*) makes the same point. It is also expressed in Leibniz' dictum: *imo extensione prius*. The whole is prior to the smallest extensive quantity. The elements of a whole are not independent entities; each is a Dedekindian section signifying the whole, and having the meaning of the "form of order" – that is, each being a symbol of the whole. In this way the whole truly consists in its parts, and the ideal directly becomes the real. This is because the whole functions immanently in a self-realizing way in the parts. When the type, or form, of order is merely external, as in spatial relationships among material phenomena, it does not perform any unifying function whatever in respect to the elements. In this case there is no true set. Consequently a material force can be described as an internal relationship in things themselves. Cantor states that a set, that is, *Ordnungstypus theta*, not only is "self-dense" but is a set including all the limit points of a "fundamental series" (*fundamental Reihe*). That is, the whole constitutes a "perfect set" (*perfekte Menge*). Cantor's meaning is that the type of order itself functions independently in itself without support from another force or power. A set is not perfect when its limit points are not completely imma-nent within that set. In a perfect set the whole constitutes one autonomous, infinite function sufficient in itself – one complete act with its own internal development. We can also characterize it as a system with its own teleological structure, and as self-activating.

In this sense a material force is not perfectly self-contained or "closed" (*abgeschlossen*). What is "closed" in these terms must be organic. A mental act established on the unity of consciousness can be described as truly self-contained. Something understood by means of another cannot be said to be autonomous in itself, or to possess its own teleological structure. An example of this would be an empirical content studied in the perspective of physics. The unity of the empirical content is attained by postulating some unknowable x behind it, while prescinding from the subjective experience itself. The physical system cannot be said to exhibit its own teleological structure. In other words, it does not qualify as a set possessing limit points within itself. Thus when various experiences involving light are not understood in themselves, but in terms of some physical continuity such as ether behind them, the manifestations of light cannot be regarded as real in themselves, but merely as signs of another reality. When the physical reality postulated to exist behind them is also not understood internally in itself, we are left with various external and accidental configurations, none of which exhibit any internal teleological structure. Physical continuity is one that is externally given, one that has been postulated, rather than one that exists in and of itself. Biological phenomena are usually thought of in teleological terms. But because biological life is so dependent upon external stimuli and support, it is easily subsumable under the mechanistic hypothesis. The mark of a true teleological process is that of immanent goal determination. Only mental phenomena exhibit such a structure, and are truly "closed" and "perfect" sets.

Let us bring our initial discussion of red colors into this frame of reference. When the series of colors is continuous in the strict sense, it constitutes a reality of a higher order than if it were a series of discrete sensations. It becomes a kind of autonomous function such as an act of visual perception. A set (*Menge*) can be a simple set having thickness (*Mächtigkeit*), and it can be an ordered set (*geordnete Menge*), the elements of which exhibit a particular kind of order. Cardinal numbers exhibit no order at all; they constitute a world of representations, a world of simple meanings. When subsumed within an order, they become a world of ordinal numbers while retaining their thickness and also their original character of cardinal numbers. The world of ordinal numbers thus established takes on autonomy and reality of its own. However, the reality of the elements is still bound up with the unifying system. The type of order is merely an ideal relationship among the elements, somewhat analogous to the reality of individual atoms in the physical world. For this reason cardinal and ordinal numbers can be treated in the same way as finite numbers. In contrast to this, when the thickness is infinite, as in the case of transfinite numbers, a type of order other than that of cardinal numbers is constituted, and this type of order possesses a reality in itself. In a transfinite number system the ordinal numbers require special treatment; the ideal becomes real, and so do the relationships among them.

Applied to our example, when sensations of red colors are arranged in accordance with their degree of saturation on a continuum, these degrees constitute the form of order of the set. When this series is considered to be infinite, the type of order has reality. The possibility of infinite transition signifies that the type (*Typus*) possesses power. It signifies that the elements are the representations of a reality in their background, and that the elements are nothing more than its determinations. Here the type of red becomes a force and act. This power cannot properly be called material or spiritual; but at any rate it signifies that the representation red can be expressed infinitely. Cantor's concept of transfinite number, or limit number, expresses this kind of power. Let the elements of a set, that is, the "extension" of a concept, be the content of the act, or the intentional object; and let the type of the set, that is, the "intension" of a concept, be the act itself, or the subjective quality. Then when the former reaches the point of falling into contradiction, the latter has already appeared as content of an act of an even higher order – that is, as the intentional object of an even higher order. That the ideas of time and space fall into antinomies (*Antinomie*) indirectly indicates that they are structural forms.

A set is a series of these kind of limit numbers; it is a series of powers. The continuity of Cantor's perfect set must be an active function at every point. Here we make the transition from the concept of a mere act, the product of abstract thinking, to the concept of power; and from the concepts of static (*Statik*) to those of dynamics (*Dynamik*). When a series of red colors is infinitely divisible, and each of its points constitutes a limit point, the experience of red here is not classifiable as a series of discrete sensations according to the method of associationist psychology. It rather constitutes an autonomous power. However, for this set to be completely self-contained, it must have an immanent unity in the strict sense – that is, be a perfect internal continuity. A self-contained set is one possessing its own immanent purpose, and is established in and through itself. Only a mental act meets these criteria.

If the type of order organizing a system of ordinal numbers is the a priori of that system, then the mental act, which functions as the act of the type of order itself, constitutes a union point of a priori established on the a priori of all a priori. In this sense, no matter how passive a mental phenomena is it differs in kind from a material phenomenon. Of course, the physicist may think of translating Cantor's concept of set into that of material force. Newton developed the concept of "fluxion" to serve as the basis of a mathematics of continuous force. But the notion of material force is a postulate made to explain a given piece of experience. It does not qualify in terms of the above criteria of immanent teleology and autonomy. In a system of dynamics, that is, in the world of pure physics, a material force may exhibit necessity and purpose in itself, and may constitute an autonomous set. But it remains a mere hypothesis as an explanation of the facts of perceptual experience. The very fact that a material force can be divided quantitatively indicates that it does not have a true internal unity. Living things, and things which exhibit mentality, cannot be so divided. Leibniz' dictum *imo extensione prius* is a critique of the concept of material force. Direct perceptual experience is a continuous thing in itself. When we try to include it within a system of thought, it becomes discontinuous in the form of Poincaré's formula of contradiction: $A = B$, $B = C$, $A < C$. To correct this, we postulate some further physical force or reality. But the perceptual experience loses its own autonomy when included within a system of thought based on a different a priori. Like the material world, it then comes to acquire its reality from the outside, to lose its autonomy, and hence is no longer a perfect whole in itself.

To summarize, when a set (*Menge*) includes its limit points, that is, its differential coefficients (*Ableitungen*), this set can be said to be "closed" (*abgeschlossen*) or self-contained. When a set includes its limit points in this way we can describe this set as already constituting an autonomous act. When each of the points of the set is a limit point, we say it is "self-dense" (*insichdicht*). That which is self-dense in this way is a living thing in all its points; it is a set of acts. However, even though it is self-dense, if it does not wholly include its limit points – that is, if a set and its differential coefficients do not coincide – it cannot be called a perfect set (*perfecte Menge*). Only when it is a perfect set can it be described as an autonomous act of consciousness.

(End of Section 1)

[The first few pages of the next section (Sect. 2) are repetitious of the preceding argument, and are thus omitted. The translated passage begins from NZK, Vol. II, p. 113, where Nishida takes the topic in another direction.]

In the second edition of his *Die neue Malerei*, Ludwig Coellen contends that the modern Impressionist school endeavors to capture the impression of the moment, in contrast to traditional schools which paint fixed objects formed by a combination of present perceptions and past memories. He characterizes Impressionism as a reaction against the *subjectivism* of the past, and as the product of a new *objektivisches Lebensgefühl*. Painters of this school such as Courbet, Monet, and Liebermann exemplified this new objectivism by means of a new *Technik*. They employed light as *das die Bildenheit schaffende Medium*. In Impressionism, all things are bathed in light, and have their existence only as *Lichtmasse* of specific qualities. In the Impressionist's world, there is no room for memory and thought to enter. Van Gogh carried through this objective-pantheistic style of intuition (*objektivistisch-pantheistische Anschauungsweise*) and deepened the Impressionists' perception of nature as a relationship of surfaces (*Oberflächenzusammenhang*) to the point of expressing nature's relationships of living forces (*lebendiger Kräftezusammenhang*) in his art. In van Gogh space becomes a dynamic organism (*dynamischer Organismus*) in which ordinary entities are entirely engulfed and the myriad things become so many *Kräftesymbolik*.

If the new Impressionism indeed possesses such a significance, we may say that modern art circumvents the traditional adulteration of perceptual and intellectual a priori, and makes the world of pure perception its object. Conrad Fiedler, in his *Schriften über Kunst*, makes the same point. He argues that art opens up the prospect of an infinite world different in kind from the infinity of thought. Just as the mathematician becomes pure thought by entering into the world of infinite numbers, so too the artist, by absorbing himself in pure plastic activity (*Gestaltungstätigkeit*), which is its own extensive continuity of what Fiedler calls the *vorstellendes Bewusstsein*, enters into an autonomous world of artistic infinity. Perception becomes finite and discontinues when viewed in the standpoint of thought. Whatever does not stand in its own order is finite; it is a dead thing.

I would like to comment in passing on an aspect of Meinong's theory of objects. Meinong describes the relationship among perceptual qualities as an objective world based on experiential content, apart from the standpoint of natural scientific existence. But the objective world of the artist, to which I have alluded, is different in kind. We have to distinguish here between the relationships among colors in themselves, which Meinong compares to geometric relations, and what I term the objective world of pure vision, which the artist perceives as an infinite continuity. The distinction I am making is analogous to that obtaining between mathematical and physical truths. The latter is the unity of the former in a certain sense. Painting and music do not express the impersonal, universal relations among colors and sounds; they express specific aesthetic nuances by their unities.

It goes without saying that the meanings expressed in painting or music are not conceptual ones. It is a mistake to conceive of the purpose of art to be that of imitation of nature, or as a tool of intellectual enlightenment. The significance of the new art, to cite Coellen again, lies in the fact that it circumvents the confusion of conceptual and perceptual a priori, and immerses itself within a pure objectivity in which everything becomes pantheistic. But this pure aesthetic objectivity is not the same as the objectivity exhibited in the naturalistic schools. This accounts for the evolution of modern art from naturalistic to impressionistic styles, and for the development from earlier to later Impressionist schools. Romantic Impressionism broke the outer shell of things, and attained to an intuition of infinity underneath. It went from the earlier intuition of nature as a *pantheistische gefasste Einheit* to a perception of the spiritual unity of the objective universe. According to Coellen, the pioneer of this tendency was F. Hodler. The early Impressionists saw things as merely a *Lichteinheit des Bildraumes*; Hodler viewed them as a *bewusste Einheit*, he saw the spiritual meaning of things. Holdler saw things as spiritual symbols. Gaugin and Matisse then deepened the Impressionistic trend already manifest in van Gogh, and perfected the style of Romantic Impressionism. They gave rhythmic power to the *dynamische Farbe* of van Gogh. They transformed van Gogh's *dynamischer Organismus* into pure *Lyrismus*. They were able truly to grasp nature in itself and spiritualize it. They sought the essence of things in immediate feeling. Not in mere feeling, but in the *lebendige Aktivität* that feeling arouses. Coellen contends that the *Kubismus* of Picasso and others emerged from this same aesthetic intention. Modern art thus advanced in the direction of objective purification, but thereby attained to a new revelation of subjectivity. This new subjectivity is not conceptual, but one that opens up in the midst of intuition itself.

Art is said to manifest individuality. Individuality is the life of art. Apart from it there is no true art. But what is individuality? An individual thing or entity must be one and not two; in other words, it must occupy a unique position in a system of relationships. However, this is not a sufficient criterion in itself. An autonomous entity must contain within itself an infinity of relations which differentiate it from everything else. As Leibniz states, each monad contains its relations with the entire universe. For merely homogeneous material atoms to qualify as individual entities they must have their own identities, with their own histories. At the least they must be regarded as having infinite relations which distinguish them from every other thing in time and space.

The more particular a thing is, the more it must have its own unique relationship to the totality – that is, the more it must be based on what Leibniz calls pre-established harmony (*harmonie préetablie*). This is the reason why Leibniz says that the predicate must exist in the grammatical subject of a true proposition (*praedictatum inesse subjecto verae propositionis*). The individual must contain infinite predicates within itself. The concept of Adam contains all that has ever happened and will ever happen to Adam. In a universal truth, on the contrary, the content contained within the grammatical subject is finite. The more it is universal, the more its content is finite. If its content becomes infinite, then there is a shift in standpoints – namely a shift from the particular (*la notion specifique*) to the individual (*la notion individuelle*). Here lies the distinction between eternal truths (*vérités éternelles*) and truths of fact (*vérités de fait*). And this shift, as in the limit concept of mathematics, implies a distinction not merely in degree but in kind. It involves a jump in standpoints. The object of cognition passes from content to act. The "type of order" becomes real. As in Cantor's *Ordnungstypus theta*, every point becomes a limit point, and the individual entity becomes a continuity of acts.

The grammatical subject in a proposition of eternal truth is based on a certain determined a priori, whereas the grammatical subject in a truth of fact is based on the will, the a priori of all a priori. Just as Leibniz holds that the free will (*les decrets libres*) of God grounds the world of the possibles (*les mondes possibles*), so too there is will at the ground of contingent truths (*vérités contigentes*). An individual entity is an internal unity of the will. Only in saying this can we make sense of Leibniz' position that "each substance contains in its own nature the law of the continuation of the series of its own operations (*legem continuationis seriei suarum operationum*) and everything that has and will happen to it (*tout ce qui lui est arrivé et arrivera*)."

For a certain element, as an autonomous entity, to contain the meaning of the whole within itself, or have a unique relationship to the totality, it must be dynamic in a sense analogous to Cohen's concept of a "productive point" (*erzeugende Punkt*). An individual entity that includes the universal within the particular must be one developmental act, in which the particular elements are symbols of the whole. In opposition to Descartes, Leibniz argued the same point in respect to the reality of the physical world. Reality is not static but dynamic; the essence of matter does not lay in its extension but in its activity. However, an individual entity which in this sense truly contains the whole within its parts must be something spiritual, as in the case of the Leibnizian monad. A material phenomenon does not possess internal unity within itself. It is unified by means of some hidden thing, and is grounded externally. Thus it cannot be said to be an individual which truly contains the whole within its parts and is self-determining. According to Leibniz' way of thinking, that which is truly individual must be spiritual. That cognition of an individual is always contingent exemplifies the point that the will cannot become an object of cognition – that is, it cannot be reflected upon. This is the reason that an individual has its own unattainable depth.

A material phenomenon is described as contingent even though the various qualities that constitute it are necessary in their own respective standpoints. Its contingency lies in the union of these contents, while the point of union remains unknown. However, this kind of unity in the background of a material phenomenon is something that has already been objectified. It possesses necessity in the sense that it has its place within a given system of cognitive objects. In the physical world itself, in the strict sense, there is no "thisness." Just as Leibniz held that there is the free will of God in the background of the world of the possibles, so too does the unity of our direct experiences become the foundation of contingent truths. The will is the essence of this unity. It is also for this reason that we are able to think that a factual truth possesses its own certainty beyond that of reason. That the will is described as subjective and as voluntary is itself the result of objectifying the individual will in thought. Truly creative free will becomes purely objective as it transcends conceptual consciousness – that is, when it achieves the fusion of subject and object. At this point true individuality, the true self, is revealed. True individuality transcends conceptual knowledge and grasps stubborn factual truth. We can factor a factual truth in innumerable ways, yet it remains a limit inexhaustible in the domain of conceptual knowledge.

Now I think that the above ideas apply to the true individuality that is the goal of art. By attaining to a new kind of objective purity modern Impressionism has revealed a new subjectivity. Truly disinterested aesthetic feeling, that is, pure feeling, accompanies the unity of this kind of transconceptual individuality. The content of aesthetic consciousness expresses the content of this kind of individuality. There must be a dynamic act at the ground of the pure object. The act is that in which the a priori itself has become real; it is its concrete basis.

Thus when we are conscious of a certain kind of color as an infinite set, we are not conscious of it as a content of consciousness, but as an act, as a type of order. When a color becomes an infinite continuity of many directions, the one qualitative universal termed color becomes autonomous in itself, and becomes one act. The characterization of the artist as one who has "pure seeing" (*reines Sehen*) refers to this as well. Similarly we can think of sound as also constituting one continuous system as pure aural perception. The acts of visual or aural perception described in empirical psychology are projections of pure visual and aural perception after the latter have been cast into objects of cognition, that is, into the natural world. But our concrete experience is comprised of more than mere visual or aural perception so described by empirical psychology. It is a continuity of infinite acts, the unity of which is the living personality. A living personality is the limit of an infinite series of these kinds of acts. The individuality that is the hallmark of artistic creation is one instance of this kind of personal unity.

For example, when a painting truly expresses individuality, it reflects one unique form of the greater life of its maker. True individuality cannot be separated from a greater personality in this sense. It has its place within the life of the greater personal life of the artist, and thus it is not subjective. The subjective has merged into the objective. In its individuality it includes a relationship with the whole universe in its own perspective. But to include the other within itself it must transcend itself. Needless to say, the Impressionist school, in transforming the myriad things into light and seeing them as a unity of light (*Lichteinheit*), is not merely expressing the relationships among universal qualities of color and light. The artist is attempting to express a unique reality. This is not a conceptual reality that becomes the object of cognition; it is a pre-conceptual, intuitive reality which cannot be expressed in concepts.

True individuality does not appear in the objective world of cognition. It appears at the time we have destroyed conceptual subjectivity. It appears when we have transcended the conceptual realm. The world of color or light in this way comes to express a transconceptual reality. In the horizon of personal unity, that is, in the objective world of absolute will, we can express a unique individuality by reflecting infinite nuances. This becomes possible because the act itself of color or light, that is, of visual perception, transcends the standpoint of the act itself. Color, at the limit of infinite continuity, becomes act, and the act becomes personality at the limit of its infinite continuity. That is, it becomes the act of absolute will. Just as every point in a continuum contains the meaning of the whole, so too a form of pure consciousness, at any point in a set, can include the meaning of the whole personality within it. In the plastic activity (*Gestaltungstätigkeit*) of pure visual perception, the whole is included within the experience based on one act. That we are able to think that pure visual perception spontaneously accompanies behavior is because it transcends the objective world of cognition and enters into the world of symbols where soul and body are one.

I wish to say a final word here concerning the distinction between an individual entity and a true individual. With regard to the former, Leibniz states that the concept of an atom contains within itself all its relations with other entities by a preestablished harmony, and thus constitutes itself as a unique entity in the universe. But a true individual must be a living thing in itself such as we see in works of art. It must have life, and its own autonomous value. Temporal and spatial uniqueness does not constitute a true individual. It must have free will. A true individual is not something merely passive, like a clothes rack or a rice bag. It must have a certain creativity. To be creative entails that it has the character of being unique in the transcognitive world, that is, in the world of affectivity and the will. Even purely theoretical learning reflects the individuality of a certain ethnic group, of a certain individual in this sense. The same is true in painting and sculpture.

(End of Section 2)

[Chapter 6 continued, Sect. 4, pp. 136–40. (Sect. 3 is repetitious of the themes developed in Sect. 2.)]

The concrete immediacy of experience, that is, true reality, is personal. It is the internal union of infinite acts. I call this unity of infinite acts the standpoint of absolute will. In the conceptual standpoint, the will is an unattainable limit point. As the mystical philosophers have said, God transcends all categories. However, while we must speak in this manner in the standpoint of thought, there is only this pure reality with its free calculus of becoming in the immediacy of experience. There is nothing simpler or clearer than this. Reason is the negative unity of absolute will. Reason functions on the one hand as the cognition of reality, and on the other hand as the moral will that is creative of reality. The moral world does not exist within the natural world; rather, the natural world is established on the basis of the moral world. Knowledge merely follows after the will, and orders its traces.

In the latter sense reason is the unifying act of the whole person. In the former sense reason is only one personal act that functions in the domain of abstract objects, that is, in the domain of logic and mathematics. But since the end of the parts is in the whole, when reason returns to its own concrete foundation, then what Kant calls the empirical world, which is the synthesis of *Kategorien* and *Wahrnehmung*, appears. At the next level, when the axioms of logic, mathematics and geometry are conjoined with empirical content, the objective world of dynamics is established between them. That is, when reason itself becomes activating, and passes from abstract to concrete standpoints, the world of dynamics appears. Or again, as reason becomes activating will, the objective world of forces arises. Kant's objective world of synthetic principles (*synthetische Grundsätze*) is established in this way. As reason itself takes the form of the act it becomes a form that unifies other infinite acts. What is described in geometrical terms as a line or form becomes what is called a vector in dynamics. When the form of "time" (*Schema "Zeit"*) is added to the categories, everything becomes active; that is, the world of dynamics is established. When reason itself becomes activating – that is, becomes the will – and unifies experience through the forms of dynamics, the empirical world is established. Taking this direction as far as it can go we have the material world.

The spiritual world, on the contrary, is attained by reversing the direction of thinking and thus returning to the original condition of personal experience. The spiritual world is the world of reality established by the form of *Aktualität*, which is the direct union of acts. The data of experimental psychology and of history are already incapable of unification in terms of the laws of causation defined in the natural sciences. They must be united by means of a causal law of internal unity. However, when the standpoint of thought is completely abandoned and we enter deeply into the standpoint of the internal unity of the whole person, we transcend the realm of cognitive objects altogether and attain the realm of moral behavior. We no longer reproduce the other from a certain standpoint of thinking, we become the personal act itself. At this level, object and act are one. In moral behavior we are both thing and mind; overcoming the duality of immanent and transcendent, we become one reality. At this consummation of the union of acts, all forms of external unity are absent. There is only personal behavior in itself. There is neither affirmation nor negation; there is only the concrete act of absolute will. This realm of absolute subjectivity is the standpoint of religion. In religious intuition, everything is "one line of steel for ten thousand miles." In religious intuition, there is neither inner nor outer, neither I nor other.

My critique of thought can be extended to other merely partial and one-sided aspects of personal experience, such as visual and aural perception. These latter modes of consciousness are autonomous and free acts of the person. In the concrete, they have their own objective worlds. Abstractly, there are the objective worlds of color and sound; but when these acts reflect their contents in the internal act of the whole person, we have the world of aesthetic creation. In its own one-sided way the world of aesthetic creation is a religious standpoint. As Schopenhauer states, the artist becomes a religiose at the moment of inspiration.

We attain to a concrete standpoint by transcending a previous, abstract standpoint. The representation of color becomes act at the limit of its infinite series. The act of visual perception is a unity of an infinite series of color tones, of degrees of brightness and of saturation. A person is a unity of an infinite set of these kind of acts. A representation becomes an autonomous mental act at the limit of its own infinite set; in like manner a set of infinite acts becomes the free person at its limit. Each kind of mental act has its own intentional object. The empirical world is the objective world of the will, which is the act underlying all acts — that is, the unity of the whole person. But from the standpoint of the moral will the empirical world is merely an infinite world as the *Typus* of free will. The moral world, which is the object of free will, is the limit point of the infinite totality of the natural world.

As in Hegel's "concept", personal experience is such that each of its parts includes the whole. Similarly, a partial form of experience such as an act of visual or aural perception can directly reflect the concrete personal experience of the fusion of subject and object. We see this in music and art. These forms of experience have their own standpoints and intentionalities. However, in the final analysis, art is only a partially concrete standpoint. The true relationships among the concrete standpoints of the various perspectives and intentionalities of experience can be clarified only in the completely personal horizons of philosophy, morality, and religion. Culture advances from the abstract to the concrete along with the development of philosophy, morality, and religion. A new culture appears as the limit of the development of the old culture. Our true eternal life is not to be found in the lives of Tithonus and Ahasver, but in the Christ who, though young, was nailed to the cross.

*First English translation by David Dilworth and V. H. Viglielmo of Kitarō Nishida's work *Problems of Consciousness*, which appeared in the original Japanese in 1920: The first passage is from Chapter 3, Section 3, p. 61 ff. of *Nishida Kitarō Zenshū* (*NKZ*, the Complete Works of Kitarō Nishida, Vol. 3). The final translation and editing of the above passages were completed by David Dilworth.

DAVID A. DILWORTH

THE CONCRETE WORLD OF ACTION IN NISHIDA'S LATER THOUGHT

Nishida's *Fundamental Problems of Philosophy* (1934) was the ninth consecutive volume of philosophical writing in his career.[1] The work is itself comprised of two volumes forming one piece: *The World of Action* (1933) and *The Dialectical World* (1934). The fact that Nishida never again attempted to formulate a total system supports his own evaluation that in this work he clarified the fundamental structure of his thought. In this essay we shall be concerned with outlining an essential theme of that work – the "concrete world of action." It is a theme in which Nishida can be observed recasting the central topic of his thought, namely, the "place of true self-consciousness," in terms of his dialectic of the "topos of Nothing ness" (*mu no basho*). In so doing Nishida pulled together many of the threads of ideas developed in his earlier writings, particularly his long-standing definition of self-consciousness in dynamic terms, or what he called "the a priori of absolute will."

In this work, however, Nishida's earlier transcendental voluntarism is absent. The emphasis now is on the concrete world of action as a dialectical field of social-historical determination. The theme of *mu no basho* as the concrete world of action becomes the basis of Nishida's critique of Hegelian and Marxian dialectics in this and his other later writings. It grew out of a sympathetic reading of Bergson and Dilthey, among other contemporary philosophers, but culminated in notions of person and community that are opposed to relativistic theories of the life-world. In this way, Nishida's thought continued to exhibit its distinctive character of being an original philosophical structure that is at the same time a sustained dialogue with Western philosophies, especially with contemporary styles of phenomenology.

1. NISHIDA'S CRITIQUE OF WESTERN ONTOLOGIES

In Nishida's understanding, Greek or classical metaphysics of the West had been essentially a metaphysics of "being" or "form."[2] In contrast to

Nitta/Tatematsu (eds.), Analecta Husserliana, Vol. VIII, 249–270. All Rights Reserved.
Copyright © 1978 by D. Reidel Publishing Company, Dordrecht, Holland.

this, the scientific and philosophical revolutions in post-medieval Western philosophy have centered around the concept of "beings in motion." But Nishida's position was a critique of both "being" and "beings in motion" as abstractions derived from the concrete immediacy of the present beyond or prior to the subject-object dichotomy.[3] From this standpoint, the concept of "acting beings," and especially "personal actions" in the real spatial-temporal, social-historical world is the fundamental one. For "personal action" presupposes the concrete fusion of the individual and environment, particular and universal, and subject and object in the dialectical field of the social-historical world. Therefore, he affirmed, the new metaphysical direction should be towards defining a metaphysics of the "self," i.e. the acting, personal self, and its real ontological dimensions, in essentially social and historical terms. Nishida was claiming that neither classical philosophers of "being" or "form" nor post-medieval philosophers of "beings in motion" could fully measure up to his requirement.

As Whitehead once remarked,[4] the task of philosophy is to explain how the abstract forms derive from the concrete facts. The concrete is given in the immediacy of experience. Philosophy is not explanatory of this concrete immediacy, but of the abstractions derived from it. It seeks the "forms" in the "facts." The facts are always more than their forms. Nishida was attempting just this critique of abstractions. He contended that previous philosophy has never escaped being "rationalistic in essence." As a result of this orientation, the "self" has always been conceived in individualistic terms in that the concept of the other, the Thou, has always been *deduced from* the false point of departure of the dichotomy and opposition between *self* and *things*. The Thou has thus never been seen as an indispensable prerequisite for the very existence of the I. Nishida applies this critique to a variety of Western philosophical positions, ranging from Greek ontologies to contemporary positivistic, and Marxist, world-views.[5]

The standpoint of personal action presupposes the social-historical world *in which* the I-Thou relation *takes place*. In this sense the Cartesian *cogito ergo sum* was a false point of departure for modern philosophy. Nishida argues that it should be amended to "I act, therefore I am," with the implied dimensions of the social-historical world in which action *takes place*. "Place" (*basho*), of course, is a special term of Buddhist connotation in Nishida's philosophy. It is meant to indicate a non-thetic, and consequently non-substantive, concept of consciousness. The phrase *mu no basho* can literally be translated as the "nihilic place" of consciousness, somewhat reminiscent of Sartre's notion of *le néant*. Nishida is here developing this definition of consciousness in broad metaphysical terms.

Nishida goes on to say that there is nothing remarkable about understanding the "self" in purely social-historical terms. Sociology, and many of the other learned disciplines, are methodologically committed to this way of thinking. But they are unconcerned with first principles, in the sense of "explaining the reasons why the acting self as a free person arises therefrom," and generally stop short of elaborating a "world-view and view of human life."[6] Not only do the special disciplines not investigate the concept of the "self determining world" which, in Nishida's analysis, such an orientation ultimately implies; they exhibit the further tendency of presupposing a "scientific standpoint" whose ultimate presuppositions retain the notion of society or history as intellectual *objects*, i.e. *noema*. To the extent that this is so, the sociologically oriented disciplines are not talking about the truly active self in the real world of personal actions, no more than Marxism whose materialistic definition of "ideology" does not avoid defining social life in exclusively noematic terms.

Now, the rationalistic standpoint manifests itself in this tendency to conceive subjectivity and objectivity, self and world, *noesis* and *noema*, as mere opposites.[7] To Nishida, these are mere abstractions incapable of withstanding sustained analysis. Social and historical components of the real world are illustrated in every instance of personal action. "We exist as the self-determinations of such a world and have our individual tasks therein."[8]

But in order philosophically to conceive of the real world of personal action there must be a logic of the active self. Greek logic is essentially a logic of the subject of predication based on the Aristotelian definition of substance as subject that cannot become predicate. But, to Nishida, neither Greek logic nor Aristotelian metaphysics can then adequately define the world of action, i.e. the world of personal selves, in a truly social-historical perspective. Modern physics took a new point of departure in Galileo's rejection of many aspects of the Aristotelian tradition. (But the Cartesian idea of substance was still medieval.) The ensuing philosphical revolution then reached a climax in Kant, who attempted to give a transcendental foundation to the new concept of empirical scientific reality. Yet Kant's transcendental logic is still not a logic of the social-historical world. Kant's concept of empirical scientific reality is also an objective world of the intellectual self. According to Nishida, it was Hegel who first attempted to develop a dialectical logic of *practice*, i.e. of action which includes subject and object. Hegel thus pioneered a logic of social and historical reality. But Hegel's dialectical logic is still a logic of the subject and the *noema* – in that sense it is a variation on Greek philosophy too. Because of this, Marxist dialectics could turn Hegelian philosophy on its head – but remain another variation on Greek philosophy while doing so. For it too remains "rationalistic," and "the meaning of true social and historical reality, i.e. truly concrete reality, has not yet been adequately formulated."[9]

This brief interpretation of Western philosophical ideas is far from revolutionary, but it takes on an added significance when we realize that Nishida had been dealing with the problem of the rationalistic point of departure of philosophy – and the underlying subject-object dichotomy – since his maiden work in 1911, and especially since his turn to a more powerful metaphysical base in his concept of the *topos* (*basho*) of "true Nothingness" since 1927. In other terms, he had been dealing with variations on the theme of what he calls a "truly dialectic logic" for many years, and was accordingly in a position to develop his own conception of personal action in the social-historical world in a dialogue with Western thought.

Nishida could thus write that we should not become entrapped in the presuppositions of classical positivism simply out of the fear of falling back into the errors of classical rationalistic metaphysics. For the alternatives are not exclusively disjunctive. We must rather proceed to develop a "truly concrete logic." He cites both Windelband and Dilthey as two thinkers who attempt to clarify the significance of social-historical reality different from the presuppositions of natural scientific reality in classical positivism. Modern Life-philosophy and *Geisteswissenschaft* are especially influenced by Dilthey. But according to Nishida, Dilthey still conceives of the historical world as an object of cognition rather than as the world of personal action. Even Heidegger's Existentialism, which partly derives from Dilthey, is a "world of the understanding." It is a species of "phenomenology," but falls short of being "dialectical."[10]

As we can see, Nishida's frame of reference here is basically that of modern German philosophy. He shows no knowledge of such "process-philosophers" as C. S. Peirce, John Dewey, or A. N. Whitehead, whose orientations were not dissimilar in trying to develop categories to deal with social-historical reality outside of the presuppositions of classical Greek logic and Hegelian or Marxist dialectics. In a certain sense Nishida parted company with the "naturalistic" universe of such process-thinkers when he took a road different from William James' reading of "pure experience" in the opening pages of *A Study of Good*.[11] His ultimate premises remain Buddhistic. Yet in some of the conclusions he draws concerning the essentially social-historical structure of the world, Nishida's thought is surprisingly reminiscent of process philosophy. Some readers may find this comparison inappropriate for a thinker so deeply committed to Buddhistic premises which, by one account at least, run counter to much of the "naturalism" and "process-philosophy" of modern Western thought. At least some traditional interpretations of Buddhism, and especially of Zen, have tended to allege anti-historical and anti-social components in that spiritual tradition. But the question remains whether these interpretations are entirely accurate, in the first place, and whether Nishida did not add something original to the Mahayana tradition, in the second place.

The closest Nishida's text explicitly gets to process-philosophy is his use and critique of Bergson for having conceived of "pure duration" in a merely temporal-linear sense. According to Nishida, Bergson's subjective interpretation of time and process is still one-sided, presupposing the subject-object dichotomy and leaning in favor of the subjective side of that dichotomy. Bergson therefore lacked a true dialectic of space and time, i.e. a truly social-historical world.[12] But this critique cannot be said to apply to such philosophers as Peirce, Dewey, or Whitehead, whose respective thought structures may be said to be "truly dialectical," in Nishida's own phrase, although not necessarily in the same terms as his own focus upon "the self-identity of absolute contradictions."

It is not necessary to pursue this point further here. Let it suffice to say that Nishida's own system of the "dialectical universal," the concrete social-historical world which is self-determining and self-creative in the immediacy of the present, seems to be an interestingly "modern" development of Mahayana Buddhist tradition. Despite signs of these ideas from *A Study of Good* on, his previous works did not fully presage this turn towards a radical social-historical "field" in which the personal actions of active selves "takes place."

A central concern of Nishida's later writings is reflected in his attempt to elaborate a "logic" that reverses the Aristotelian logic of the grammatical subject of predication to conceive of a "logic of the concrete predicate." The concrete predicate is the "concrete *topos*" in which abstract subject and predicate emerge, as in the case of judgment. He thus stresses, not a return to Plato in the sense of seeing the subject in the determined predicate or universal Idea, but the "empty" place in which all determinations take place – i.e. True Emptiness. In *The World of Action* he argues that the Spinozan concept of *substantia* is also merely an absolute subject of predication still presupposing the Aristotelian logic. Schelling's concept of *Identität* ultimately goes no further. For true self-identity must be considered as predicate "in the sense of absolute Nothingness as subject, the self-identity of absolute contradictories as subject-qua-predicate and predicate-qua-subject."[13] Such a dialectical logic is precisely to be conceived from the standpoint of the *active self* considered as a social-historical reality. But what, then, did this relatively new conception of the "active self" mean?

2. THE WORLD OF MUTUALLY DETERMINING INDIVIDUALS

To answer this question, it is crucial to see that what Nishida calls "rationalistic" means the attempt to begin from the subject-object point of departure. This is the main thread running throughout the previous cited texts illustrating his fundamental argument with Western philosophical positions, as he understood them. To take either side of this dichotomy – or, indeed, even the unity of subject and object conceived objectively – as the starting point for philosophical discourse is, in Nishida's analysis, to lose the concrete immediacy of experience itself, and thereby to deal in abstractions. It might be remarked here that Nishida seems to be superimposing another kind of rationalistic interpretation on the immediacy of experience. For his own position is deeply committed to the discovery and application of a dialectical logic, variations on the theme of which have characterized his writings since 1927. This dialectic itself is rationalistic in the sense of presupposing the possibility of ultimate formulations in terms of binary oppositions. It is a Buddhistic logic of a long metaphysical tradition which Nishida uses with some originality in *The World of Action* (1933). Thus it might be remarked that while Western categories may seem rationalistic to an Eastern philosopher, Eastern metaphysical categories may seem rationalistic to a Western observer.

The seeming "rationalism" of Nishida's position consists in its reductionism of the immediacy of experience to the components of the epistemological dialectic between relative "being" and relative "non-being," and their resolution in the Mirror of Emptiness. To Nishida, this "recollection of contradictions" yields the Buddhist insight that being is "just as it is" in the Mirror of Emptiness. It also yields the insight of the perfect transparency of all beings in the *topos* of true Nothingness – in the Kegon Buddhist phrase, "the unhindered mutual interpenetration between phenomena and phenomena." These phrases are other ways of getting at the essential message of the Mahayana tradition, namely absolute negation-qua-absolute affirmation of experience itself achieved by virtue of the dialectic of the "negation of the negation." Yet Nishida would not call this dialectic "rationalistic." He takes pains to show how, in contrast to rationalistic positions attached to the subject-object dichotomy, his own standpoint is grounded on the ultimate and immediate *irrationality* of experience. In that sense, what he calls a "truly concrete logic," or "logic of experience" is a dialectic of the very irrationality in which experience, through its own self-determinations, concretely "takes place." Further, he insists that it is not an objective or deterministic logic in any sense of the word. Therefore it is better expressed as the self-determining dialectic of the self-determining world of experience itself.

This was the import of his long essay entitled 'A Preface to Metaphysics' with which Nishida begins *The World of Action*. In this opening essay Nishida tries to illustrate the functioning of a "truly concrete logic" in various fundamental areas of the world of personal action. But we shall see that even in his analysis of the world of action, which is a "social-historical world," the Buddhistic ideas of "dependent causation" (*pratitya-samutpaca*) and "the unhindered mutual interpenetration between phenomena" pervade the text.

Nishida begins by stating his point of departure in clear and simple terms. He writes that "reality" can indeed be understood in various ways. One view takes objects of sensation to be the real, as in the "empirical" view of the natural sciences. Another view regards the "rational," i.e. objects of thought, to be the real, as in the Platonic theory of Ideas and the subsequent tradition of *logos* in the West. A third position might be a "theory of immediate, or pure experience" that holds the real to be empirical content immediate to the self. Nishida writes that this latter position, although taking a perspective prior to the subject-object distinction, is still merely "seeing the external from the inside."[14] He does not elaborate the point here, but it is evidently meant to be differentiated from his own point of departure which he spells out next. "The true self is an acting self," he affirms, and "true reality is the object of the acting self. We realize our selves by living and acting in this world." Since true reality "must be considered from the perspective of the acting self," and since "true action must be personal," it follows that the "objectively real in the deepest sense must be sought in that which opposes personal action in the sense of resisting subjective reality."[15]

Nishida conceives of the world of the acting self as one of conflict and resistance. But such conflict and resistance is precisely the concrete sphere of dialectical immediacy in which personal action takes place. His purpose, then, is not to minimize the subject-object polarity, but to philosophize from the standpoint of the concrete immediacy of action itself in which the subject-object polarity emerges in intellectual analysis.[16] To Nishida, this is the very antithesis of a rationalistic view. For action precedes knowledge, not the other way around. For example, an epistemological self determined from the world of sensation, the physical world, is not the true self, but the negation of the self. Even Fichte's voluntarism, which he once followed, is called "still rationalistic," in contrast to which Nishida's present view of the real self is described as "grounded on absolute irrationality."[17]

The question, then, is: what is personal action concretely taken? To answer this question, Nishida's 'Preface to Metaphysics' makes a series of analyses whose basic premises and conclusions are consistently dialectical and Buddhistic. I shall briefly take up several of these themes here, namely (i) Action, (ii) Personal Unity, (iii) The I-Thou Relation, and (iv) The Subjectivity of Nothingness as True Self-Identity. These themes continue to be interwoven through the ensuing long essays of *Fundamental Problems of Philosophy* entitled 'The Self and the World' (vol. one) and 'The World as Dialectical Universal' (vol. two).

(i) *Action*. Nishida begins with an analysis of the notion of the motion of acting beings in the style of Aristotle, who conceived of an underlying substance or individual as the unchanging subject of change. But he points out that this Aristotelian conception does not truly account for the contradiction of movement in time, which necessitates that the individual include its own real self-negation within itself if it is to be truly moving and changing in time. For time is essentially discontinuous; it is a perpetual perishing. A being in motion is either truly *temporally* moving or it is not. The individual must therefore be an acting *individual*, not merely a universal which undergoes accidental specification while remaining immune to time and change. But it must then be a *self-determining* individual in the flow of discontinuous time, yet at the same time a *self-negating* individual. How is this possible? To Nishida, it is only possible in the sense of *dialectical self-determination* in which the negative is seriously considered. This means, in a sense, that the individual "lives by dying," for it is a "unity of contradictions."

Hence, a logic of action must truly account for both continuity and discontinuity in time. In Nishida's key phrase used throughout his later writings, it must be a logic of "the continuity of discontinuity."[18] "Living by dying" and "continuity of discontinuity" are key Nishida expressions which take on increasingly Buddhistic nuances in the course of the text as a whole. For in the above text the true universal which mediates the process of living by dying is "the universal of Nothingness."[19] Indeed, the constantly recurring phrase "the mutual determination of individuals" is now defined as "the determination of the continuity of discontinuity mediated by absolute negation."[20] Nishida's point is that this "logic of true being, i.e. of concrete reality" can be conceived only from the idea of action itself, where the self is already plunged into the realm of dependent causation mediated by absolute negation.

(ii) *Personal Unity.* "Personal action" implies a conception of the person and of personal unity. How is this to be understood in the perspective of the concrete world of action in time? Here Nishida takes up the theme of the "personal" as the meeting point of the rational and the irrational, and of the transcendent and immanent. Desire, for example, involves the clash between both rational and irrational, transcendent and immanent aspects of the self in its temporal existence. But desire, which is "born to die and dies to be reborn again," is essentially contradictory in itself.[21] Indeed, it is another instance of the self-contradiction of the dialectical individual which "lives by dying." The analysis of "desire" also takes on increasingly Buddhistic coloration in the course of *Fundamental Problems of Philosophy*.

Furthermore, personality cannot be unique and simple. It too can only be conceived in relation to other personalities. The *self* becomes a *person* only by recognizing the personality of the other, as in Kant's statement that in ethical action we must regard the other as an end in itself. Thus true individual personality "must be mediated by absolute negation." And therefore action as the relation between individuals must be grounded on the relation between persons. The real must be grounded on the self-determination of the personal self in this sense. This is another instance of the "continuity of discontinuity."

But what does even personal unity mean if we proceed in this manner "from concrete experience instead of from abstract logic"? Here, again, the same fundamental use of the dialectic of negative mediation is employed.

The personal self is always a process of unification from the focus of the actual present, each point of which is self-determining. This kind of *quantum* idea of the person in the self-determining but discontinuous flow of the present is a "living by dying" of the self, by virtue of which personal unity is constituted.[22] "Self-consciousness" is precisely such a continuity of discontinuity in which, with freedom at every point, the *I* of the present regards the *I* of yesterday as a Thou.

(iii) *The I-Thou Relation*. From the foregoing brief summary of some threads of Nishida's argument we can see that a consistent application of the concept of "dependent causation" and the underlying logic of negation has been employed. Indeed, even the few texts cited illustrating the idea of the personal individual involve the use of the concept of dialectical self-negation (seen from the point of view of the individual) and of mutual determination (seen from the point of view of the total relationship of action and reaction). Consequently the foregoing analysis has actually yielded two distinct applications of the logic of negation in reference to the personal individual. One is the I-Thou relation *within* the individual self in which the *I* of today regards the *I* of yesterday as a *Thou* – which is an interesting way of saying that there is no "permanent ego." In that sense, there is no unique *I*, any more than there is a unique present. It too is "a self-identity of absolute contradictories" as a constant "living by dying." Therefore we see the *absolute other* in the depths of the self. For "the personal self determines itself dialectically as a dynamic individual."[23]

But further, such a unity of self-consciousness of the personal self as a continuity of discontinuity *occurs in* a social-historical world.[24] Accordingly, the same metaphysical structure is the ground of the I-Thou relation *between* persons. Nishida's present analysis presupposes the 'I-Thou Relation' chapter of *The Self-Conscious Determination of Nothingness* (1932), and was further expanded in the chapters 'The Self and the World' (1933) and 'The World as Dialectical Universal' (1934) of the present volume under consideration, *Fundamental Problems of Philosophy*. But in the present 'Introduction to Metaphysics' Nishida puts down the basic components of his final position concerning the I-Thou relation *between* persons. For example, he writes that individuals "face one another separated by absolute negation."[25] This is not mere opposition, but a "relationship of expression" and "a mutual seeing separated by absolute negation" in which the self exists "in absolute negation-qua-affirmation."[26] In this respect everything which confronts the self – the "mountains, rivers, trees, and stones" as well – has the meaning of a Thou.

In this passage, therefore, Nishida was affirming that the concrete world, in which everything is a Thou to the self, constitutes a "metaphysical society."[27] Such a metaphysical society is evidently the community of transparent immediacy of "the realm of the unhindered mutual interpenetration between phenomena," which are united expressively in the I-Thou relationship because they are grounded in absolute Nothingness. Nishida goes on to say that action means such "mutual determination through absolute negation," which in turn implies a *topos* or a field (*basho*) in which such mutual determination takes place. This *topos* is the "concrete universal" in which mutually determining individuals are truly what they are. Nishida often uses the Buddhist metaphor of a mirror to describe this. The mirror, because it is itself empty, can reflect all things just as they are. In this conception, the idea of the *topos* of Nothingness, which is the absolute ground of the concrete immediacy of the self-determining present and self-determining social-historical world, guarantees the dialectical identity of "absolute negation-qua-affirmation." But we can note how the idea of dialectical identity here comes together with the idea of "the mutual interpenetration of phenomena." It is precisely the underlying conception of the Mirror of True Emptiness which has yielded these social-historical conclusions in Nishida's analysis of the I-Thou relation.

(iv) *The Subjectivity of Nothingness as True Self-Identity*. We are now in a position to pull together some of the foregoing threads of ideas and see the relation between the world of action, i.e. of "dependent causation," and the universal Mirror of Nothingness in which such action "takes place." A Logic of Dependent Causation, as it were, has been consistently applied to the world of action which, in turn, is the world of the concrete present prior to intellectual analysis into subjects and objects. The core of this logic seems to be as follows. On the one hand, the individual is absolutely independent. It is *causa sui*. On the other hand, a merely unique individual has no meaning. Therefore we must conceive of the mutual determination of individuals both temporally and spatially. But this further presupposes a concrete *topos* (*basho*) in which individuals are mutally determining. Hence the relation between individuals is dialectical, a relation of absolute contradictories. Indeed, even the relation between the discontinuous temporal points *within* the life of the individual is dialectical in this sense. Their concrete "place" is thus a self-identity and unity of contradictions.[28]

Nishida's critique of Hegel in his 'Introduction to Metaphysics' and 'Summary and Conclusion' of *The World of Action* (1933) derives from this Buddhistic reading of the concept of dialectical negation. Nishida agrees with Hegel's critique of Spinozan substance as a mere self-identity which lacked internal dialectical determination. But Hegel's own *Phenomenology of Mind* then proceeds to redefine Spinozan substance as dialectical *subject*. To Nishida, this is still conceiving self-identity *noematically*, i.e. in the direction of the Aristotelian grammatical subject which presupposes the subject-object and subject-predicate forms of predication. In that sense it is merely unique – indeed, an *absolute Subject*. In another sense, the Hegelian absolute is still a kind of abstract *universal* which does not truly allow for the mutual determination of concrete *individuals*. It is ultimately a self-determination of a universal, i.e. the "rational-qua-real," and therefore "does not avoid being essentially Greek and accordingly cannot truly ground the irrational."

Nishida is here claiming that his own position has a true dialectical logic of the concrete world. He is further claiming that he has a conception of true (mutually determining) *individuals* whose existences are not "subsumed" by the Hegelian universal. For he takes seriously the concept of *absolute* dialectical negation.[29]

Another way of understanding this is that Nishida is implying that previous philosophers fail to take into adequate account the fact of mutual determination itself. Or when they deal with the concrete world in which "living is dying," they account for it in rationalistic terms. In this sense, Western philosophies generally continue the tradition of Greek *logos* which sees the real *above* the contradictions of spatial and temporal existence, i.e. above the flux of "dependent causation." They do not conceive of a true dialectic of contradictions in the immediate world of action itself. To Nishida, true self-identity positively involves the fact of mutually determining individuals. Therefore true self-identity must be the "universal of Nothingness,"[30] the mirror-like self-identity of contradictions. In contrast to this position, the self-negation of an absolute (Hegelian) subject would be a "mere contradiction," not the "self-identity of contradictions."

If we were to interpret this position once again in reference to Mahayana Buddhist tradition, I think we can see the following sequence of ideas. The world of "dependent causation," which destroys the concept of a unique ego or permanent substance, presupposes the ground of Nothingness or Emptiness, which in turn guarantees that sensible forms or beings are precisely as they are. The logic of ideas has progressed to the point of insight into the meaning of the famous Mahayana phrase: "sensible form, precisely as it is, is emptiness; emptiness, precisely as it is, is sensible form." But this seeming logical progression can now be restated in terms of the Hua-yen phrase: "the dharma realm of the unhindered mutual inter-penetration between phenomena and phenomena," which in turn is to be understood as the true meaning of "dependent causation." We have thus come full circle. In one sense we have gone through a process of rational deduction of ideas, but in another sense this rational process is a single insight. Nishida's concept of the "world of action" is clearly both.

If this rendering of Nishida's concept of the world of action is allowed, we are in a position to conclude that the Mahayana tradition, or at least its philosophical repossession by Nishida Kitarō, is not lacking in a concept of the *individual*. Indeed, Nishida's position seems to be claiming to have discovered the "logic" of the individual in terms of the irrationality of experience itself in which individuals are mutually determining. In his phrases, "the mutual determination of individuals" or "mutual deter-mination between individual and individual" must take place in "the deter-mination of the *topos* of Nothingness." It is the precise function of "true Nothingness" to be the ground, not of mere negation, but of the *absolute affirmation of individuals in the plural*.[31] From this position which absolutely affirms individuals in the plural Nishida went on to elaborate a concept of creative, social-historical world.

3. THE SELF-CREATIVITY OF THE WORLD

We have so far surveyed the leading ideas of two of the three long essays comprising *The World of Action* (1933), namely chapter one, 'An Introduction to Metaphysics,' and chapter three, 'Summary and Conclusion.' The middle chapter entitled 'The Self and the World' was the longest and structurally most important articulation of Nishida's argument in that work. It is repetitious of some themes of the preceding chapter, 'An Introduction to Metaphysics,' but at the same time Nishida's circular style of writing permitted him to deepen his ideas in certain directions. For example, he further elaborated such themes as the world of "desire" as illustrative of the world of personal action as a self-identity of contradictions; the concept of the "eternal present" as the ultimate ground

of irrationality identifiable with absolute Nothingness; and the concept of
"true creative evolution" as the dialectical self-determination of the world.
As we shall see, these directions exhibit Nishida's concern in 1933–34 to
reap the fruit of his Buddhistic epistemological analyses since the 'Basho'
chapter of 1927 in the sphere of a metaphysics of the social-historical
world.

The very title of this main essay, 'The Self and the World,' demonstrates
Nishida's primary concern. The Japanese title "Watakushi to sekai" may
also be translated 'The I and the World,' implying a creative contrast with
his often repeated phrase "The I and Thou." From even a few previously
cited texts we can see that Nishida conceives the I-Thou relation in the
broad sense of a "metaphysical society" in which the I is in dialogue "with
the mountains, rivers, trees, and stones" as well.[32] It is not too difficult to
read into this idea the notion of the community of all sentient beings of
traditional Buddhist thought,[33] especially when we are given to understand that this realm of interrelated beings exists in the Mirror of True
Emptiness.

As usual, we will find that Nishida's actual text is short on traditional
Buddhist references – with the exception of his cardinal metaphysical
concept of mu or Nothingness – and is long on philosophical analysis in
the language of Western philosophy. The reason for this is again that
Nishida develops his ideas somewhat originally and independently of
Buddhist scriptural traditions, especially by virtue of a generic categorial
dialogue with certain ideas and presuppositions of Western thought. The
theme of "The I and the World," understood against the background of the
modern discussion of the "I-Thou relation," in fact exemplifies this point.
It is important to notice that Nishida is thinking primarily in terms of "I-
World" rather than "I-Thou," in the sense that the latter is a limited case of
the former.

The chapter 'The I and the World' carries this point in the actual
progression of ideas. For it begins with a recapitulation of his doctrine of
the "personal self" and ends with a wider concept of the "true creative
evolution" of the self-creative world. The former doctrine is itself an
excellent summary of his position developed to this juncture, along with
adding new emphases that would continue to be explored throughout
volume two, The World as Dialectical Universal.

Nishida in fact begins his analysis by returning again to the Cartesian
doubt, which yields the conclusion that the self exists. The question then
becomes: what is the "self" which "exists"? The existential self is
"personal," which for reasons surveyed above involves the I-Thou relation

in which the I discovers itself only by recognizing the absolute other, the Thou. This conclusion returns us to the conception of the "continuity of discontinuity" and the underlying logic of the identity of absolute contradictories. Here Nishida makes the interesting observation that "moral phrases not only signify the evaluation of our person, but also of its existence." For they signify its dialectical existence, its self-negation-qua-self-affirmation. Even the continuity of discontinuity of the temporal unity within the personal self is a moral relation, a "dialogue" in which "the world of personal unity is a world of *logos*."[34] But he does not develop this point further here. He rather returns again to his fundamental metaphysical position:

The individual person is determined through the dialectical determination of absolute negation-qua-affirmation, i.e. through the self-determination of absolute Nothingness.[35]

In the course of the ensuing analysis it is clear that the moral relationships which the I-Thou relation *within* the person and *between* persons involves is an integral, but not primary, part of Nishida's doctrine. The question of the contradictions of "desire" is given far more attention than that of the moral dialogue between persons.[36] In fact, the "moral *ought*" is treated as illustrative of the same basic point as that of "desire," namely that, as a "living by dying" of the contradictions of negation and affirmation of the self, the moral relation is based on a deeper irrationality.[37] We are not merely rational beings, but social-historical beings. Hence Hegel's concept of the human community as the "true moral substance" is affirmed by Nishida over the Kantian idea of an ethical imperative. But in another respect he parts company with Hegel in that Hegel's social-historical community is grounded in an absolute Reason.[38]

Nishida's conception of "culture" as "ideology" in a sense which criticizes the presuppositions of the Marxist position is another case in point. For culture as "ideology" proceeds from a deeper ground of irrationality than the Marxist point of departure allows for.[39] Indeed, Nishida's own frequent use of the concept of "seeing the Platonic Ideas" in the social-historical process of the self-creative world is a continuous critique of Marxist premises. His position here is reminiscent of Schopenhauer, as the following text illustrates: "History, which is a dialectical process of self-determination, may be said to be the process of the transformation of the content of our life, our racial life, into the Platonic Ideas."[40] Since his earlier work *Intuition and Reflection in Self-Consciousness* (1917) we have seen Nishida's basic affinity with the reversal of

Hegelian Idealism by Schopenhauer. (Cf. Nishida's concept of "absolute will" as the irrational ground of the "true self".)

In sum, neither morality nor "ideology" are the primary focus of Nishida's analysis of the "I-World" relation. Rather he attempts to elaborate a concept of creative intelligence within the dynamic world in which the I-World relation "takes place." The concept of creative intelligence is grounded in the "self-determination of the eternal present."[41] Nishida takes over the Bergsonian concept of intelligence as "tool-making" and the concept of man as "*homo faber*" in both the present chapter and in *The World as Dialectical Universal* (1934). Conceptual knowledge functions in the realm of instinct and the satisfaction of desire. A deeper "intuition" of the irrational ground of the self-determining world pervades the thought of both Bergson and Nishida, but the latter criticizes the former's position as well for still being "rational" and "subjective" in the sense of presupposing the subject-object dichotomy.

The prime focus of Nishida's concept of the I-World relation may be said to be that of the paradox that the world of "birth-and-death," of *samsara*, is more than "mere" birth and death. For it is dialectically grounded on the self-determination of absolute Nothingness, the ultimate irrationality of experience. In this light, the "mutual determination of individuals" is not a mere struggle for existence in the noematic or "physical" world which still presupposes the subject-object point of departure. "Desires," like "time" and "morality," negate yet affirm us. They "kill yet give us life."[42] This involves the key point that our true self-identity can be conceived *neither* in the direction of the subject, or abstract spirit, *nor* in the direction of the predicate, or physical matter. True self-identity is the self-identity of absolute contradictories which genuine mutual determination implies.

In the chapter 'I and the World,' it is this ground beyond mere life and death which occupies Nishida's attention as he builds up his case for a conception af a self-creative social-historical world. He writes that "time" or the "individual" fall into "mere contradiction" if conceived in and through themselves, for they can only be truly conceived in terms of a true dialectical ground, i.e. absolute Nothingness. Innumerable individuals can be conceived from this ground, which is ultimately the "self-determination of the eternal present," and "the true world" which is "essentially irrational."[43] In other terms, the world of true reality is "the eternal present as the self-identity of absolute contradictories," which may be further conceived from the personal experience of the self in the I-Thou relation. In such a sense, there is "social-historical determination at the foundation of reality."[44] But this ground of acting things, which must possess dialectical self-negation within themselves, guarantees that the possession of absolute negation within the self means "resurrection from absolute death."[45] It means that in the *topos* of the mutual determination of individuals should be conceived as "the self-determination of absolute space, the absolute present."[46]

4. CREATION,

From the latter half of 'The I and the World' chapter of 1933, Nishida begins to elaborate a concept of "creation" and the "self-creativity of the world." He further develops this theme in volume two, *The Dialectical World*. His logic of self-determining individuals in a field of mutual determination in both temporal and spatial dimensions of existence has yielded the conclusion that "true life must be social and historical." Therefore he can further write: "Action which determines itself can also be conceived as the creativity of the self-determining environment. The self-determination of the environment in the social-historical world also directly has the meaning of self-determining individuals."[47] In this perspective, the environment is not merely the objective world of desire, but "the world of objective spirit." We possess a "common world of spirit, a world of a universal self."[48] But these Hegelian sounding ideas are ultimately grounded on a deeper irrationality than Hegel contemplated. For we have seen that this world of "dependent causation" is a world of the "mutual interpenetration between phenomena and pheomena" without a substantive ground.

Because of this radically dialectical structure of the world, our actions are social-historical in essence. For example, the self as a free person paradoxically finds itself through submerging itself in objectivity, as in the case of aesthetic creation. The artist himself does not foreknow how his creation will turn out. This is an instance of "seeing through acting," which is true of all forms of social action.[49] Nishida seems here to give a certain priority to the social, i.e. the spatial field of contemporaneous coexistence, over the historical dimension of this thought structure, since he goes on to say that "true dialectical determination" is not found in the relation of I and Thou *within* the temporal unity of the personal self, but *between* persons in dialogue in the present. True dialectical determination, he adds, is found "not between temporal things but rather spatial things, i.e. between subjectivity and objectivity."[50] Indeed, it may be pointed out that the concept of absolute Nothingness as the *absolute* ground of the dialectical negation-qua-affirmation of individuals has been defined as both the "eternal present" and "absolute space" or "place," both of which conceptions lean toward a certain primacy of spatial over temporal metaphors in Nishida's analysis. "Mirror" is of course another spatial metaphor.

This strong focus on the self-determining *present* as the mirror surface of the *Eternal Present* may derive naturally from the fact that "place" (*basho*) has been Nishida's key vocabulary item since 1927. Nishida in fact continues this nuance of the primacy of spatial over "process" dimension of dialectical determination in his concept of "the world as dialectical universal." But nevertheless Nishida insists throughout both volumes of *The Fundamental Problems of Philosophy* that the true dialectical world is "self-creative," and that his concept of the self-creativity of the world is not to be confused, for example, with either Platonic *logos* of Hegelian absolute.

Perhaps for this reason the present text deals again with the philosophy of Bergson, a philosopher Nishida always felt close to, even while criticizing some of Bergson's ideas from his own Buddhistic premises. His affinity with Bergson derives from the fact that Bergson's philosophy is also a negation of both Greek *logos* and Hegelian *absolute* insofar as these latter positions conceived of true reality as ultimately above the flow of time. His critique of Bergson is simply that "true creative evolution is a dialectical determination which is subjectivity-qua-objectivity and objectivity-qua-subjectivity."[51] In other words, in Nishida's view, Bergson's concept of "pure duration" lacks a genuine dialectical character. Being conceived in the standpoint of the subject which presupposes the subject-object point of departure, it is a concept of an infinite internal flow in time. Therefore in Bergson, the universal, i.e. the physical, spatial, and environmental, is subsumed within the determination of the individual. Bergson's position is the opposite extreme of Fichte's concept of *Tathandlung* that subsumes the individual within the universal. But "a world of pure duration which is not physical is a mere world of dreams."[52] True creative evolution must be the dialectical determination between individual and environment. Bergson's concept of "pure duration" lacks this "physical" dimension.

From this background we can see why Nishida defines "creation" itself in a sense which subsumes both Bergsonian and some traditional Western theological and philosophical ideas in a wider dialectical framework.[53] In the light of many years of his variations on the theme of the logic of the

Middle Path, it seems perfectly understandable that Nishida should here define "true creative activity as the self-identity of absolute contradictions," and as "not merely the activity of the unity of subject and object, but a self-determining *topos* (*basho*), in the sense of the self-determination of absolute Nothingness."[54] Symbolically, if not consciously, Nishida then takes to task the ideas of both Plotinus and Erigena, two of the philosophers whose thought structures have greatly contributed to the philosophical idea of creation in the West.

Nishida's essential position is that *noesis* and *noema* must be conceived dialectically, but Plotinus, according to Nishida's analysis, thinks of the One and its emanations "in the direction of the *noema*." In the opposite direction of *noesis*, emanation is "the self-determination of *basho* within itself, i.e. an infinite seeing of the self within itself. It may therefore be thought to be creation as the self-determination of absolute Nothingness."[55] This would truly go beyond the world of *logos* or of *nous*, as the Plotinian One is meant to do. But it would not be a kind of Hegelian subject, for Nishida is here criticizing the attempt to take either abstract *noema* (Plotinus)) or abstract *noesis* (Hegel) as point of departure. Nishida's basic point is that the dialectical logic of concrete action involves the mutual determination of *individuals*. "Creation," therefore, as the matrix or medium of such mutual determination, is "the function of the self-determination of absolute Nothingness."

Nishida goes on to comment on the position of Scotus Erigena, who is perhaps the most original and radical "creationist" thinker of the early Christian Neo-Platonic tradition. Nishida often cites Erigena's ideas in his early writings. At the end of the 'I and the World' chapter of *The World of Action*, he significantly returns to Erigena.[56] Nishida evidently approves of Erigena's concept of creation, which is sufficiently dialectical to suit his own Buddhistic purposes. Yet Nishida concludes by ringing a Buddhistic change even on Erigena. He writes:

The ancient mystics still stood in the perspective of subjective logic, and consequently they had to negate the world from such a standpoint. The position of the self-determination of the dialectical universal which I am enunciating does not negate the world.[57]

This observation is a challenging one.

NOTES

[1] Kitarō Nishida, *Fundamental Problems of Philosophy*, transl. by David A. Dilworth, Tokyo: Sophia University Press, 1970. The Japanese text is published under the title of *Tetsugaku no kompon mondai*, and appears as volume seven of *Nishida Kitarō zenshū*

(Complete Works of Kitarō Nishida), 19 vols., Tokyo: Iwanami Shoten, 1965. Subsequent citations in this essay will be from the Japanese edition, hereafter abbeviated to *NKZ*.

² *NZK*, VII, p. 203.

³ Ibid., pp. 83–84.

⁴ A. N. Whitehead, *Process and Reality*, New York: Harper Torchbook Edition, 1960, p. 30.

⁵ *NZK*, VII, p. 173: "I feel that philosophy has hitherto not truly philosophized in the standpoint of the active self. Consequently, it has never fundamentally considered what this real world in which we act is. Not only Greek philosophy, but also modern philosophy which centers on empirical reality, have been essentially rationalistic. Even taking sensation instead of reason, it too does not avoid being an object of the intellectual self. And as long as Marxist philosophy, which claims to take practice as its focus, defines the world from the side of the objective, it cannot truly philosophize in the perspective of the acting self. For history does not truly emerge from the objective world conceived by positivism."

⁶ Ibid., p.175.

⁷ Ibid., p. 176: "The standpoint of the active self transcends the opposition between subjectivity and objectivity and is that in which such opposition itself is grounded. Even knowing is a kind of action in this sense. In the standpoint of the active self, this world is neither merely subjective nor merely objective, neither merely a world of things nor of consciousness. It is the world of personal life from which we are born and to which we die again."

⁸ Loc. cit., p. 176.

⁹ Ibid., p. 179.

¹⁰ Ibid., pp. 179–180.

¹¹ See David A. Dilworth, "The Initial Formulations of 'Pure Experience' in Nishida Kitarō and William James," *Monumenta Nipponica* 24, 1–2 (1969), 93–111.

¹² *NKZ*, VII, p. 181.

¹³ Ibid., pp. 182–3.

¹⁴ Ibid., p. 5.

¹⁵ Ibid., p. 6.

¹⁶ Ibid., pp. 6–7: "The real resists or conflicts with the person. However, something merely such is not the real. Something which entirely transcends the person has no meaning at all. Reality must have the further meaning of determining us. Resistance or conflict already implies this. . . . The real in such a sense does not entirely transcend the person, for it always retains the meaning of determination. True reality is that which fully determines us – indeed, which determines us *from our very depths*. Accordingly, the opposition between subject and object must be sought *from within reality itself.*"

¹⁷ Ibid., p. 9.

¹⁸ Ibid., p. 16: "The individual determines itself only in relation to other individuals. The idea of a single individual has no meaning. For the individual to determine itself there must be what I call the determination of *topos* of Nothingness, (*bashoteki gentei*), i.e. a unity of absolute contradictories. The principle of particularization has the meaning of the continuity of discontinuity. . . . The mediating function of the universal which truly mediates the individual must have the significance of an absolute negation-qua-affirmation, i.e. of a continuity of discontinuity."

¹⁹ Ibid., p. 17.

²⁰ Ibid., p. 18.

²¹ Ibid., p. 19.

²² Ibid., p. 23: "Continuous and discontinuous, or rational and irrational determinations are

the two directions of concrete being, i.e. of the self-determination of the dynamic individual. Concrete being, i.e. the truly self-determining individual, may be conceived as always determined by the universal and be its limit point. But at the same time, the individual transcends and determines the universal as well. It lives by dying. .... The self as something truly temporal must be understood through such a logic."

23 Ibid., pp. 38–39.

24 Ibid., pp. 40–41: "We always have our individual personalities as beings who exist in a social and historical world which can be regarded as the self-determination of the present. An infinite number of individuals can be determined in the standpoint of the self-identity of absolute contradictions which is both a one-qua-many and many-qua-one, i.e. in the sense of the mutual determination of individuals."

25 Ibid., p. 59.

26 Ibid., p. 59.

27 Loc. cit.

28 Ibid., pp. 19–24.

29 Ibid., pp. 36–7: "A dialectical subject in the above sense must be absolute Nothingness. Moreover, mere non-being is nothing at all. Dialectical movement cannot emerge from such a thing. It must be absolute negation-qua-affirmation, i.e. absolute Nothingness-qua-being. Or again, it must be the unity of absolute contradictories. It must be a self-identity of aspects which touch yet do not know each other. From such a perspective, dying is living and living is dying. There is an infinite process of negation-qua-affirmation. Dialectical determination touches such an absolute self-identity at each step of its process."

30 Ibid., p. 50.

31 Ibid., pp. 54–55: "A true individual must be absolutely independent. Therefore the mutual determination of individuals must be one between absolutely independent entities, and it must be a unity of absolute contradictories. Moreover, as long as such a mutual determination can be conceived, it must be considered to be the determination of a universal. The self-determining universal in such a sense must be a self-identity of absolute contradictories, i.e. an absolute affirmation-qua-negation. *Individuals exist in worlds which differ absolutely. Each individual has its own world. There is no universal which subsumes the I and the Thou.* For the I and the Thou to be what they are in the mutual determination of I and Thou, there must be *mutual determination through absolute negation.* ... I and Thou face each other and determine one another in the absolute aspect of being-qua-non-being, death-qua-life. They do so as the self-identity of absolute contradictories, i.e. as the self-determination of absolute self-identity."

32 Ibid., p. 59.

33 The phrase is in fact an old Zen phrase.

34 Ibid., p. 86.

35 Ibid., p. 87.

36 Ibid., pp. 103–25.

37 Ibid., pp. 101–02.

38 Ibid., p. 103.

39 Ibid., p. 160.

40 Ibid., p. 170.

41 Ibid., p. 88

42 Ibid., p. 87.

43 Ibid., p. 90.

[44] Ibid., p. 92.

[45] Ibid., p. 93.

[46] Ibid., p. 94: "Each and every point must touch the absolute, i.e. must be resurrection from absolute death. True dialectic must have such a meaning, i.e. it must be the determination of the *topos* of Nothingness. A [mere] processive dialectic is still not a dialectic of absolute negation. For the mutual determination of individuals should be conceived from the self-determination of absolute space, the absolute present."

[47] Ibid., p. 134.

[48] Loc. cit.

[49] Ibid., p. 136. "Seeing through acting," and "active intuition" (or "action-intuition"), are technical phrases in Nishida's writings.

[50] Ibid., p. 137.

[51] Ibid., p. 146.

[52] Ibid., p. 147.

[53] Ibid., p. 156: "Creation may be conceived from the side of the subject or from the side of the object. Bergson's concept of creative evolution conceives creative activity from within. The *élan vital* is pure duration which may be thought to be the very interior of our Self. When the contrary idea of a God creating the world is entertained, creation is conceived as coming from without. However, the true meaning of creation is found at the point where such contradictory meanings of creation are one, i.e. when there is an identity between absolute immanence and absolute transcendence."

[54] Ibid., p. 157.

[55] Loc. cit.

[56] Ibid., p. 172: "As Scotus Erigena has said, God who both creates and is not created, yet is infinitely creative, must also be God who neither creates nor is created. On the basis of such a perspective, that which is infinitely creative sees itself within itself, reflects itself within itself. Our objective worlds emerge within it."

[57] Loc. cit.

SELECTED BIBLIOGRAPHY OF THE MAJOR PHENOMENOLOGICAL WORKS TRANSLATED INTO JAPANESE AND OF THE MAJOR PHENOMENOLOGICAL WRITINGS BY JAPANESE AUTHORS

I. JAPANESE TRANSLATIONS OF THE WRITINGS OF E. HUSSERL

(Translator, Japanese Title, Original Title, Publisher or Name of the Journal and Year of Publication. The sign indicates a long vowel.)

1915

Itō, Kichinosuke: 'Gaku to shite no Tetsugaku' (Philosophie als strenge Wissenschaft) in the journal *Tetsugaku Zasshi* (= *Journal of Philosophy*, Tōkyō University) Nr. 343–346, 1915 (abbreviated translation and commentary).

1923

Watsuji, Tetsurō: 'Franz Brentano no Omoide' (Erinnerungen an Franz Brentano), in the journal *Shisō* (= *Thoughts*, publisher: Iwanami), Nr. 18, 1923.
(Unknown translator): 'Saishin-sono Mondai to Hōhō' (Erneuerung – Ihr Problem und Ihre Methode', in the journal *Kaizō* (= *Transformation*, publisher: Kaizō) Series 5, Nr. 3, 1923.

1924

(Unknown translator): 'Koin Rinri Mondai no Saishin' (Erneuerung als individualethisches Problem), in the journal *Kaizō*, Series 6, Nr. 2, 1924.

1932

Kitō, Eiichi: *Junsei Genshōgaku oyobi genshōgakuteki Tetsugaku Kan* (*Ideen zu einer reinen Phänomenologie und phänomenologischen Philosophie*), publisher: Shunjūsha, 1932.

1933

Terada, Yakichi: *Sanjutsu Tetsugaku* (*Philosophie der Arithmetik*), publisher: Monasusha, 1933 [very poor translation].

Nitta/Tatematsu (eds.), Analecta Husserliana, Vol. VIII, 271–287. All Rights Reserved.
Copyright © 1978 by D. Reidel Publishing Company, Dordrecht, Holland.

1934

Ikegami, Kenzō: *Junsui Genshōgaku oyobi genshōgakuteki Tetsugaku Kōan* (*Ideen zu einer reinen Phänomenologie und phänomenologischen Philosophie*), the first half; publisher: Iwanami, 1934.

1941

Ikegami, Kenzō: *Junsui Genshōgaku oyobi genshōgakuteki Tetsugaku* (*Ideen zu einer reinen Phänomenologie und phänomenologischen Philosophie*), the second half; publisher: Iwanami, 1941.

1954

Yamamoto, Mañjiro: *Genshōgaku Josetsu - Dekarutoteki Seisatsu* (*Eine Einleitung in die Phänomenologie - Cartesianische Meditationen*), publisher: Sōbunsha, 1954.

1965

Tatematsu, Hirotaka: *Genshōgaku no Rinen (Idee der Phänomenologie)*, publisher; Misuzu, 1965.

1967

Tatematsu, Hirotaka: *Naiteki Jikan-Ishiki no Genshōgaku* (*Verlesungen zur Phänomenologie des inneren Zeitbewusstseins*), publisher: Misuzu, 1967.
Tatematsu, Hirotaka: *Ronrigaku Kenkyū* (*Logische Untersuchungen*, Bd. I), Vol. I, Publisher: Misuzu, 1968.

1969

Satake, Tetsuo: *Genmitsuna Gaku to shiteno Tetsugaku* (*Philosophie als strenge Wissenschaft*), publisher: Iwanami, 1969.

1970

H. Tatematsu, Y. Matsui and H. Akamatsu: *Ronrigaku Kenkyū* (*Logische Untersuchungen,* Vol. II/1, I and II. *Untersuchungen*), Vol. 2, publisher: Misuzu, 1970.
Koike, Minoru: *Genmitsuna Gaku to shiteno Tetsugaku* (*Philosophie als strenge Wissenschaft*), publisher: Chūōkōronsha, 1970.
Funahashi, Hiromu: *Dekarutoteki Seisatsu* (*Cartersianische Meditationen*), publisher: Chūōkōronsha, 1970.

1974

H. Tatematsu and Y. Matsui: *Ronrigaku Kenkyū* (*Logische Untersuchungen*, Vol. II/ 1, III, IV, and V. *Untersuchungen*), Vol. 3, publisher: Misuzu, 1974.

Hosoya, Tsuneo and Kida, Gen: *Yōroppa Shogaku no Kiki to chōetsuronteki Genshōgaku* (*Die Krisis der europäischen Wissenschaften und die transzendentale Phänomenologie*), publisher: Chūōkōrcnsha, 1974.

1975

Hasegawa, Hiroshi: *Keiken to Handan* (*Erfahrung und Urteil*), publisher: Kawade-shobō Shinsha, 1975.

1976

H. Tatematsu: *Ronrigaku Kenkyū* (*Logische Untersuchungen* Vol. II/2), Vol. 4, publisher: Misuzu, 1976.

H. Tatematsu: *Anthology of Husserl*, publisher: Heibonsha, 1976.

New translations of the three-volume *Ideen* (Watanabe, Jiro and Tatematsu) and of *Formale und transzendentale Logik* (Tatematsu and Noe, Keiichi) are now in progress. The first half of *Ideen*, Vol. I, will be published next year by the publisher Misuzu.

II. JAPANESE LITERATURE ON THE PHENOMENOLOGY
OF E. HUSSERL (A SELECTION)

1911

Nishida, Kitarō: 'Ninshikiron ni okeru jun ronriha no shuchō ni tsuite' (On the Theses of the Pure-Logic Schools in Epistemology), in *Geibun* 2, Nos. 8 and 9, 1911.

1916

Nishida, Kitarō: 'Gendai no Tetsugaku' (Contemporary Philosophy), in *Tetsugaku-kenkyū* (*Philosophical Investivations*), Kyōto University, No. 1, 1916.

1923

Kihira, Masami: 'Genshōgaku ni tsuite' (On Phenomenology), in *Tetsugaku Zasshi* (= *Journal of Philosophy*, Tōkyō University) Nr. 436, 1923.

Honda, Kenzō 'Kahei-Riron no genshōgakuteki Kōsatsu' (Phenomenological Observations on the Money Theory), in the monthly journal *Shisō* (= *Thoughts*, publisher; Iwanami) of July and September, 1923.

1925

Tanabe, Hajime: 'Ninshikiron to Genshōgaku' (*Epistemology and Phenomenology*), in *Kōza* (which is a series of philosophical treatises) Vols. 24 and 25, publisher: Ōmura-shoten, 1925.

Satake, Tetsuo: '*Genshōgakuteki Kangen*' (*The Phenomenological Reductions*), in *Kōza*, Vols. 35 and 36, 1925.

Ikegami, Kenzō: 'Taishō-Ishiki to Hanchū no Mondai' (The Problem of Object-consciousness and of Categories), in *Tetsugaku Zasshi*, Nr. 459 and 460, 1925.

1926

Yamanouchi, Tokuryū: *Genshōgakuha no Tetsugaku* (*The Philosophy of the Phenomenological School*), in *Tetsugaku Kōza* (a series of philosophical treatises), publisner: Kindaisha, 1926.

Ikegami, Kenzō: 'Junsui Ishiki to ronriteki Ninshiki' (Pure Consciousness and Logical Knowledge), in *Tetsugaku Zasshi*, Nr. 473, 1926.

1927

Miki, Kiyoshi: 'Kaishakugakuteki Genshōgaku no Kisogainen' (The Basic Concepts of Hermeneutical Phenomenology', in *Shisō*, Nr. 63, 1927.

Ikegami, Kenzō: 'Genshōgakuteki Hontairon ni kansuru ippanteki Kōsatsu' (General View of Phenomenological Ontology), in *Shisō*, Nr. 63, 1927.

1928

Satake, Tetsuo: 'Zengenshōgaku to Genshōgaku' (Pre-phenomenology and Phenomenology), in *Tetsugaku Zasshi*, Nr. 491, 1928.

Hayami, Keiji: 'Ninshiki no genshōgakuteki Kaimei ni tsuite' (On the Phenomenological Enlightenment of Knowledge), in *Tetsugaku Kenkyū* (= *Philosophical Research*, Kyōto University), Nr. 148, 1928.

1929

Satō, Keiji: 'Ishiki no Genshōgaku' (Phenomenology of Consciousness), in the monthly journal *Risō* (= *The Ideal*), Nr. 8, publisher: Risōsha, 1929.

Honda, Kenzo: 'Genshōgaku to Benshōhō' (Phenomenology and Dialectics), in *Shisō*, Nr. 89, 1929.

Takahashi, Satomi: 'Husserl no Genshōgaku, tokuni sono genshōgakuteki Kangen' (Husserl's Phenomenology and Particularly its Phenomenological Reductions), in *Risō*, Nr. 12, 1929.

1930

Takahashi, Satomi: 'Husserl ni okeru Jikan to Ishikiryū' (Time and the Stream of Consciousness in Husserl), in *Tetsugaku Nenpō* (= *Yearbook of Philosophy*), Vol. 1, publisher: Daiichi-shobō, 1930.

Satō, Keiji: *Genshōgaku Gairon* (*Introduction to Phenomenology*), Waseda University Press, 1930.

Yamanouchi, Tokuryū: *Genshōgaku Josetsu* (*Presentation of Phenomenology*), publisher: Iwanami, 1930.

Takahashi, Satomi: 'Genshōgakuteki Kangen no Kanōsei' (The Possibility of Phenomenological Reductions), in *Tetsugaku Zasshi*, Nr. 523 and 524, 1930.

1931

Takahashi, Satomi: *Husserl no Genshōgaku* (*The Phenomenology of Husserl*), publisher: Daiichi-shobō, 1931.

Mutai, Risaku: 'Genshōgaku ni okeru Chōetsu no Mondai' (The Problem of Transcendence in Phenomenology), in *Tomonaga Hakase Kanreki Kinen Ronbunshū* (*Festschrift for the 60th birthday of Dr. Tomonaga*), publisher: Iwanami, 1931.

Kitō, Eiichi: 'Genshōgakuteki Kannenron to Kangenteki Hōhō' (Phenomenological Idealism and its Reductive Method), in *Tetsugaku Zasshi*, Nr. 535 and 536, 1931.

1932

Satake, Tetsuo: 'Genshōgaku ni okeru Jisshōshugi no Mondai' (The Problem of Positivism in Phenomenology), in *Tetsugaku Zasshi*, Nr. 540-547, 1932.

Makita, Kazuo: 'Husserl ni okeru kyōtsūhukanteki Kangen' (Intersubjective Reduction in Husserl), in *Risō*, Nr. 36, 1932.

1933

Mutai, Risaku: 'Taishōron to Genshōgaku' (Object Theory and Phenomenology) in *Iwanami Kōza Tetsugaku*, Vol. 7, 1933.

Sakasaki, Akira: 'Husserl Tetsugaku no sho Sō' (Various Phases of Husserl's Philosophy), in *Risō*, Nr. 41, 1933.

Hosoya, Tsuneo: 'Genshōgaku ni okeru Nichijōsei no Mondai' (The Problem of the Ordinary in Phenomenology), in *Risō*, Nr. 41, 1933.

Ōzeki, Schōichi: 'Genshōgaku towa nanika' (What is Phenomenology?), in *Risō*, Nr. 41, 1933.

Satake, Tetsuo: 'Genshōgakuteki Kangen no Kenkyū' (Studies of Phenomenological Reduction), in *Tetsugaku Zasshi*, Nr. 573 and 574, 1934.

1935

Hosoya, Tsuneo: 'Genshōgaku no Genmitsusei to Kongensei' (The Strictness and Radicality of Phenomenology), in *Risō*, Nr. 54, 1935.

Makita, Kazuo: 'Keitaishinrigaku to Genshōgaku' *Gestalt Psychology and Pheno-
menology*), in *Tetsugaku Ronshū* (= *Series of Philosophical Treatises*), Vol. 2, 1935.
Kishimoto, Masao: 'Ishiki no mittsu no Shikōsei' (Three Intentionalities of Consciousness),
in *Tetsugaku Zasshi*, Nr. 579, 1935.
Ōkuma, Taiji: 'Seishinbyōgaku to Genshōgaku' (Psychopathology and Phenomenology) in
Shisō, Nr. 161, 1935.

1936

Hosoya, Tsuneo: *Ninshiki Genshōgaku Josetsu* (*Introduction to the Phenomenology of
Knowledge*), publisher: Iwanami, 1936.
Katsube, Kenzō: 'Genshōgakuteki Hōhō to kaishakugakuteki Hōhō (The Pheno-
menological and Hermeneutical Method), in *Risō*, Nr. 63, 1936.
Ōnishi, Katsuyoshi: *Genshōgakuha no Bigaku* (*The Aesthetics of the Phenomenological
School*), publisher: Iwanami, 1937.

1937

Ōzeki, Shōichi: *Genshōgaku* (*Phenomenology*), publisher: Risōsha, 1937.
Wakayama, Chōkan: 'Husserl's "Die Krisis der europäischen Wissenschaften und die
transzendentale Phänomenologie"' (review) in *Tetsugaku Zasshi*, Nr. 607, 1937.
Shitahodo, Yūkichi: *Husserl*, publisher: Kōbundō, 1937.

1938

Mutai, Risaku: 'Edmund Husserl,' in *Shisō*, Nr. 194, 1938.
Inoue, Tetsujirō: 'E. Husserl shi o tsuiokusu' ('Recollections of E. Husserl'), in *Tetsuguku
Zasshi*, Nr. 620, 1938.
Kuwaki, Genyoku: 'Husserl no Sokumen' (One Side of Husserl), in *Tetsugaku Zasshi*,
Nr. 620, 1938.

1939

Shitahodo, Yūkichi: 'Genshōgaku to Benshōhō' (Phenomenology and Dialectics), in *Risō*,
Nr. 103, 1939.

1940

Mutai, Risaku: *Genshōgaku Kenkyū* (*Studies of Phenomenology*), publisher: Kōbundō,
1940.

1943

Satake, Tetsuo: *Genshōgaku* (*Phenomenology*), publisher: Asakura-shoten, 1943.

1946

Satake, Tetsuo: 'Senkenteki Genshōgaku no Hōga' (The Germ of Transcendental Phenomenology), in *Tetsugaku Zasshi*, Nr. 696, 1946.

1949

Satake, Tetsuo: *Husserl no Genshōgaku* (*Husserl's Phenomenology*), publisher: Shunjūsha, 1949.

1955

Ōzeki, Schōichi: *Genshōgaku Gaisetsu* (*Introduction to Phenomenology*), publisher: Sakurai-shoten, 1949.
Satake, Tetsuo: *Genshōgaku Gairon* (*Introduction to Phenomenology*), publisher: Ishizaki-shoten, 1954.
Nitta, Yoshihiro: 'Husserl no Sekai Bunsekiron ni tsuite' (On Husserl's Analysis of the World), in the journal *Bunka* (= *Culture*), 18th year, Nr. 3, 1955.

1959

Kida, Gen: 'Husserl ni okeru Jikan-Ishiki to Ishiki no Jikansei' (Time-consciousness and the Temporality of Consciousness in Husserl), in *Bunka*, 23rd year, Nr. 1, 1959.
Katō, Yasujoshi: 'Genshōgakuteki Hōhō no Tenkai' (The Development of the Phenomenological Method), in *Risō*, Nr. 319, 1959.

1963

Mizuno, Kazuhisa: 'Husserl no Monado-ron' (Husserl's Monadology), in *Tetsugaku Kenkyū*, Nr. 469, 1960.
Satō, Keiji: 'Genshōgakuteki Hōhō' (The phenomenological method), in *Risō*, Nr. 341, 1961.
Katō, Seiji: 'Husserl no Seisekai ni tsuite' (On the Life-world in Husserl), in the annual report *Tetsugaku*, Nr. 13, 1963.
Yamamoto, Manjirō: *Seimeikai Gainen o chūshin tosuru Husserl kōki-Shisō no Tenkai* (*The Development of Husserl's Later Thought Around the Concept of the Life-world*), publisher: Senbuncō, 1963.
Tanabe, Hajime: 'Genshōgaku ni okeru atarashiki Tenkō' (New Departure in Phenomenology), in *Tanabe Hajime Zenshū* (= Collected Works of Tanabe), Vol. 4, 1963, publisher: Chikuma-shobō.

1964

Tanabe, Hajime: 'Genshōgaku no Hatten' (The Development of Phenomenology), in *Tanabe Hajime Zenshū*, Vol. 15, publisher: Chikuma-shobō, 1964.

1965

Katō, Seiji: 'Shikōteki Mondai toshiteno Rekishi' (History as a Problem of Intentionality), in *Risō*, Nr. 380, 1965.
Tsunetoshi, Sōzaburō: 'Genshōgaku no Shinri' (The Truth of Phenomenology), in *Tetsugaku Kenkyū*. Nr. 498, 1965.

1967

Takahashi, Nobuaki: 'Husserl kara Sartre e' (From Husserl to Sartre), in the journal *Jitsuzonshugi* (= *Existentialism*), *Nr. 39, 1967.*

1968

Nitta, Yoshihiro: *Genshōgaku towa nanika* (*What is Phenomenology*), publisher: Kinokuniya, 1968.
Nitta, Yoshihiro: Genshōgaku no Sekai (The World of Phenomenology), in *Jitsuzonshugi*, Nr. 44, 1968.

1969

Nitta, Yoshihiro: 'Husserl ni okeru Hōhō no Mondai' (The Problem of Method in Husserl), in *Shisō*, Nr. 536, 1969.
Kitō, Eiichi: 'Genshōgaku no Tenbō to Kadai' (Prospectus and Task [or Agenda] of Phenomenology), in *Risō*, Nr. 437, 1969.
Tatematsu, Hirotaka: 'Husserl no Gengo-Riron' (Husserl's Theory of Language) in *Tetsugaka Zasshi*, Nr. 756, 1969.
Tatematsu, Hirotaka: 'Genshōgaku no Hōhō' (The Method of Phenomenology), in *Risō*, Nr. 437, 1969.
Ichikawa, Hiroshi: 'Shintai no Genshōgaku' (Phenomenology of the Organism), in *Risō*, Nr. 437, 1969.

1970

Kida, Gen: *Genshōgaku* (*Phenomenology 1970*), publisher: Iwanami, 1970.
Tatematsu, Hirotaka: 'Shikōsei' (Intentionality), in *Genshōgaku Kenkyū*, 1970.
Imamichi, Tomonobu: 'Genshōgakuteki Hōhō no jissaiteki Shihyō' (The Practical Index of the Phenomenological Method), in *Tetsugaku Zasshi*, Nr. 757, 1970.
Tatematsu, Hirotaka: 'Zenki Husserl Tetsugaku no Hatten to Keisei' (The Development and Formation of Husserl's Earlier Philosophy), in *Tetsugaku Zasshi*, Nr. 757, 1970.
Watanabe, Jirō: 'Genshōgakuteki Kangen ni tsuite' (On Phenomenological Reductions), in *Tetsugaku Zasshi*, Nr. 757, 1970.

1972

Takiura, Shizuo: *Sōzō no Genshōgaku* (*Phenomenology of the Imagination*), publisher: Kinokuniya, 1972.

1975

Tatematsu, Hirotaka: 'Genshōgaku towa nani ka' (What is Phenomenology), in *Genshōgaku Tokushū* (omnibus volume for phenomenology), publisher: Jōkyō-shuppan, 1975.

Nitta, Yoshihiro: 'Husserl no "Aruyo no Taiwa" Sōkō o megutte' (On Husserl's Manuscript 'Ein Nachtgespräch'), in *Genshōgaku Tokushū*, 1975.

Watanabe, Jirō: 'Heidegger to Genshōgaku' (Heidegger and Phenomenology), in *Genshōgaku Tokushū*, 1975.

Kuroda, Wataru: 'Genshōgakū to Bunpō' (Phenomenology and Grammar), in *Tetsugaku*, Nr. 25, 1975.

Kumagai, Tadao: 'Genshōgaku ni tsuite' (On Phenomenology), in *Tetsugaku*, Nr. 25, 1975.

Hiromatsu, Wataru: 'Genshōgaku to busshōkateki Sakusi' (Phenomenology and its Objectifying Deception), in *Tetsugaku*, Nr. 25, 1975.

III. JAPANESE TRANSLATIONS OF THE WRITINGS OF MAX SCHELER

1935

T. Kaba and K. Satō: *Tetsugakuteki Ningengaku* (*Philosophische Weltanschauung*), Risō-sha, 1935.

1937

Y. Ōshima: *Uchū niokeru Ningen no Ichi* (*Die Stellung des Menschen im Kosmos*), Daiichi-shobō, 1937.

1942

S. Terajima: *Tetsugakuteki Sekaikan* (*Philosophische Weltanschauung*), Sōgen-sha, 1942.

1960

K. Shinoda: *Ai to Ninshiki* (*Liebe und Erkenntnis*), Risō-sha, 1960.

Scheler Chosakushū (Selected Works of Scheler), *Hakusui-sha, 1976–* (in 15 volumes):

1976

1. D. Yoshizawa: *Rinrigaku ni okeru Keishikishugi to jisshitsuteki Kachirinrigaku* (*Der Formalismus in der Ethik und die materiale Wertethik*), Part 1, 1976.
2. D. Yoshizawa and N. Okada: the same Part 2, 1976.
3. Y. Ogura: the same, Part 3 (in progress).

1977

4. S. Hayashida *et al.*: *Kachi no Tento* (*Vom Umsturz der Werte*), Part 1, 1977.
5. H. Ōtani *et al.*: the same, Part 2, 1977.
6. H. Kamei and K. Kashiwabara: *Ningen ni okeru Eien narumono* (*Vom Ewigen im Menschen*), Part 1, 1977.

In Progress

7. S. Ogura: the same, Part 2 (in progress).
8. S. Aoki and S. Kobayashi: *Dōjō no Honshitsu to Keishiki* (*Wesen und Formen der Sympathie*) (in progress).
9. M. Iijima *et al.*: *Shakaigaku oyobi Sekaikangaku Ronshū* (*Schriften zur Soziologie und Weltanschauungslehre*), Part 1 (in progress).
10. M. Iijima *et al.*: the same, Part 2 (in progress).
11. O. Hamai: *Chisiki-keitai to Shakai* (*Die Wissensformen und die Gesellschaft*), Part 1 (in progress).
12. M. Hiro and Y. Tsumura: the same, Part 2 (in progress).
13. H. Kamei *et al.*: *Uchū ni okeru Ningen no Chii, Tetsugakuteki Sekaikan* (*Die Stellung des Menschen im Kosmos, Philosophische Weltanschauung*) (in progress).
14. Y. Ogura *et al.*: *Shoki Ronbunshū* (*Frühe Schriften*) (in progress).
15. Y. Hamada *et al.*: *Ikōshū* (*Schriften aus dem Nachlass*) (in progress).

IV. JAPANESE LITERATURE ABOUT MAX SCHELER
(A SMALL SELECTION)

1934

Sudō, Kōtarō: *Jisshitsuteki Kachi-rinrigaku no Rinen* (*The Idea of the Material Ethic of Value*), 1934, Seibidō.

1936

Matsushita, Sansei: *Ningengaku towa nanzoya – Scheler no Ningengaku* (*What is Anthropology – Scheler's Anthropology*), 1936, Dōbunkan.

1937

Tanaka, Hiroshi: *Max Scheler*, 1937, Kōbundō.

1955

Ishizeki, Keizō: *Jisshitsuteki Kachi-rinrigaku no Kenkyū* (*Studien zur materialen Wertethik*), 1955, Maeno-shoten.

1969

Ogura, Sadahide: *Max Scheler*, 1969; Ban Shobō.

V. JAPANESE TRANSLATIONS OF THE WRITINGS OF M. HEIDEGGER

Various translations of Sein und Zeit

1939

S. Terajima: *Sonzai to Jikan*, Part 1, 1939; Part 2, 1940; Mikasa-shobō.

1960

T. Kuwaki: *Sonzai to Jikan*, Part 1, 1960; Part 2, 1961; Part 3, 1963, Iwanami-shoten.
K. Matsuo: *Sonzai to Jikan*, Part 1, 1960; Part 2, 1966, Keisō-shobō.

1963

S. Hosoya, H. Kamei and H. Funahashi: *Sonzai to Jikan*, Part 1, 1963; Part 2, 1964, Risō-sha.

1967

K. Tsujimura: *U to Toki*, 1967, Kawade-shobō-shinsha.
T. Hara and J. Watanabe: *Sonzai to Jikan*, 1971, Chūōkōron-sha.

Translations of Other Works

1936

K. Satō: *Kant to Keijijōgaku no Mondai* (*Kand und das Problem der Metaphysik*), 1938, Mikasa-shobō.

1958

T. Kuwaki: *Hyūmanizumu ni tsuite* (*Über den Humanismus*), 1958, kadokawa-shoten.

1962

K. Tsujimura and H. Buchner: *Konkyoritsu* (*Der Satz vom Grund*), 1962, Sōbun-sha.

282 SELECTED BIBLIOGRAPHY

1977

M. Sonoda: *Nietzsche*, Part 1, 1976; Part 2, 1976; Part 3, 1977, Hakusui-sha.

Heidegger Senshū (Selected Works of Heidegger), Risō-sha, 1952—
(Up to 1977, 25 volumes have appeared.)

1952

1. S. Ōe: *Keijijōgaku towa nani ka (Was ist Metaphysik?)*, 1952.

1954

2. S. Hosoya: *Nietzsche no Kotoba "Kami wa shiseri," Hegel no "Keiken" Gainen (Nietzsches Wort "Gott ist tot," Hegels Begriff der Erfahrung)*, 1954.

1955

3. Tezuka, Saitō, Tsuchida and Takeuchi: *Hölderlin no Shi no Kaishaku (Erläuterungen zu Hölderlins Dichtung)*, 1955.

1957

4. M. Tanaka: *Anaximandros no Kotoba (Der Spruch des Anaximander)*, 1957.

1958

5. T. Tezuka and H. Takahashi: *Toboshiki Jidai no Shijin (Wozu Dichter)*, 1958.

1960

6. K. Tsujimura: *Shii no Keiken yori (Aus der Erfahrung des Denkens)*, 1960.
7. T. Hara: *Tetsugaku towa nanika (Was ist das – Die Philosophie?)*, 1960.
8. M. Kōsaka and K. Tsujimura: *No no Michi, Hebel (Feldweg, Hebel)*, 1960.
9. E. Kawahara: *Keijijōgaku Nyūmon (Einführung in die Metaphysik)*, 1960.
10. S. Ōe: *Dōitsusei to Saisei (Identität und Differenz)*, 1960.

1961

11. S. Kiba: *Shinri no Honshitsu ni tsuite, Platon no Shinriron (Vom Wesen der Wahrheit, Platons Lehre von der Wahreit)*, 1961.
12. E. Kikuchi: *Geijutsu-sakuhin no Hajimari (Der Ursprung des Kunstwerkes)*, 1961.

1962

13. T. Kuwaki: *Sekaizō no Jidai* (*Die Zeit des Weltbildes*), 1962.

1963

14. M. Miki: *Shi to Kotoba* (*Die Sprache in Gedicht, Das Wort*), 1963.
15. K. Tsujimura: *Houge* (*Gelassenheit*), 1963.
16. S. Hosoya, H. Kamei and H. Hunahashi: *Sonzai to Jikan* (*Sein und Zeit*), Part 1, 1963 (already mentioned above).

1964

17. The same, Part 2, 1964

1965

18. T. Kojima and Armbruster: *Gijutsu* (*Die Technik und die Kehre*), 1965.

1967

19. S. Kiba: *Kant to Keijijōgaku no Mondai* (*Kant und das Problem der Metaphysik*), 1967.

1968

20. K. Tsujimura: *U ni tsuite no Kant no Tēse* (*Kants These über das Sein*), 1968.
21. T. Tezuka: *Kotoba ni tsuite no Taiwa* (*Aus einem Gespräch von der Sprache*), 1968.

1970

22. A. Kakihara: *U no Toi e* (*Zur Seinsfrage*), 1970.

1974

23. K. Sasaki: *Hyūmanizumu ni tsuite* (*Über den Humanismus*), 1974.

1976

24. S. Hosoya: *Nietzsche*, Part 1, 1976.

1977

25. S. Hosoya: *Nietzsche*, Part 2, 1977.

In Progress

26. S. Hosoya: *Nietzsche*, Part 3, in progress.

VI. JAPANESE LITERATURE ABOUT M. HEIDEGGER
(A SMALL SELECTION)

1935

Kitō, Eiichi: *Heidegger no Sonzairon (Heidegger's Ontology)*, 1935, Tōyō-Shuppansha.

1939

Kuki, Shūzō: *Ningen to Jitsuzon (Man and Existence)*, 1939, Iwanami.

1943

Nakagawa, Hideyuki: *Heidegger Kenkyū (Heidegger Studies)*, 1943, Koyama-shoten.

1950

Miyake, Gouichi: *Heidegger no Tetsugaku (The Philosophy of Heidegger)*, Kōbundō, 1950.

1958

Kaneko, Takezō (ed.): *Heidegger no Shisō (Heidegger's Thought)*, 1958, Kōbundō.
Hara, Tasuku: *Heidegger*, 1958, Keisō-shobō.

1962

Watanabe, Jirō: *Heidegger no Jitsuzon Shisō (Heidegger's Existential Thought)*, 1962, Keisō-shobō.
Watanabe, Jirō: *Heidegger no Sonzai Shisō (The Idea of Being in Heidegger)*, 1962, Keisō-shobō.

1971

Tsujimura, Kōichi: *Heidegger Ronkō (Studies on Heidegger)*, 1971, Sōbun-sha.

VII. JAPANESE TRANSLATIONS OF THE WRITINGS OF
M. MERLAU-PONTY

S. Takiura and G. Kida: *Kodō no Kōzō* (*La structure du comportement*), 1964, Misuzu-shobō.

1965

K. Morimoto: *Hyūmanizumu to Teroru* (*Humanisme et terreur*), 1965, Gendaishichō-sha.

1966

S. Takiura and G. Kida: *Me to Seihin* (*L'Oeil et L'Esprit, Éloge de la philosophie*), 1966, Misuzu.

1967

Y. Takeuchi and S. Kogi: *Chikaku no Genshōgaku* (*Phénoménologie de la perception*), Part 1, 1967, Misuzu.

1969

Y. Takeuchi, S. Takiura, G. Kida *et al.*: *Sinyu* (*Signes*), Part 1, 1969, Misuzu.

1970

The same translators: the same (*Signes*). Part 2, 1970, Misuzu.
T. Eito: *Imi to Muimi* (*Sens et Non-sens*), 1970, Kokubun-sha.

1972

S. Takiura, G. Kida, S. Tajima and H. Ichikawa: *Benshōhō no Boken* (*Les aventures de la dialectique*), 1972, Misuzu.

1974

Y. Takeuchi, G. Kida, T. Miyamoto: *Chikaku no Genshōgaku* (*Phénoménologie de la perception*), Part 2, 1974, Misuzu.

Several additional smaller essays have been translated into Japanese; e.g., N. Takahashi: *Gengo no Genshōgaku ni tsuite* (*Sur la phénoménologie du langage*), 1969, Serika-shobō.

Very many Japanese essays about Merleau-Ponty have appeared in various journals and bulletins, but no large treatise has yet been published in book form.

VIII. JAPANESE TRANSLATIONS OF THE WRITINGS OF J.-P. SARTRE

Sartre Zenshū (*Oeuvres complètes de J.-P. Sartre*), Jinbun-shoin, 1950— . (Up to 1977, 36 volumes have appeared, including the following philosophical works.)

1955

Y. Hirai: *Sōzōryoku no Mondai* (*L'imaginaire*), (as 12th vol.), 1955.
T. Ibuki: *Jitsuzonshugi towa nanika* (*L'existentialisme est un humanisme*), (13th vol.), 1955.

1956

S. Matsunami: *Sonzai to Mu* (*L'être et le néant*), Part 1 (18th vol.), 1956.

1957

Y. Hirai and Y. Takeuchi: *Tetsugaku Ronbunshū* (*L'imagination, La transcendance de l'ego, Esquisse d'une théorie des émotions*), (23rd vol.), 1957.

1958

S. Matsunami: *Sonzai to Mu*, Part 2 (19th vol.), 1958.

1960

S. Matsunami: *Sonzai to Mu*, Part 3 (20th vol.), 1960.

1962

Y. Hirai: *Hōhō no Mondai* (*Question de méthode*), (25th vol.), 1962.
Y. Takeuchi and I. Yanaibara: *Benshōhōteki Risei Hihan* (*Critique de la raison dialectique*), Part 1 (26th vol.), 1962.

1965

Y. Hirai and K. Morimoto: *Benshōhōteki Risei Hihan*, Part 2 (27th vol.), 1965.

IX. JAPANESE LITERATURE ABOUT J.-P. SARTRE
(A SMALL SELECTION)

1954

Kaneko, Takezō, ed.: *Sartre no Tetsugaku* (*Philosophy of Sartre*), 1954, Kōbundō.

1966

Matsunami, Shinzaburō: *Sartre*, 1966, Keisō-shobō

1967

Yanaibara, Isaku: *Sartre*, 1967, Chūōkōron-sha.
Takeuchi, Yoshirō: *Sartre to Marx-shugi* (*Sartre and Marxism*), 1965, Kinokuniya-shoten
Takeuchi, Yoshirō: *Sartre Tetsugaku Josetsu* (*Introduction to the Philosophy of Sartre*), 1966, Morita-shoten

1971

Suzuki, Michichiko, ed.: *Sartre to sono Jidai* (*Sartre and His Time*), 1971, Jinbun-shoin

Nagoya   HIROTAKA TATEMATSU

INDEX OF NAMES

289

ANALECTA HUSSERLIANA

The Yearbook of Phenomenological Research

Editor

ANNA-TERESA TYMIENIECKA